ESG Reporting Manual

500+ Legal Tips and Tricks to Improve Your ESG Reporting

RIVER PUBLISHERS SERIES IN ENERGY SUSTAINABILITY AND EFFICIENCY

Series Editor

PEDRAM ASEF
Lecturer (Asst. Prof.) in Automotive Engineering,
University of Hertfordshire,
UK

The "River Publishers Series in Sustainability and Efficiency" is a series of comprehensive academic and professional books which focus on theory and applications in sustainable and efficient energy solutions. The books serve as a multi-disciplinary resource linking sustainable energy and society, fulfilling the rapidly growing worldwide interest in energy solutions. All fields of possible sustainable energy solutions and applications are addressed, not only from a technical point of view, but also from economic, social, political, and financial aspects. Books published in the series include research monographs, edited volumes, handbooks and textbooks. They provide professionals, researchers, educators, and advanced students in the field with an invaluable insight into the latest research and developments.

Topics covered in the series include, but are not limited to:

- Sustainable energy development and management;
- Alternate and renewable energies;
- Energy conservation;
- Energy efficiency;
- Carbon reduction;
- Environment.

For a list of other books in this series, visit www.riverpublishers.com

ESG Reporting Manual
500+ Legal Tips and Tricks to Improve Your ESG Reporting

Robin Boustead

River Publishers

Routledge
Taylor & Francis Group
NEW YORK AND LONDON

Published 2025 by River Publishers
River Publishers
Alsbjergvej 10, 9260 Gistrup, Denmark
www.riverpublishers.com

Distributed exclusively by Routledge
605 Third Avenue, New York, NY 10017, USA
4 Park Square, Milton Park, Abingdon, Oxon OX14 4RN

ESG Reporting Manual: 500+ Legal Tips and Tricks to Improve Your ESG Reporting / Robin Boustead.

Routledge is an imprint of the Taylor & Francis Group, an informa business

ISBN 978-87-7004-721-0 (hardback)
ISBN 978-87-7004-791-3 (paperback)
ISBN 978-87-7004-723-4 (online)
ISBN 978-8-770-04722-7 (ebook master)

Disclaimer

This publication has been prepared for general guidance on matters of interest only, and does not constitute professional advice. You should not act upon the information contained in this publication without obtaining specific professional advice. The methodologies, indicators, and metrics included in this publication rely on users providing information that is relevant and faithfully represents the substance of phenomena that it purports to represent, is comparable and comprehensible, complete, neutral, verifiable, and free from error. No responsibility is accepted for the relevance, faithful representation, understandability, or meaningfulness of the use and application of indicators, methodologies, and metrics outlined in this publication by the reader. No representation or warranty (express or implied) is given as to the accuracy or completeness of the information contained in this publication, and, to the extent permitted by law, the author, publisher, employees, and agents do not accept or assume any liability, responsibility, or duty of care for any consequences of you or anyone else acting, or refraining to act, in reliance on the information contained in this publication or for any decision based on it.

Decisions that you make based on information in this publication are at your own responsibility and risk. We (editors-in-chief, authors, and publisher) have endeavoured to ensure that the information is correct, complete, and up-to-date at the time of going to press. Nevertheless, this publication may contain inaccuracies or sources that may no longer be available. When hyperlinks or references to third-party websites are used, this does not mean that we recommend products or services offered via these internet sites. The use of these hyperlinks is entirely at your own risk. We do not accept any responsibility or liability with regard to the content, use, or availability of such internet sites. The veracity, correctness, reasonableness, reliability, and completeness of information on such internet sites have not been verified by us. It is not permitted to reproduce, forward, distribute, make public, or make available to third parties this publication for a fee, digitally or materially. We do not accept any liability for direct or indirect damage resulting from the use of this publication. You may print and/or download this publication for your own personal use, in accordance with the Creative Commons License for this work. Danish copyright applies to this publication and the terms of use. The Danish court has exclusive jurisdiction over any disputes that may arise in connection with this website.

Impact Promise

The entire text of **ESG Reporting Manual:** ***500+ Legal Tips and Tricks to Improve your ESG Reporting*** is focused on reducing negative impacts and developing/enhancing positive environmental and social impacts. The text has been written on a computer supplied with green energy and the supporting website is hosted on a green server. It is a book that walks its own talk and is designed to encourage businesses and organisations to take realistic steps to making the world a better place.

This logo indicates that the content or text or graphic is part of our extensive range of support materials, for example downloadable excel files, indicator and metric tools, graphics and more on our support website: https://esgreportpro.com

Contents

Acknowledgements

This book is the result of some very dedicated work by some exceptional individuals, most notably Sudan Aryal, whose contribution in both overall methodologies and selection of specific metrics comparisons was invaluable.

A special thank you goes to Andrea Orsag from MissionC NL for her insight to Nudie Jeans and CSRD insights in general. Another big thank you for encouragement and insight goes to Robin Dufek from Sustainova for pushing me for clarifications and explanations that have helped to make the descriptions much more 'hands-on'.

Another very special thank you goes to the River Publishing team, Ana Koren, Junko Nakajima, and Subrahmanyan for their belief in the project and hard work to embrace additional workloads in their already hectic schedule.

I also owe all the foreword writers a huge note of thanks. There is no single ESG authority, so their combined willingness to view the text in the light of their own fields of expertise and experience is very deeply appreciated.

Finally, my deepest appreciation and gratitude must go to my partner, Sabine Peetz, who has been my rock and unwavering support over two years of research and writing - and then she made time to help proof the text!

Key Sources

This book draws on Annex I to the Commission Delegated Regulation (EU) 2023/2772 supplementing Directive 2013/34/EU of the European Parliament and of the Council as regards sustainability reporting standards and as published in the Official Journal of the European Union on 22 December 2023. Explanatory and interpretive information relies on EFRAG's published Draft European Sustainability Reporting Standard (ESRS) Disclosures: ESRS 1, ESRS 2, E1, E2, E3, E4, E5, S1, S2, S3, S4, G1, and Annexes, as well as related guidance and online materials (seminars, public consultation forums, etc.).

Details regarding the EU Taxonomy draw on the Commission Delegated Regulation (EU) 2021/2139 of 4 June 2021 supplementing Regulation (EU) 2020/852 of the European Parliament and of the Council by establishing the technical screening criteria for determining the conditions under which an economic activity qualifies as contributing substantially to climate change mitigation or climate change adaptation and for determining whether that economic activity causes no significant harm to any of the other environmental objectives.

Information regarding the ISSB/IFRS S1 and S2 disclosures draws on ESRS–ISSB Standards Interoperability Guidance, April 2024. Global reporting Initiative (GRI) indictor comparisons are referenced to *GRI 1, 2, 3: 11, 12, 13, 14, 306, 101, 201, 202, 203, 204, 205, 206, 207,301, 302, 303, 304, 305, 306, 308,401, 402, 403, 404, 405, 406, 407, 408, 409, 410, 411, 413, 414, 415, 416, 417, and 418 as of 2023.*

Additional resources of use to the reader that are referenced in this book:

- Assessing the Materiality of Nature-Related Financial Risks for the UK. Green Finance Institute, 2024.
- Authentic Sustainability Assessment: A User Manual for the Sustainable Development Performance Indicators, UNRISD, 2022.
- Circular Transition Indicators V4.0 Metrics for business, by business. WBCSD, 2022.

- Discussion paper on Conducting Advanced Scenario Analysis, TNFD, 2023.
- Final Report: Recommendations of the Task Force on Climate-related Financial Disclosures, TCFD, 2017.
- [draft] GRI-ESRS Interoperability Index, GRI & EFRAG, 2023.
- Technical Guidance for Calculating Scope 1, 2 and 3 Emissions, Greenhouse Gas Protocol, 2011.
- The Enhancement and Standardization of Climate-Related Disclosures for Investors, Securities and Exchange Commission, 2024.
- The TNFD Nature-related Risk & Opportunity Management and Disclosure Framework, TNFD, 2022.
- TPT Disclosure Framework, Transition Plan Taskforce, 2023.
- [draft] Voluntary ESRS for Non-Listed Small- and Medium-Sized Enterprises (VSME ESRS), EFRAG, 2024.

Forewords

In today's financial landscape, ESG reporting is increasingly a business necessity. With growing regulatory pressures and heightened stakeholder expectations, the demand for transparent, accountable, and comprehensive ESG disclosure has never been greater. Investors, regulators, and consumers alike are no longer satisfied with vague commitments to sustainability; they are demanding detailed, verifiable data on how companies are managing their environmental and social impacts, as well as how they are governed. The evolving regulatory landscape, from the European Union's CSRD to the global rise of ISSB standards, has solidified ESG reporting as a key pillar of business strategy.

This ESG Reporting Manual: 500+ Legal Tips and Tricks to Improve Your ESG Reporting is a vital tool for any organization navigating these complex waters. Designed with both seasoned sustainability professionals and those newer to ESG in mind, it offers a practical, step-by-step guide to achieving compliance while driving business performance. Written by a business owner with decades of experience in the field, this manual is packed with legal insights, best practices, and actionable strategies to help your organization not only meet its regulatory obligations but also unlock the value that comes with being a sustainability leader.

At the heart of this manual is the idea that ESG reporting is more than a compliance exercise. It is a mechanism for fostering trust with stakeholders, mitigating risks, and identifying new opportunities for growth. The book emphasizes that effective ESG reporting requires a holistic approach—one that integrates sustainability into every facet of business operations, from governance structures to supply chains. By following the guidance laid out in these pages, organizations can ensure their reporting is both rigorous and aligned with global standards, thereby positioning themselves to thrive in an increasingly green economy.

This manual walks its own talk, produced using green energy and hosted on a sustainable server, underscoring the very principles it advocates. Whether you are just beginning your ESG journey or looking to refine your existing practices, this comprehensive guide will empower you to deliver

ESG reports that stand up to scrutiny and contribute to a more sustainable, resilient future for both your organization and the planet.

David Carlin, Founder
D. A. Carlin and Company, LLC.
Accelerating sustainable finance and strategic success
www.dacarlin.com

A company's continued success and market resilience are increasingly dependent on its performance in Environmental, Social, and Governance (ESG) criteria. The demand for more rigorous and transparent ESG reporting has never been more urgent driven by three sets of stakeholders: 1) investors, owners, bankers and insurers, as stewards of capital and enterprise; 2) legislators and regulators as guardians of compliance and standards and; 3) consumers with increased expectation for corporate accountability. Against this backdrop, the "*ESG Reporting Manual*" emerges as an essential guide for corporations navigating the ESG waters.

Sustainability is no longer a peripheral aspect of business strategy. Physical, transitional and financial risks are core topics to business management in a world faced with environmental and societal challenges and where resilience is the name of the game. As Environmental, Social, and Governance issues become central, businesses are now tasked with integrating both financial and nonfinancial metrics into their strategies to remain viable in a rapidly changing global landscape

The *ESG Reporting Manual* adeptly sets the stage for organizations to rise to these challenges through comprehensive strategies, actionable insights, and a clear roadmap towards robust ESG reporting. It is a needed guide to navigating the sea of acronyms and sustainability-related jargon and to understand the 'what', 'why' and 'how' of ESG laws, regulations, standards and frameworks all the way to the process and practices of ESG reporting.

Expertly written, the ESG Reporting Manual caters to a broad spectrum of business or organizational needs—from designing efficient ESG systems and engaging stakeholders to mastering the detailed nuances of ESG disclosures. It offers over 500 tips, which enhance the reader's ability to transfer the ESG knowledge into practices.

In addition, businesses today are under significant scrutiny concerning their operational impact on the environment and society. However, the fear of 'greenwashing' or social-washing looms large. This manual provides practical steps to mitigate these risks and to transform the reporting endeavours into opportunities for business innovation and leadership in sustainability.

For anyone involved in ESG reporting—whether you are initiating a reporting system, refining your existing processes, or aiming to educate your corporate teams—this manual is an indispensable resource. With this knowledge, you are prepared to navigate the complexities of the ESG landscape with confidence and emerge as a leader in your industry and transform your business by fostering resiliency in a challenging world.

Dr. Willy Legrand
Professor
IU International University of Applied Sciences
Germany

In today's rapidly evolving business landscape, the integration of Environmental, Social, and Governance (ESG) reporting has become mainstream especially with the European Union's Corporate Sustainability Reporting Directive (CSRD) that mandates that companies disclose detailed ESG information.

The opportunities presented by ESG reporting are vast and transformative. Companies that embrace ESG principles can unlock new avenues for growth and innovation. By prioritizing sustainability, businesses can tap into emerging markets, attract forward-thinking investors, and build stronger, more resilient brands. ESG reporting if done right can also foster a culture of responsibility and purpose, aligning corporate goals with the broader societal good.

However, ESG reporting is not without its challenges. The primary hurdle is the lack of standardized metrics and frameworks, which can lead to inconsistencies and difficulties in benchmarking performance. Also, integrating ESG considerations into existing business models requires a significant shift in mindset and operations, often needing substantial investment in time and resources.

The risks associated with ESG reporting are equally important. Companies that fail to accurately report their ESG performance may face reputational damage, legal repercussions, and loss of investor confidence. ESG reporting has the risk of becoming yet another tool of greenwashing (the practice of making misleading claims about sustainability efforts) and to avoid taking real actions for sustainability if not used the right way. Greenwashing poses a significant threat to the credibility of ESG initiatives. It is therefore crucial for businesses to approach ESG reporting with integrity and transparency, ensuring that their actions genuinely reflect their commitments.

This ESG Reporting Manual has all the tips and tricks you should know when producing the reports for your organisation. It easily guides you through every definition, possible step and scenarios that you might encounter in your process. It's by far the most comprehensive guide to the process I have encountered so far.

To conclude, ESG reporting if done right represents a powerful tool for companies to navigate the complexities of modern business while contributing to a more sustainable and equitable world. By embracing the opportunities, addressing the challenges, and mitigating the risks, businesses can pave the way for a brighter future for all - and if you want to do it right – let this book be your main guide in the process!

Elin Bergman
Circular Impact Officer Cradlenet
Co-founder & Managing Partner Nordic Circular Hotspot
www.cradlenet.se
www.nordiccircularhotspot.org

ESG factors are increasingly important for achieving long-term success in today's business world. Regulations evolve at a rapid pace and companies feel the pressure to adapt. Keeping up with ESG reporting is challenging, and this is where The ESG Reporting Manual comes in—timely and essential for organizations looking to thrive and improve.

Written by Robin Boustead, a highly accomplished business owner, environmental advocate, and mountaineer, this manual offers a clear and practical guide to ESG reporting. Robin combines his expertise in business with a strong commitment to environmental stewardship, drawing on the determination he's developed from climbing some of the world's most challenging peaks. ESG reporting is about more than just meeting regulatory requirements. It requires commitment to sustainability and ethical leadership and for a company that focuses on the right things, the right reports will follow.

What really sets this book apart is its hands-on, accessible approach. Robin takes an overwhelming topic and breaks it down into manageable steps. Through this approach he makes it so much easier for companies, both large and small, to implement effective ESG practices. Whether you're new to the field of ESG reporting or you have years of experience, this manual provides a straightforward and actionable roadmap.

ESG is a powerful way to future-proof businesses and make companies become and remain appealing for customers, talent, financing and investors. Robin has written a book that helps you make this happen and it's a privilege for me to be part of introducing it to you. I'm confident that The ESG

Reporting Manual will educate and empower you to meet the challenges of future-proofing your business and building a more responsible and resilient future.

Petter Binde
Author of "ESG SPRINT: 7 Steps to Get Your Business on Track for a Sustainable Future"

ESG – When people hear this term many get excited, while many get worried. These days we talk about Environmental, Social and Governance factors as something to chase and achieve, whereas in a thriving economy they truly should be an integrated part of company's strategy. In other words: common sense.

While resources and extra effort are needed to accomplish an insightful ESG reporting, I do believe it has a big potential to move from being a "ticking a box" exercise, to becoming an insightful guide on how the overall organizational strategy can be aligned, adjusted and transformed to stir the organization into becoming a force for good. This ESG Reporting Manual book written by Robin Boustead provides in incredible detail handy "how to" tips, guidance and examples for crucial activities from stakeholder engagement, materiality assessment and data collection, covering all important focus areas such as Climate Change, Water, Biodiversity, Circular Economy, Workforce, Communities and Business Conduct. It is a useful information source not only for all working on ESG reporting but also for those keen to shape a futureproof business strategy that incorporates all the ESG factors into business as usual.

Andrea Orsag
Founder: MissionC

Throughout my career in sustainable business practices, I've been on the front lines of the ESG evolution, witnessing first-hand how organizations of all sizes—from global corporations to small enterprises—struggle with ESG reporting. The challenges are real: complex data collection, the ever-present fear of greenwashing accusations, and the difficulty of aligning ESG efforts with business goals. On top of that, businesses must contend with constantly shifting standards and rising stakeholder expectations. Yet, despite these hurdles, I've also seen how embracing ESG reporting can transform businesses, sparking innovation, enhancing reputations, and, most importantly, creating lasting value. Today, ESG reporting isn't just about ticking compliance boxes—it's an essential pillar of business resilience and strategy in an increasingly uncertain world.

Robin Boustead's ESG Reporting Manual couldn't have come at a more critical time. As the world grapples with intensifying environmental crises and greater social responsibility, businesses must embed ESG into their DNA to stay relevant and competitive. Robin's guide stands out because it doesn't just focus on compliance—it goes further, offering over 500 legal tips and actionable insights that help companies enhance their performance within the boundaries of our planet. What I truly appreciate about this manual is its practicality. Boustead offers a step-by-step, approachable guide to demystifying ESG reporting.

As Robin wisely puts it, "Don't let confusion hold you back any longer – let the ESG Reporting Manual guide you towards confident compliance and a profitable future of responsible and resilient growth." This manual is much more than a guide—it's a roadmap to build trust, foster innovation, and turn challenges into opportunities. From identifying material impacts to implementing robust disclosure practices, this book offers everything a business needs to navigate the complexities of ESG reporting with confidence.

For those committed to sustainability—whether just beginning or refining their ESG approach—this manual is an indispensable resource, designed to guide you toward measurable progress and success.

Jaime Amoedo
Executive Director
The ESG Institute

ESG is simultaneously controversial and important. Whilst many of its critics call it 'woke capitalism', its advocates say it's about good business — that there is no trade-off between long-term profits and ESG.

There is another type of ESG critic. They have no problems with the principles underpinning the idea but misgivings about the practicality. They say ESG can be used as a kind of smokescreen — painted over in green colours, in an attempt to portray an organisation in a way that does not reflect reality.

The challenge often boils down to how we measure ESG.

I first met Robin when he appeared as a guest on the ESG Show — the theme was how to get started. His passion and deep knowledge were immediately obvious. And for Robin, the data ESG can create helps an organisation make better decisions.

There are many reasons why an organisation adopts ESG — because of regulations, investors, customers, to aid recruitment and retention of staff or maybe because they see ESG as good business. Whatever the motivation, measuring is the key.

If we can measure ESG effectively, warts-an-all, we can begin to turn ESG into an opportunity that can help guide an organisation towards success as this century of extraordinary challenges and opportunities continues.

See a parallel with disruptive technology. Organisations late to embrace disruptive technology are more likely to instead embrace the corporate graveyard instead. The same risks and opportunities apply to ESG, first mover advantage is critical.

This book is thus an essential guide — it can support the ESG journey much like a map can support a physical journey, a constant companion to the ESG practitioner.

For the organisation applying ESG this book can create certainty where once there was doubt and clarity where once there was fog, but above all, help organisations manage risk. Business risk is a constant companion, and the ESG Reporting Manual is an important part of the tool kit required to address this.

Michael Baxter
Editor in Chief Techopian.com
Host and founder ESG Show.

List of Graphics

List of Tables

List of Abbreviations

ACA	Absolute contraction approach
AGM	Annual general meeting
ALE	Annualised loss expectancy
AQI	Air quality index
ATG	Average tax gap
CapEx	Capital expenditures
CDP	Carbon Disclosure Project
CDR	Commission Delegated Regulation
CDSB	Climate Disclosure Standards Board
CEAOB	Committee of European Auditing Oversight Bodies
CEPA	Classification of Environmental Protection Activities
CPUE	Catch-per-unit-effort
CRHI	Coral reef health index
CSDDD	Corporate Sustainability Due Diligence Directive
CSR	Corporate Social Responsibility
CSRD	Corporate Sustainability Reporting Standard
DA	Delegated Act
DE&I	Diversity, equity, and inclusion
DEI	Diversity, equity, and inclusion
DP	Data point
DPO	Days payable outstanding
DSO	Days sales outstanding
EAR	Exposure at risk
EBT	Earnings before tax
EFRAG	European Financial Reporting Advisory
EPR	Extended producer responsibility
ESAP	European Single Access Point
ESG	Environmental, social, and governance
ESMA	European Securities and Markets Authority
ESOP	Employee Stock Ownership Plan
ESRS	European Sustainability Reporting Standard
ETR	Effective tax rate

EWC	European Works Council
FCCC	Framework Convention on Climate Change
FTE	Full-time equivalent
GDP	Gross domestic product
GHG	Greenhouse gas
GHG Protocol	Greenhouse Gas Protocol
GHGe	Greenhouse gas emissions
GIS	Geographic information systems
GRI	Global Reporting Initiative
GSTC	Global Sustainability Tourism Council
IASP	Independent assurance service providers
IFRS	International Financial Reporting Standards
IPCC	International Panel on Climate Change
IRO	Impacts, risks, and opportunities
ISSB	International Sustainability Standards Board
KPI	Key performance indicator
LEAP	Locate, Evaluate, Assess, Prepare
LEF	Loss event frequency
LEM	Loss event magnitude
LSMEs	Listed small- and medium-sized enterprises
MDR	Minimum disclosure requirements
MDR	Minimum general disclosures
MSS	Managers selected by staff
MSY	Maximum sustainable yield
MWP	Members/workers/producers
NCAs	National Competent Authorities
NFRD	Non-Financial Reporting Directive
NGFS	Network for Greening the Financial System
NGO	Non-governmental organisations
NPV	Net present value
NUTS	Nomenclature of Territorial Units of Statistics
ODP	Ozone-depleting potential
ODS	Ozone-Depleting Substances
OECD	Organisation for European Co-operation and Development
OpEx	Operational expenditures
OTH	Other purposes or programs
P&L	Profit and loss
PIE	Public interest enterprise
PRE	Proportion of renewable energy
R&D	Research and development

REC	Renewable energy consumed
RIO	Reinvestment in the organisation
RoI	Return on investment
RTS	Regulatory Technical Standards
SAI	Social Accountability International
SASB	Sustainability Accounting Standards Board
SBTi	Science-Based Targets initiative
SCE	Societas Cooperativa Europaea
SDGs	Sustainable Development Goals
SE	Societas Europaea
SEC	Securities Exchange Commission
SEEA	System of Environmental Economic Accounting
SFDR	Sustainable Finance Disclosure Regulation
SHR	Shareholders
SME	Small and medium enterprise
SOP	Standard operating procedure
STR	Statutory tax rate
T&D	Transmission and distribution
TCFD	Taskforce of Climate-based Financial Disclosures
TEC	Total energy consumption
THT	Total hours of anti-corruption training
TME	Total number of managers employed
TNFD	Taskforce on Nature-based Financial Disclosures
TPT	Transition Plan Taskforce
TSP	Total surplus/profits distributed
UNGC	United Nations Global Compact
VOC	Volatile organic compound
VRF	Value Reporting Foundation
WASH	Water, sanitation, and hygiene
WQI	Water quality index
WRA	Water Risk Atlas
WRF	Water risk filter
WWF	World Wildlife Forum

PART 1

Designing Your ESG Reporting System

1

Introduction to ESG Reporting

In this chapter: A brief overview of ESG reporting systems from around the world. A comparative summary of the major systems including the CSRD (ESRS), IFRS, GRI, SEC Climate Reporting, and the UK SDR. How to choose between them (if required) and should you want to engage with more than one or more reporting systems, then which one should you begin with and how to structure your time and budgets.

In today's rapidly evolving business landscape, stakeholders are increasingly prioritising sustainability and responsible business practices. The general public, investors, media, and politicians are voicing growing concern over the impacts that commercial activities are having both on our planet and society. Recent introduction of environmental, social, and governance (ESG) legislation in many countries is set to formalise reporting processes and procedures as much as international accounting practices did in the 1990s. ESG reporting is an enhanced and more transparent approach to Corporate Social Responsibility (CSR) reporting that was often used to drive competitive advantages. We now see standardised, statutory ESG reporting evolving to become as important as annual financial reports and driving organisational strategy.

Rising to this new reporting obligation is a time of uncertainty and stress for organisations, carrying large potential risks for inaccurate statements or greenwashing. Many organisations feel confused about what is required in their reporting, and even how to begin the reporting process is a major source of concern. Further anxiety is caused by not wanting to be seen as a laggard among peers, so sector-based insights are in very high demand but extremely hard to find. ESG reporting for organisations poses a profound challenge to accurately assess an organisation's impacts and ensure that there is no risk of exposure to greenwashing or social-washing allegations.

This book provides you with the necessary tools to effectively assess and communicate your ESG efforts. Such transparency not only enhances corporate reputation and trust but also attracts investors, customers, and top talent who value sustainability. In a world where ESG considerations are shaping investment decisions and influencing consumer behaviour, mastering ESG reporting is no longer just a choice but a strategic imperative for long-term success and impact.

Organisations must now rapidly master a new skillset, annual ESG report production, as well as incorporate continuous improvement systems that prioritise minimising negative impacts and establish a pattern of positive outcomes across all operations. Being able to produce an ESG report once is not the sole objective, rather it is how to integrate ESG systems into ongoing processes and procedures, and then how to further improve them over time. The legislation adopted by governments aims to see ESG reporting 'ripple' through value chains and across sectors, leading to a greener and more socially responsible world. This is perhaps best expounded by the European Green Deal, which sets out goals and pathways for the continent to become both planet 'friendly' and more socially 'just'.

Organisations require insights and ideas about how they can effectively begin their ESG reporting journey that then leads to a comprehensive approach to managing their impacts on climate change, pollution, water and marine resources, biodiversity, waste management, enhanced workforce management throughout the value chain, and community, consumer, and end-user impacts. **ESG Reporting Manual:** ***500+ Legal Tips and Tricks to Improve your ESG Reporting*** is an essential deep dive into mastering ESG reporting. It allows you to unlock the secrets to creating impactful reports that showcase your company's commitment to sustainability, social responsibility, and good governance. It also guides learning of how to navigate complex frameworks, engage stakeholders effectively, and drive positive change. It is a book designed to elevate ESG reporting for any size of organisation and put you in the driving seat.

In summary, the complexity of compliance with ESG legislation is causing confusion around the world. Organisations are unsure how they can fulfil their ESG reporting obligations without exposing themselves to allegations of green or social-washing and yet continue to deliver shareholder value. This book is written by a business owner who has decades of experience and will guide you through creating ESG reports in a practical step-by-step format.

1.1 ESG Reporting Relevancy

Environmental, social, and governance (ESG) reporting standards have become increasingly important in the global business landscape. These standards provide a framework for companies to report on their performance in areas related to sustainability, social responsibility, and corporate governance. ESG reporting allows stakeholders, including investors, customers, employees, and the public, to assess a company's impact beyond financial performance. By adhering to global ESG reporting standards, companies can enhance their credibility, attract investment, and build trust with stakeholders. Moreover, transparent ESG reporting can help organisations identify risks and opportunities, drive innovation, and improve overall performance.

But are global ESG standards and reporting systems really important or is this just another administration exercise?

Let us begin by approaching the sticky question of corporate responsibility. Many argue that the purpose of a corporation is solely to provide a financial return to shareholders. How they do so is up to the corporation, and shareholders will vote with their investments if they approve of the actions taken to provide that return. So, for many, the concept of accountability for actions is simply couched in financial impact terms; this is discussed further in Chapter 4: Single and Double Materiality Assessment. Over the years, we have seen this approach result in some devastatingly bad outcomes for the environment and communities, and it is obvious that to rely on organisations to 'do the right thing all the time' is at best naïve. In response to some of the worst impacts, sustainability reporting systems were established, for example, the Exxon Valdez oil spill in 1997 is cited as the trigger for creating the GRI. So, is ESG reporting an incursion on the corporation's right to do what it pleases to create shareholder value, or is it a fair and reasonable response to potentially avoidable disasters? This philosophical question has largely been muted in Europe by the introduction of the CSRD, which has the clear agenda of rippling through global value chains thus reducing negative impacts, especially climate change. However, the question remains for each organisation – is ESG reporting simply an administrative exercise or somehow more significant? Let us reflect on the introduction of international accounting standards in the 1990s, which were greeted with much the same antipathy and have now become a vital source of information. Whether you are in favour with ESG standards and principles or not, it would appear that they are here to stay in the same way as accounting standards took hold of the global investment marketplace. This is because:

1. Standards provide a common language and framework for companies to report on their performance. This allows for better comparability between companies and industries, making it easier for investors to assess their risks and opportunities related to ESG.

2. Standards and reporting systems help companies identify areas where they can improve their practices, so in many respects, ESG is about doing 'better business' thus negating (partially) the naivety of the market. By setting clear and measurable targets, companies can track their progress and make more informed decisions about their sustainability strategies.
3. ESG reporting can also have a positive impact on a company's reputation and brand. As consumers become more conscious of the social and environmental impact of their purchases, companies that demonstrate a commitment to sustainability and responsible business practices are more likely to attract and retain customers.
4. And finally, and perhaps most critically, there is a growing trend of ESG investing, where investors consider a company's ESG performance alongside traditional financial metrics when making investment decisions. This has led to an increased demand for ESG data and reporting, making it even more important for companies to adhere to global ESG standards and reporting systems.

However, the 'administration exercise' criticism of ESG reporting lingers, with some arguing that it can be overly complex and subjective, and that companies may use it as a marketing tool rather than a genuine effort to improve their ESG practices. There are also concerns about the lack of standardisation and consistency in ESG reporting, which can make it difficult for investors to compare and evaluate companies.

In conclusion, global ESG standards and reporting systems are an important tool for companies to measure and report on their ESG performance. They provide a common framework for reporting, help companies identify areas for improvement, and can have a positive impact on a company's reputation and brand. However, there is still room for improvement in terms of standardisation and consistency in ESG reporting, and it is important for companies to use these standards and systems in a genuine effort to improve their ESG practices rather than just for marketing purposes.

1.2 Choosing the Right ESG Standard

Over the last 20 years, various organisations have developed ESG reporting standards and frameworks to help companies disclose relevant information in a consistent and comparable manner, and importantly, from specific reporting perspectives. Each of these standards has different focuses and applications within the realm of ESG reporting and can appear confusing at first.

Frameworks mostly focus on greenhouse gas management and reporting, for example (see summary Graphic 1), the Greenhouse Gas (GHG) Protocol, Science Based Targets initiative (SBTi), Carbon Disclosure Project (CDP), Taskforce for Carbon-based Financial Disclosure (TCFD), and the UK-focused Transition Plan Taskforce (TPT). All frameworks have the same foundation in international climate research stemming from the International Panel on Climate Change (IPCC), so they are similar with some slight variations in methodology or geographic or industry focus. There are other frameworks relevant to ESG reporting, for example, the Taskforce on Nature-based Financial Disclosures (TNFD) and more are launching to fill the gaps in relevant assessment methodologies for ESG impacts. Choosing between frameworks can be complicated and time consuming, but for the majority of organisations, the SBTi offers sector-focused GHG emission frameworks and will be the basis for much of Chapter 5, Climate Change.

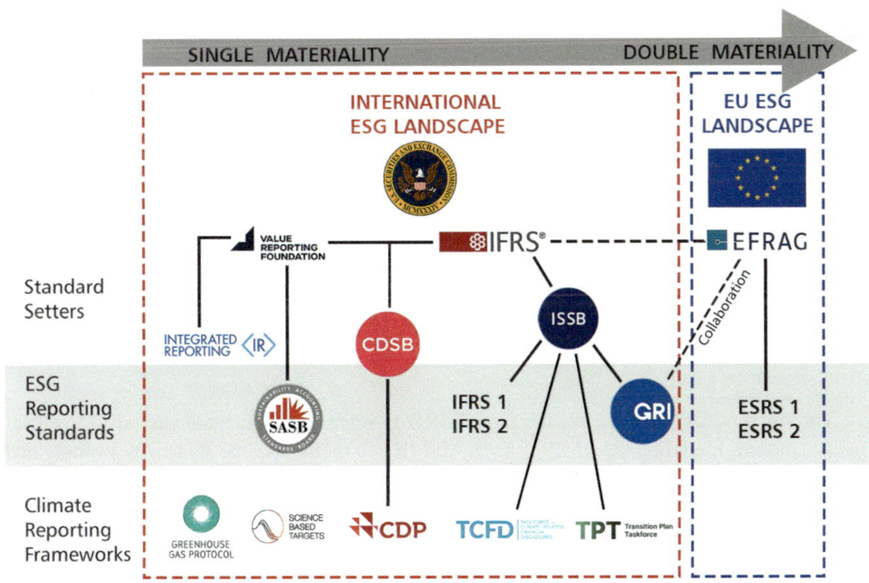

Graphic 1 Summary ESG standards and frameworks landscape in 2024. Source: Author.

Regarding standards, it gets somewhat more complicated as the landscape is almost constantly changing. The International Sustainability Standards Board (ISSB) has become one of the most important standard and framework setters by publishing the International Financial Reporting Standards S1 and S2 (IFRS, established 2001) in 2023. The ISSB consolidates a range of frameworks and standards, including the Sustainability Accounting Standards Board (SASB, established 2011), which provides industry-specific sustainability accounting standards, the Global Reporting Initiative (GRI, established 1997), which focuses on voluntary sector-based sustainability reporting, and from 2024, the TCFD. Also linked to the ISSB through the IFRS is the Climate Disclosure Standards Board (CDSB), which manages the CDP framework and the Value Reporting Foundation (VRF), which is an independent non-profit managing the Sustainability Accounting Standards Board as well as Integrated Reporting, which has been absorbed into the IFRS.

In mid-2023, Europe passed into law the Corporate Sustainability Reporting Directive[1] (CSRD, replaces the NFRD) aimed at enhancing sustainability reporting requirements through the European Sustainability Reporting Standard (ESRS). The CSRD/ESRS is the focus of this ESG reporting manual as it is (by far) the most comprehensive of the ESG standards and is also linked to the Sustainable Finance Disclosure Regulation (SFDR), Pillar 3 (the third accord from the European Banking Authority), and various EU Commission regulations (referred to Commission Delegated Regulations, CDRs) and EU Climate Laws, which are all currently in force across Europe and also referenced through this manual.

Additionally, the European Union Deforestation Regulation[2] (EUDR) demands that organisations trading in coffee, cocoa, palm oil, cattle, rubber, soya and wood must report that their value chains are engaged with deforestation-free processes. Importers inside the EU must comply with the rules, which will affect raw commodities as well as processed goods liked chocolate, leather, tyres or furniture[3].

---

[1] Throughout this book we use the term CSRD to refer to the delegated act and all subsequent standards including the ESRS, LSME, VSME and sector specific standards promulgated by EFRAG.

[2] For details of the regulations see https://environment.ec.europa.eu/topics/forests/deforestation/regulation-deforestation-free-products_en

[3] For more information and guidance for reporting see https://www.intracen.org/news-and-events/news/new-step-by-step-guide-on-eudr

In addition to these global standards, the United Nations Sustainable Development Goals (SDGs) also serve as a framework for companies to report on their contributions to global sustainability goals. To a greater or lesser extent, they all provide guidelines for reporting on a range of ESG factors, such as carbon emissions, diversity and inclusion, human rights, and board diversity.

One of the major sources of confusion in ESG reporting is the rapid development of new standards in different countries. For example, in early 2024, the US Securities and Exchange Commission (SEC) (see summary Graphic 2) introduced a climate reporting standard and has since been embroiled in legal challenges. The United Kingdom is in the process of developing a national sustainability reporting standard based on the IFRS S1 and S2 called the UK Sustainability Reporting Standards[4] (draft due in 2024) and related Sustainability Disclosure Requirements[5] (UK SDR). Similarly, Japan, Australia, and Switzerland, to name only a few countries, are also developing their own standards.

Arguably, the most important difference between the various standards is the concept of single or double materiality, which is discussed in detail in Chapter 4: Single and Double Materiality Assessment. Simply put, materiality is a 'test' applied by organisations to assess the severity and/or relevance of an impact or action during the performance of operations on either the environment or finances.

In 2023, the GRI was the most widely used standard for sustainability reporting, with over 10,000 organisations in over 100 countries using it to report on their ESG performance. However, this is likely to rapidly change as the CSRD comes into effect in 2024 where an estimated 11,000 organisations will need to prepare ESG reports based on the ESRS by March 2025. Moving forward, the CSRD will embrace more organisations as the qualification boundaries decrease, as can be seen from Graphic 3:

---

[4] For further information, see: https://www.gov.uk/guidance/uk-sustainability-disclosure-standards

[5] For further information, see: Financial Conduct Authority (FCA), 2023, *Policy Statement PS23-16: Sustainability Disclosure Requirements (SDR) and investment labels.*

Table 1 ESG laws, regulations, and frameworks summary in 2024.

	Laws and regulations		
	European Union		**USA**
	CSRD	**SFRD**	**SEC**
	Corporate Sustainability Reporting Directive: European Sustainability Standards	Sustainable Finance Disclosure Regulation	Securities Exchange Commission: Enhancement and standardisation of climate-related disclosures
Estimated companies reporting in 2024*	11,000+	15,000+	Not applicable yet in 2024
GHG emissions approach	Mandatory Scope 1, 2, and 3	Mandatory Scope 1, 2, and 3	Mandatory Scope 1 and 2 Scope 3: only if deemed financially material and/ or the organisation has a publicly stated goal that includes Scope 3
GHG disclosure information	Double materiality assessment to cover 220 data points and report all metrics, including risks and opportunities, value chain, intangible resources, and comprehensive transition targets	Nine GHG indicators including intangible resources and investment decision-making in line with EU taxonomy	Only GHG scopes, other metrics are optional
Materiality approach	Double materiality	Double materiality	Single materiality
Scenario analysis	Dependent on risks identified and must include alignment with 1.5 C scenario. Must report on forward-looking targets corresponding to EU 2050 net-zero target	Must report on forward-looking targets corresponding to EU 2050 net-zero target	Not required unless the organisation has already established the practice but is recognised as a 'reasonable' way to inform investors about resiliency
Social disclosure information	High priority, 400 data points required for own workforce, value chain workforce, communities, and end-users/ consumers	Simplified key data points relating to human rights, due diligence, and governance	Not included
Third-party assurance required?	Yes, and report to be made public and machine readable	Yes, and report to be made public	Yes

Source: Author.

Frameworks and standards		
Global		
ISSB/IFRS	**GRI**	**SASB**
International Sustainability Standards Board	Global Reporting Initiative	Sustainability Accounting Standards Board
5000+	10,000+	5000+
Mandatory Scope 1, 2, and 3	Mandatory Scope 1, 2, and 3 (although omission can be justified)	Voluntary Scope 1, 2, and 3
Must report progress on all GHG metrics and indicators	Double materiality assessment to identify appropriate metrics to be reported over time	Industry-specific indicators may include GHG scopes, which are generally optional
Single materiality Included in resiliency analysis and aligned with 1.5 C scenario	Double materiality Included in strategic plans, capital requirements, transition plans, risk and resilience assessments	Double materiality Only if relevant to forward-looking forecasts, otherwise not included
High priority, 400 data points required for own workforce, value chain workforce, communities, and end-users/ consumers	High priority, 400 data points required for own workforce, value chain workforce, communities, and end-users/ consumers	Industry-specific indicators may include a range of social and human capital indicators
Under review but should have a strong internal governance process and be made public	Under review but should have a strong internal governance process and be made public	Not required, but should be public

Table 2 ESG regulation status by jurisdiction, April 2024.

	Sustainable taxonomies	Corporate climate disclosures	Corporate ESG disclosures	ESG fund requirements
Europe				
European Union	✓	✓	✓	✓
Russia	✓	✓		
Switzerland		✓		
Turkey			✓	
United Kingdom	✓	✓	✓	✓
Americas				
Argentina			✓	✓
Brazil	✓	✓	✓	✓
Canada		✓	✓	
Chile	✓		✓	✓
Colombia			✓	
Mexico			✓	
Peru			✓	
United States		✓	✓	✓
Asia Pacific				
Australia			✓	
Hong Kong			✓	✓
India				✓
Indonesia				✓
Japan		✓		✓
Malaysia			✓	
New Zealand		✓		
Philippines			✓	✓
Singapore		✓		✓
Taiwan			✓	✓
Thailand			✓	
Vietnam			✓	
Middle East, Africa				
Egypt			✓	
Nigeria			✓	

Source: https://www.sustainablefitch.com/banks/esg-regulations-reporting-standards-tracker-march-2024-highlights-18-03-2024

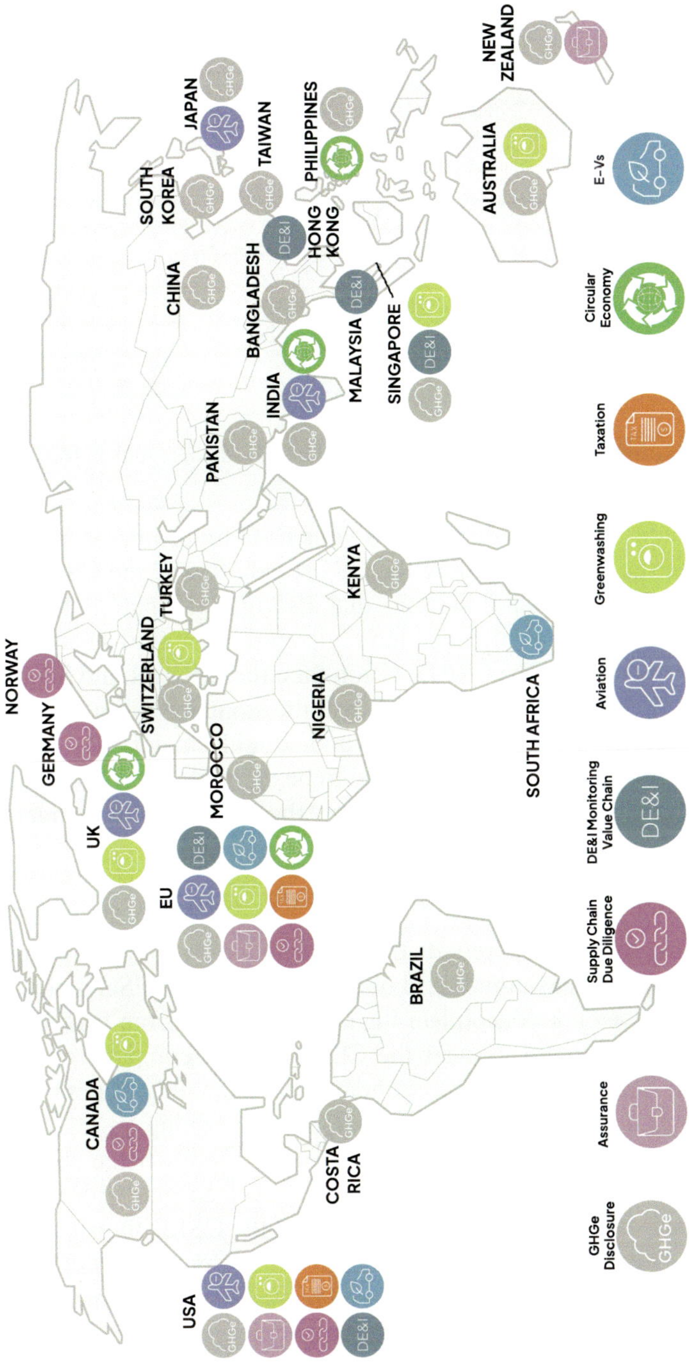

Graphic 2 Major ESG-related regulations by topic (June 2024). Source: Author.

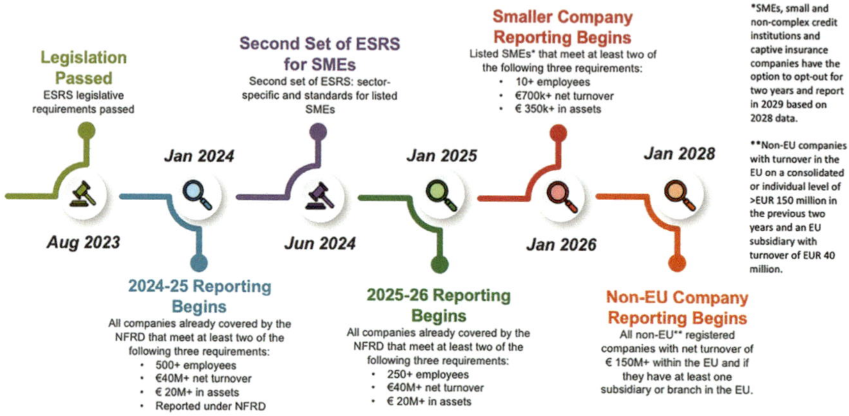

Graphic 3 CSRD/ESRS timeline until 2028. Source: Author.

Which standard should you apply to your ESG reporting? It is likely that we will see the IFRS (including the GRI standard) and the CSRD become the two dominant standard systems that also underlie national or regional initiatives. If you are an organisation registered in, or with a heavy reliance on Europe, then apply the CSRD. If not, then the IFRS S1 and S2 are likely to be more appropriate. Having said that, the standard 'level' of the CSRD is considered higher due to the rigour of double materiality testing and range of reporting indicators that need to be applied. So, if you are unsure, then err on the side of caution and follow the CSRD/ESRS standard.

1.3 ESG Reporting for Consolidated Entities

Consolidation of financial reports through parent or holding companies may mean that your organisation is still obliged to submit an ESG report (if your organisation does not consolidate financial reporting from diverse entities, then skip to section 1.4). Under the CSRD, parent companies must either include subsidiaries in their consolidated sustainability statement, which probably means using a reporting-exemption, or issuing an individual sustainability statement in their own management report. If the sustainability statement is consolidated at parent-company level, such as with an EU holding company, and the company does not meet the CSRD threshold, then the subsidiary is exempted from publishing a stand-alone report. This exemption does not apply to public interest enterprises (PIEs) that are large, listed companies on an EU regulated market. For those, whether they are consolidated by their parent company or not, a sustainability statement in their own management report is still required.

Furthermore, if there are significant differences between the sustainability risks or impacts of the group and those of its subsidiaries, the parent company should provide, in the consolidated sustainability statement, an adequate understanding of the risks or impacts of its subsidiaries – including information on their due diligence processes where appropriate.

Exempted EU-based subsidiaries are subsidiaries that will be consolidated; their sustainability information will be incorporated in the report of their parent company. In this case, they will still need to:

a) report the name and registered office of the parent company that is reporting on the sustainability information at group level;

b) publish the weblinks to the consolidated management report of their parent company, with clear information and instructions on how to access the consolidated management report, especially if it is not available online;

c) clearly state that they are exempted from sustainability reporting, in their management report.

A form of the subsidiary exemption is currently applicable on a temporary basis to EU subsidiaries or subgroups, with a parent company headquartered outside the EU. Until 2030, these companies may be exempted from individual reporting requirements if they are incorporated into a report using ESRS and 'artificial consolidation', meaning that the EU subsidiaries and subgroups are consolidated. The entity responsible for consolidating other EU-based subsidiaries is the one with the highest turnover within the EU. The purpose of this transitional provision is to reduce the burden for the non-EU groups that do not have a single EU-based holding entity that controls the group's EU entities.

The ultimate parent companies that are not based in the EU may also consolidate their EU subsidiaries in a global sustainability statement. However, such consolidation exempts the EU subsidiaries from the obligation to produce their own sustainability statements only if the global consolidated statement is drawn up in accordance with the ESRS, or in an equivalent manner.

In summary:[6]

- For non-EU parent groups with large subsidiaries or branches in the EU that, when consolidated, meet the CSRD threshold, they must submit a full ESRS sustainability statement in 2026 on a consolidated basis and from 2030 on a subsidiary-basis.
- For non-EU parent groups with small-, medium-, or listed subsidiary, then the subsidiary must submit full ESRS (or equivalent standard) sustainability statement from 2025.
- For non-EU parent groups with net turnover >€150m for two consecutive years, a full ESRS (or equivalent standard) sustainability statement from 2028 must be submitted.

1.4 Solving the Resource Dilemma: Voluntary vs. Statutory in Five Questions

Let us take a dive into a big decision – voluntary vs. statutory ESG reporting – what is the rub and how can your organisation come out with the best of both worlds?

First, let me put a bunch of sustainability reporting standards and frameworks into the same bucket – B-Corp, GSTC, GRI, ISO, SASB, ISSB, and IFRS – and call them voluntary standards (there are many more that can be added but the list is not the point here). Some are highly sector specific and others are general. Some have been around for decades and others are relatively new. What is important is that they are not statutory. It is a choice to engage with them, and many companies choose not to.

Then we have statutory standards, and this group is dominated by new entrants, most notably the CSRD/ESRS, SFRD, and the SEC. But more are coming and the need to fulfil them is causing quite a bit of concern, even panic, around the world. How much will they cost and can you even do it? If you are already part of a voluntary reporting program, what is the extra cost? Is this the end of the voluntary program that you have worked so hard on?

Size matters:

If your organisation has more than 500 employees and is based in Europe, then you must comply with the CSRD/ESRS from this year. There are a couple

[6] For further details, see Commission Delegated Regulation (EU) 2023/2772 supplementing Directive 2013/34/EU of the European Parliament and of the Council as regards sustainability reporting standards and as published in the Official Journal of the European Union on 22 December 2023.

of other qualifying criteria to check as well, but if you had to report under the (now defunct) NFRD, you have to fulfil the CSRD. By 2025–2026, the number of staff limit will decrease to 250 and/or €40M+ net turnover and/or € 20M+ in assets. It is important to note that the CSRD also includes a statutory standard for Listed SMEs, a voluntary standard for non-listed SMEs, and specific sector standards by 2027.

But here is the first rub. Even if you do not qualify as a reporting entity in your own right, you are probably doing business with someone who does, and that catches you as part of their value chain. Reporting companies must consider upstream and downstream impacts, including issues like workers' rights and waste management. For now, let us assume that you have less than 500 employees and you are asking yourself one of these two questions:

1. We are likely to have to report in the next year or two as we have 250 staff, so it makes sense to start preparing now but what is the best lead-in?

2. We have put a lot of effort into a voluntary program, but now we are afraid we are going to lose all that work. Is there a way to do both?

Of course, there is a third question, 'How deep is the hole I have to stick my head in to ignore ESG reporting?', but let us not go there.

Question 1: Flexibility vs. Compliance

The first decision is critical as it will set the tone for all subsequent questions, *'Looking across your entire organisation, what is the required scope and detail of your policies and extent of your due diligence?'* If you answer to either of those issues, 'Yes, I think this is going to get complicated', then statutory reporting should be your focus to start with. Statutory reporting mandates compliance with specific regulations, limiting flexibility but ensuring adherence to a standardised set of reporting requirements. So, the more complicated your organisation, the less likely a voluntary reporting system is going to provide a more competitive cost–benefit outcome at the outset. Answering this question should probably involve a consultant who has an idea of the range of challenges you may face in various sectors.

If your answer is along the lines of, 'Our business units are aligned around a few core competencies that drive market access and/or specific product/service deliverables', then you have options on the table. This is because your policy and due diligence landscape is quite narrow and voluntary reporting is applicable in this context. Voluntary reporting can offer flexibility in choosing reporting frameworks and indicators aligned with the organisation's specific sustainability goals.

Question 2: Stakeholder Engagement

Now we come to one of the really big deciding factors – the extent of your value chain and associated level of stakeholder engagement required. Voluntary reporting allows organisations to engage with stakeholders to determine material ESG issues and reporting priorities. However, most voluntary standards prioritise financial impact perspectives meaning that they may not fully capture all stakeholder concerns or industry-specific nuances. However, it is important to note that there is no reason why you cannot take a double materiality approach and include physical impacts – just about any reporting system will embrace that idea.

On the other hand, statutory reporting demands double materiality from the start and this causes more organisational stress than anything else. Additionally, approaching your value chain for information that many will feel is sensitive and/or a waste of time and energy is not easy and will require extensive rapport-building.

To help guide, you ask yourself, *'Can I easily identify a member of senior management in 80% of my value chain partners?'* If you can say yes, then you have a tightly knit chain and access to data will be much easier than if you say no. If you do say no, then a voluntary reporting approach will help you build those relationships through information sharing.

Question 3: Granularity and Detailing

Now there are a group of considerations that are largely interrelated but are treated in different ways by voluntary reporting systems: Transparency, Credibility, Data Quality, Assurance, Risk Management, and Legal Compliance.

I mentioned before that your value chain partners might not buy in to your sustainability approach and you will need to make a serious effort to convince them. One of the first hurdles is overcoming transparency-reticence; this is best achieved by showcasing credibility. It is the old carrot and stick approach, but it works if done sensitively. Remember that there are more things that you need than just transparency – ensuring data quality and consistency are vital in both voluntary and statutory systems as both will require some sort of assurance, probably by auditors. Assurance then raises some sticky questions about risk (and opportunity) management as well as our legal friends demanding compliance measures… all of which are deeper and more serious in statutory systems.

So how to differentiate here? The best approach is to look outwards at your competitors and the broader sector-approach, *'To what extent is data reporting granularity the standard in your industry?'* For example, sectors with high adoption of ISO standards will always tend towards statutory reporting as the norm, and again, some advise from an experienced consultant in your sector can be handy here.

On a final note, whether you are tending towards voluntary or statutory reporting, you should take your GHG transition planning very seriously, possibly adding an additional voluntary reporting framework (for example, the SBTi or the Greenhouse Gas Protocol) to enhance credibility. For any company operating in or near nature, also add the Taskforce for Nature-related Financial Disclosures (TNFD) or similar for the same reason.

Question 4: Support Required

ESG reporting is a cost and time burden. Voluntary reporting frameworks know that and have developed a variety of tools to help you encourage and cajole your staff and value chain into positive action. Statutory reporting systems will involve additional compliance costs to your time and any capital investments (for example, ESG data management systems). So, if you think your organisation could do with some hand-holding at times, then voluntary reporting is probably a good idea. Either way you may well need to access sector-specific skills and resources through a dedicated consultancy.

There is also a related question here about your long-term strategy. Voluntary reporting can support the integration of sustainability into the organisation's core strategy, fostering a culture of continuous improvement and innovation. In contrast, statutory reporting tends to focus on providing a structured framework for measuring progress towards sustainability goals, aligning reporting with long-term strategic objectives. Ultimately, they are aiming for the same outcome using slightly different pathways.

To help navigate question 4, ask yourself, *'How do we expect to be improving our sustainable business in 5-years; through internal innovations or by external regulation?'* The internal innovator will have the market advantage, but the external regulation approach will probably have a cost advantage.

Question 5: Business Development

Perhaps unsurprisingly, we have come down to strategy and business development – the source of your competitive advantage and market positioning. Over the last 20 years, we have seen early adopters embrace voluntary reporting as a way to differentiate their organisations, attracting socially responsible investors, and enhancing brand reputations. But statutory reporting might rain on their parade as a commitment to sustainability becomes compulsory, thus rippling through market positioning and investor perceptions.

One of the strengths of voluntary reporting systems is their flexibility across international operations and jurisdictions. Harmonising reporting practices can streamline processes and enhance consistency in ESG disclosures, facilitating global stakeholder engagement and compliance. If you do not have a large amount of leverage within your value chain, then voluntary reporting might be the only viable option to begin. In such a situation, you might also need some extra time to see how regulations evolve and if any trends emerge within your sector(s). So, our question here is, *'To what extent can we influence our value chain towards sustainability and if not, could we consider a low-cost market position rather than a differentiated position?'* There is a balancing act between voluntary and statutory reporting that can help organisations adapt to changing expectations, address emerging ESG issues, and stay ahead of regulatory developments. Or it could sink them.

Take the time to carefully consider each of the above questions to identify your priorities and then make informed decisions between voluntary sustainability reporting and statutory sustainability reporting standards. Aligning your operational practices with sustainability goals, stakeholder expectations, and regulatory obligations is very likely to become the key core competence of the future.

1.5 Understanding Your ESG Report Audience

The natural next step when considering your approach to ESG reporting is to get to grips with your ESG report audience. Understanding the various perspectives of stakeholders will play an important role from your initial materiality consultations, through the data acquisition and analysis processes, to final report preparation and ongoing continuous improvements. Graphic 3 is a summary review and comparison of the major groups and their engagement with the ESG reporting process. There are ten stakeholder types that we will partition into four groups and they will all have something to say at some point.

During your ESG report development process, there is a greater detail-orientation by management and value chain partners and very little involvement of regulators. However, external audiences become highly engaged with detail once your report is published.

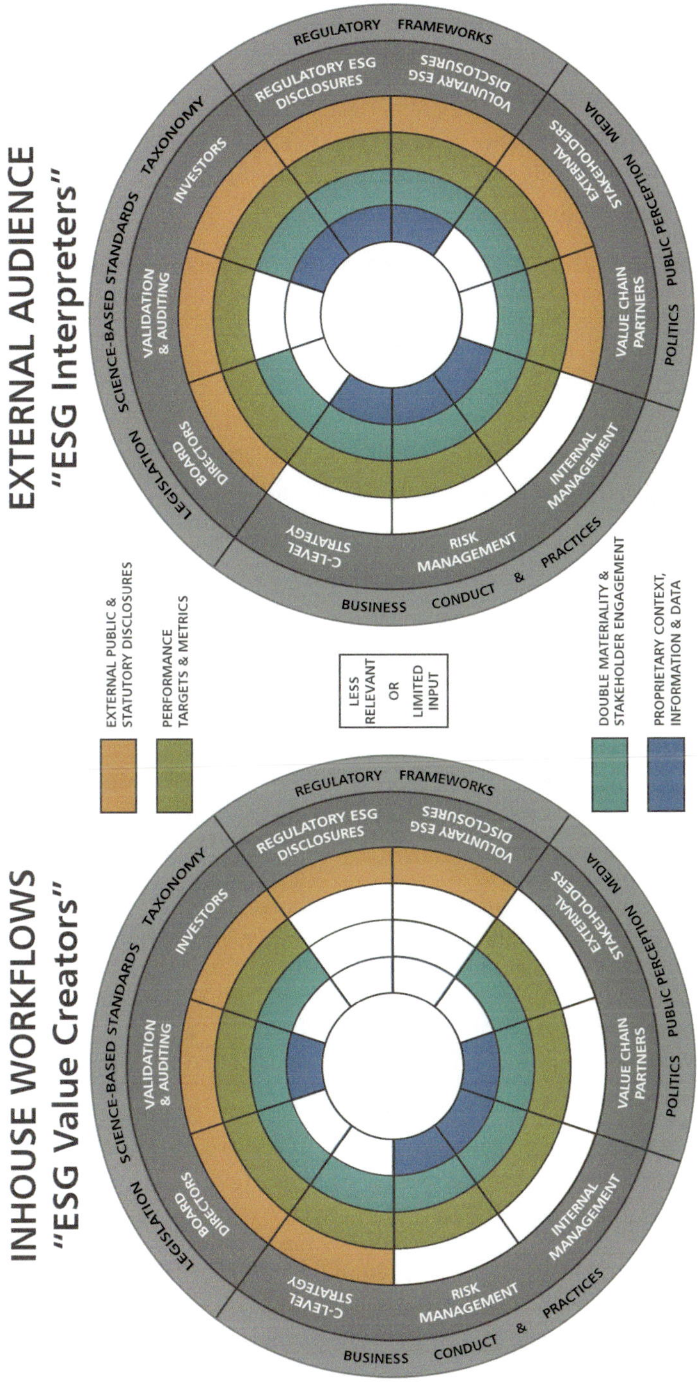

Graphic 4 ESG report audiences and their interests. Source: Author.

PHASE 1: When creating your ESG report, there is relatively little concern about public disclosures and only a few stakeholders will actually be involved in data collation.

- **Organisational performance**: The Board of Directors, auditors, and investors tend to be focused on ensuring compliance with standard or framework methodologies.
- **Business conduct and practices**: Your senior and internal management staff tend to be absorbed with trying to find relevant data; bigger picture issues tend to be a low priority.
- **Public perception and media**: Aside from making a declaration that your organisation intends to produce the most accurate ESG report possible, most external energy is dedicated to convincing value chain partners to dig into their organisations for relevant data.
- **Regulators**: Their only concern is adoption of their standard and how organisations are positioning their ESG reporting process in the public domain.

PHASE 2: Now you have done the hard work, published your report, and everyone wants to grab a copy. But they have different reasons for doing so:

- **Organisational performance**: The Board of Directors are now interested in how they can pitch to investors and minimise risk both the organisation as well as themselves. Investors are now relating shareholder value and performance for you and alternative investments.
- **Business conduct and practices**: For your internal management, preparing the next report is already underway, with your corporate level now taking an interest in the data.
- **Public perception and media**: Publicity will spotlight your strengths and weaknesses, which your value chain will certainly be looking at, as well as your competitors.
- **Regulators**: The production of your ESG report is also a test of the regulators ability to formulate robust ESG reporting standards, so they now take a very detailed interest.

The important lesson here is that your report generation process should be treated differently from the report preparation process, and that both should be covered by due diligence and rigorous compliance. This is often a tricky issue for companies as the marketing department almost inevitably wants to produce the public report, but as the sustainability director, it is your job to make sure they do not forget the full range of stakeholder perspectives.

1.6 How to Begin – Your Green Team

Complying with any ESG reporting standard requires a substantial effort from multiple staff members and should eventually involve most of your organisation. For large organisations, it is common to create a dedicated sustainability team who have access to senior management across all functional areas. At the very least, a sustainability manager needs to be directly responsible for all ESG matters and who has the authority to petition for information and expect to receive it in a timely manner.

Whether you need to prepare for a statutory ESG report or voluntarily reporting as part of a business strategy or that you are a member of a value chain of a company that must report, it makes sense to begin with nominating an internal Green Team who are fully and publicly supported by top management.

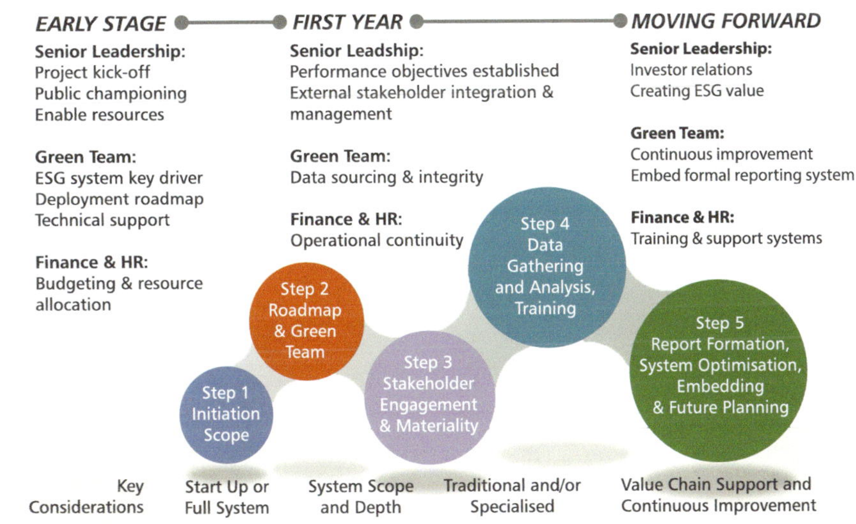

Graphic 5 ESG reporting roadmap. Source: Author.

Green Teams (or similar roles) are key drivers responsible for coordinating planning, executing, monitoring and submitting your ESG report (see

Graphic 5), as well as subsequent ESG performance improvements. The Green Team must have at least one person with executive authority (normally at board level), management representatives from all major functional areas and who all possess strong critical-thinking skills. The Green Team needs to set the foundation of the ESG reporting system and includes overcoming the initial challenge of data gathering, aggregation, and analysis, see Graphic 6 for a typical rollout timeline.

The first major recommendation the Green Team normally makes is the choice of ESG reporting data management system or dedicated software – this is covered in more detail in Chapter 4: Data Collection and Analysis. It is worth considering a number of alternatives that should be easy to use and intuitive for categorising relevant information for each of the reporting criteria and specific indicators. You may well need a system that is compliant with other standards and reporting criteria, especially if you have international subsidiaries or partnerships. It is also critical that you have access to skilled guidance on how to implement the reporting system and embed it wherever possible to ensure efficiencies.

Simultaneously, two important processes must begin; human resources must update all workforce records and ensure an effective diversity, equity, and inclusion (DEI) system is in place, and senior management develops a clear ESG policy and ensures that it is reflected in all organisational policy and governance documents. Both of these tasks typically take multiple months to complete, so it is essential that they begin as soon as possible after the formation of the Green Team, and they will form the foundation of your approach to governance, strategy, impacts, risks, and opportunities, which are covered in detail in Chapter 5: General Disclosures.

Graphic 6 Starting your ESG reporting process. Source: Author.

The next step for the Green Team will be to build your greenhouse gas (GHG) inventory, which is covered in detail in Chapter 6: Climate Change and involves identifying, categorising, and quantifying each source and reliance scale of each GHG emission associated within your organisation. This includes all internal (Scopes 1 and 2) and external (Scope 3) emissions as well as defining all risks and opportunities associated with each emission. While building the GHG inventory, you will need to identify a GHG baseline year for which you have comprehensive records and is since 2015. The goal is to comprehensively monitor and report your organisations interface with climate change through a scenario and resilience analysis that allows you to set target emission levels and establish a transparent transition plan to reach net zero emissions before 2050.

Your journey into ESG reporting has now genuinely begun, with many more reporting groups and criteria to review, which are covered in Chapters 7–16. Once you have identified all relevant reporting criteria, collected and analysed the data, assessed and reviewed your conclusions, it is time to prepare your ESG report, which is covered in Chapter 16: ESG Report Preparation.

1.7 Integrated Due Diligence Process

Sustainability and due diligence are inherently interlinked, in that one cannot exist without the other. The EU taxonomy requires compliance assessment with minimum safeguards as part of the taxonomy-alignment process for organisations. An economic activity can only be counted as taxonomy-aligned if is carried out in compliance within the minimum safeguards, which are defined in article 18 of the taxonomy and further explained by the EU Platform on Sustainable Finance. Non-compliance with minimum safeguards is determined if one of the following criteria applies to the company:

- The organisation has not established an adequate sustainability due diligence process, as outlined in the UN Guiding Principles for Business and Human Rights and the OECD Guidelines for Multinational Enterprises; or
- There are clear indications that the organisation does not adequately implement human rights due diligence. For example, it does not take appropriate action to address identified potential and actual impact (this could be, but does not need to be, confirmed by final court rulings, OECD National Contact Point cases).

The EU taxonomy minimum safeguards refer to the same process of sustainability due diligence as the CSRD/ESRS (see 5.2.6, GOV-4: Statement on Sustainability Due Diligence), and the proposed Corporate Sustainability Due Diligence Directive (CSDDD), and overlap with the OECD Due Diligence Guidance for Responsible Business Conduct.[7] The CSRD/ESRS requires organisations to provide 'a description of the due diligence process implemented with regard to sustainability matters', by which organisations must identify, prevent, mitigate, and account for material actual and potential negative impacts on the environment and people connected with their operations. Due diligence therefore begins with responsible business conduct, strategy, policies and management systems, and extends throughout all functions and activities. So, it is imperative that your Green Team consider how to integrate due diligence either in parallel or combination throughout each of the following stages of ESG reporting covered in this book.

INFORMATION BOX

Corporate Sustainability Due Diligence Directive (CSDDD)[8]

After its adoption by the European Parliament on 24 April and the EU Member States' vote on 25 July 2024, the CSDDD is now in force. Financial institutions are included in the scope but are exempt from conducting due diligence on their clients, subject to a review clause within two years. The Directive contains provisions to facilitate compliance and limit the burden on companies, both in scope and in the value chain, and apply as follows:

- Large EU limited liability companies & partnerships: +/- 6,000 companies - >1000 employees and >EUR 450 million turnover (net) worldwide.
- Large non–EU companies: +/- 900 companies - > EUR 450 million turnover (net) in EU.

[7] For further information, see: OECD, 2018, Due Diligence Guidance for Responsible Business Conduct, and OECD, 2023, ACHIEVING ALIGNMENT: Syncing EU due diligence legislation with the updated OECD Guidelines.

[8] For further information see: https://commission.europa.eu/business-economy-euro/doing-business-eu/sustainability-due-diligence-responsible-business/corporate-sustainability-due-diligence_en

- SMEs: Micro companies and SMEs are not covered by the proposed rules. However, the Directive provides supporting and protective measures for SMEs, which could be indirectly affected as business partners in value chains.

The scope of your due diligence process extends from internal management to all stakeholders, and is ongoing.

- *Internal scope* embraces mindsets, behaviours, and practices necessary to minimise impacts on human rights and the environment that should be grounded and are expressed through strategy coherence with human rights and environmental commitments, allocation of responsibilities, integration in decision-making, budget allocations and oversight, and effective information flows to the senior management and the board of the company.
- *External scope* involves engaging with stakeholders to identify, address, and remediate actual and potential impacts, and encourages open expression of their views and experiences.

The due diligence process should first identify instances where your organisation is, or is at risk of being, involved with negative impacts on the environment or people, and then assess the nature and severity of actual or potential impacts (context, causes, severity, etc.) by categorising them as: caused by, contribute to, or directly linked. For each impact, then specify actions to cease, prevent, and mitigate adverse impacts that are tracked to ensure effectiveness. Finally, the due diligence process requires clear communication on human rights impacts and any remediation, including the establishment of a grievance mechanism.

In each case, the *severity* of an impact is determined by how grave the harm is or would be, how widespread the harm is or would be, and how hard it is or would be to put the harm right. Any one of these factors may be sufficient to judge the impact to be severe. Where impacts are assessed to be potential in nature, a description of their likelihood is also required.

Tracking and communicating the effectiveness of your organisation's efforts is expected to be based on a mix of qualitative and quantitative indicators and draw on internal as well as external feedback, including from affected stakeholders. Ultimately, tracking effectiveness should be focused on evaluating whether the company's activities are leading to better outcomes for stakeholders.

Table 3 Due diligence impact categories and actions.

<table>
<tr><th>Caused by your organisation</th><th>Contributed to by your organisation</th><th>Directly linked to your organisation</th></tr>
<tr><td>Causing means that your organisation's actions (or omissions), on their own, result in an impact</td><td>Contributing means that your organisation facilitates or incentivises a third party to cause an impact, or that its actions (or omissions), combined with those of others, result in the impact</td><td>Direct linkage means that, while your organisation does not cause or contribute to it, an impact is nevertheless directly linked to operations, products, or services through a business relationship (which may be a direct or remote relationship in the value chain)</td></tr>
<tr><td>If your organisation causes or may cause an impact, it should take action to cease or prevent the impact</td><td>If your organisation contributes, or may contribute, to the impact, it should take action to cease or prevent its contribution, and use its leverage to mitigate any remaining impact</td><td rowspan="2">If an impact is directly linked to your organisation's operations, products, or services, it should, first and foremost, use leverage with third parties to seek to prevent or mitigate the impact. The more complex or systemic the issue, or the more remote in your value chain, the more likely that exercising leverage will require some form of collaboration with others, whether industry peers or other public, private, international, or civil society organisations. It may also require forms of collective action</td></tr>
<tr><td colspan="2">If your organisation has caused or contributed to an adverse impact, it should provide for, or cooperate in securing remedy for those harmed, whether by itself or in collaboration with others</td></tr>
</table>

Source: Author.

TIPS AND TRICKS

Minimum safeguards:

Pursuant to the EU taxonomy, SFDR, and ESRS, organisations should disclose how and to what extent they comply with minimum safeguards and international documents referring to human and labour rights such as:

- The UN Guiding Principles on Business and Human Rights, covering corporate responsibility to respect human rights and human rights due diligence.
- The Organisation for Economic Co-operation and Development (OECD) Guidelines for Multinational Enterprises, a set of recommendations for responsible business conduct in a global context. The OECD Guidelines incorporate UN Guiding Principles and extend the human rights due diligence concept to employment and industrial relations, environment, bribery, consumer interests, science and technology, and competition and taxation).
- The eight fundamental conventions of the International Labour Organisation, legal instruments covering fundamental principles and rights at work (such as freedom of association, elimination of all forms of forced or compulsory labour, effective abolition of child labour, and the elimination of discrimination in respect of employment and occupation). The UN Guiding Principles and the OECD Guidelines also refer to these international human rights.
- The International Bill of Human Rights.

1.8 ESG Report Costs

There are basically two major cost-centres associated with creating an ESG report: those of staff hours (including your Green Team) and your ESG data management system, and neither are inexpensive. There are also indirect costs, for example, value chain staff time that need to source data, especially when it is not convenient to handle. Plus, there are transitioning costs, which can become substantial capital outlays, so it is wise to integrate your finance and accounting department into your Green Team or at least have a close rapport.

There are many software solutions on the market, offering specific tools like GHG emission calculations to fully integrated ESG data management and report generation systems. Deployment and licensing software costs vary from a few thousand to hundred Euros to over 100,000 Euro per annum. However, ESG reporting does have a range of benefits, from identifying operational cost savings to driving innovation, so you might find yourself having to 'sell' the idea of software to the finance folks, and to help you, the following Graphics 7 and 8 contain 50 examples of ESG-based initiatives with positive return on investment, albeit over time:

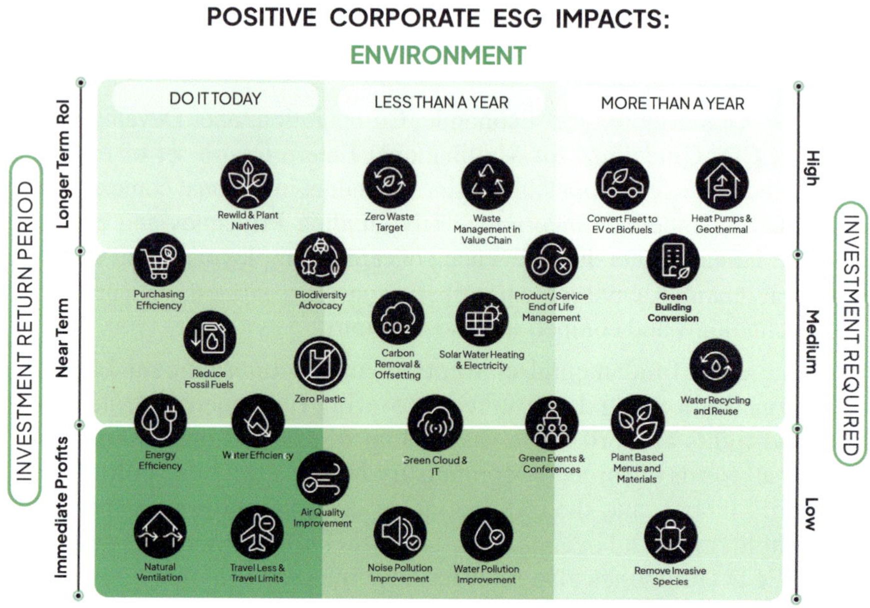

Graphic 7 25 positive corporate ESG impacts: Environment. Source: Author.

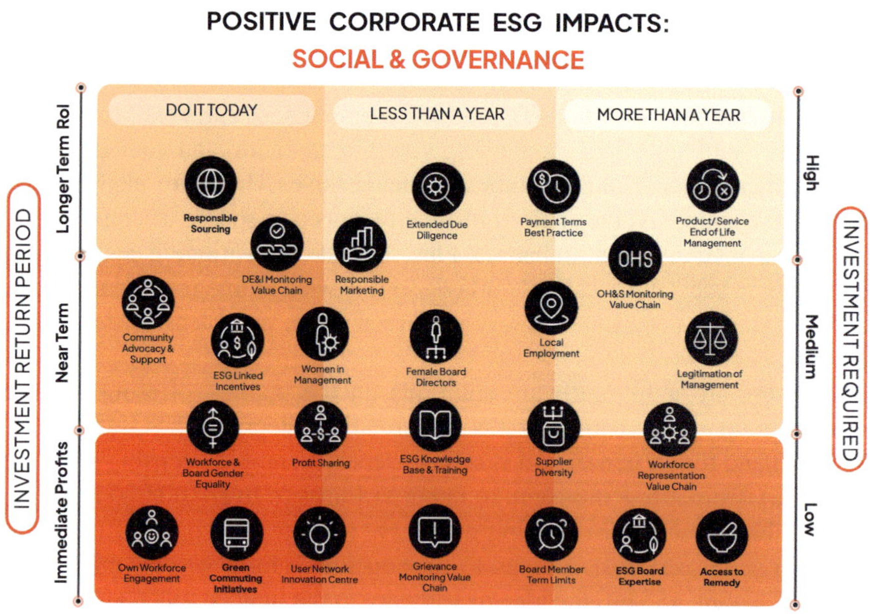

Graphic 8 25 positive corporate ESG impacts: Social and governance. Source: Author.

Table 4 Positive corporate ESG impacts and RoI: Environment.

Sustainability indicator	Example action	Investment return period	Investment required	Implementation period	Long-term impact
GHG management, emissions intensity, and GHG transition plan	Energy efficiency	Near term	Low	Today	High
	Fossil fuels use reduction	Immediate	Moderate	Today	High
	Carbon removal and offsetting	Near term	Moderate	<1 year	Low
	Convert fleet to EV or biofuels	Longer term	High	>1 year	Moderate
	Green cloud & IT	Near term	Low	<1 year	Low
	Plant-based menus and materials	Immediate	Low	<1 year	Low
	Green events and conferences	Immediate	Low	<1 year	Low
	Travel less and travel limits	Immediate	Low	Today	Moderate
	Natural ventilation	Immediate	Low	Today	Low
Renewables %	Heat pumps and geothermal	Longer term	High	>1 year	Moderate
	Solar water heating and electricity	Near term	Moderate	<1 year	Moderate
Air quality improvement	Green buildings including links to renewables %	Near term	Low	Today	Low
Green building conversion		Near term	High	>1 year	Moderate

(Continued)

Table 4 (*Continued*).

Sustainability indicator	Example action	Investment return period	Investment required	Implementation period	Long-term impact
Noise pollution improvement		Immediate	Low	<1 year	Low
Water pollution improvement		Near term	Low	<1 year	Moderate
Water recycling and reuse		Near term	Moderate	>1 year	Moderate
Water efficiency (consumption reduction)		Immediate	Low	Today	Moderate
Biodiversity improvements	Biodiversity advocacy	Near term	Moderate	<1 year	Moderate
	Remove invasive species	Immediate	Low	>1 year	Moderate
Invasive species management	Rewild and plant natives	Immediate	High	>1 year	Moderate
Materiality circularity rate	Zero waste target	Near term	Moderate	<1 year	High
Resource efficiency & hazardous waste management	Value chain waste management plan	Longer term	High	<1 year	High
	Purchasing efficiency	Near term	Low	Today	Low
Product/service lifetime management	Product lifetime management with customers	Near term	Moderate	>1 year	Moderate
	Zero plastic	Immediate	Moderate	<1 year	High

Source: Author.

Table 5 Positive corporate ESG impacts and RoI: Social and governance.

Sustainability indicator	Example action	Investment return period	Investment required	Implementation period	Long-term impact
CEO-worker pay ratio	Profit sharing	Immediate	Moderate	>1year	Moderate
Gender pay gap	Workforce and board gender equity	Immediate	Moderate	Today	High
Female board directors	Female board directors	Near term	Moderate	<1year	Moderate
Women in management	Women in management	Near term	Low	<1year	Moderate
Workforce representation in house and value chain	Own workforce engagement	Immediate	Low	Today	Moderate
	Workforce representation value chain	Near term	Moderate	>1year	Moderate
	Access to remedy	Immediate	Low	>1year	Low
Community investment	Community advocacy and support	Near term	Moderate	Today	Moderate
Local sourcing and employment	Supplier diversity	Near term	Low	>1year	Low
	Local employment	Near term	Moderate	<1year	Low
Average hours of training	ESG knowledge base and training	Near term	Moderate	<1year	High
Disability inclusion	DE&I monitoring in house & value chain	Longer term	Moderate	Today	Moderate

Table 5 (*Continued*).

Sustainability indicator	Example action	Investment return period	Investment required	Implementation period	Long-term impact
Local economic impact	Responsible sourcing	Near term	High	Today	Moderate
Stakeholder grievance	Grievance monitoring value chain	Near term	Low	<1year	Low
Product Safety and Assurance	Value chain OH&S monitoring	Longer term	Moderate	>1year	High
Customer satisfaction	User network innovation Centre	Near term	Low	Today	Moderate
	Responsible marketing	Near term	Moderate	<1year	
Sustainable design	Product/service end of life management	Longer term	High	>1year	High
Responsible sourcing	Green commuting initiatives	Immediate	Low	Today	Low
Employee turnover	ESG-linked incentives	Near term	Moderate	Today	Moderate
Days payable outstanding	Payment terms best practice	Near term	High	>1year	Moderate
Term limits for directors	Board member terms limits	Immediate	Low	<1year	Low
ESG board focus	ESG board expertise	Near term	Moderate	>1year	Low
	Legitimation of management	Longer term	Moderate	>1year	Moderate
	Extended due diligence	Longer term	High	<1year	High

Source: Author.

1.9 Qualitative Information Characteristics

The combination of corporate disasters and market naivety has led legislators and ESG reporting standard setters to take a rigid approach to reporting content. The type, scope, and detail of information required are clearly defined in an effort to standardise approaches and make them as transparent as possible. Comparability is one of the criteria required and provides a good example of how a single piece of information can be used and interpreted in many different ways. For example, an organisation must report greenhouse gas emissions (GHGe) intensity, which it would use internally to monitor the progress of its net zero transition plan, by customers to assess the emissions associated of competing products, by suppliers to pitch lower emission inputs, by competitors to differentiate their operations, by financiers to assess eligibility for lower interest rate 'ESG loans', by investors to assess the organisation's ability to minimise risk, by regulators to monitor industry adaption progress, and by ratings companies to compare GHG management implementation effectiveness, thus leading to enhanced shareholder value. GHGe intensity, therefore, has a very specific calculation formula and presentation format so that values can be easily compared between reporting organisations. Information comparability from one reporting entity to another is critical as it can enhance transparency, accountability, risk management, performance improvement, regulatory compliance, investor attractiveness, and access to capital. The same is true for relevancy, faithful representation, verifiability and understandability, and meaningfulness – each will be discussed below. It is important to note that the approach taken by each ESG reporting standard is largely similar but with some semantic differences.

Qualitative information makes up the majority of an ESG report and due to its nature, the influence of legal interpretations, and the agenda of marketing departments, meaning can vary considerably between very similar expressions. Accordingly, ESG standards come with some extensive meaning definitions and requirements that will keep lawyers and ESG consultants busy for years to come.

Comparability and consistency

As we have just seen, sustainability information is comparable when it can be likened with information provided by other organisations, particularly those with similar activities or operating in the same industry sector, and can be related with information provided by the organisation in previous period(s). The principle is to compare like with like, and an organisation must make demonstrable efforts that it has done so. Typical points for comparison are targets, a baseline, an industry benchmark, information from another organisation or from an internationally recognised organisation or association, etc.

Consistency is related to, but not the same as, comparability. Consistency refers to adhering to the same use, approach, or method for the same sustainability topic from period to period. Consistency supports and is integral to achieving comparability.

It is important to note that comparability is not uniformity or sameness. Similar information should look alike and different components of sustainability topics should look different; in other words, emphasis or redaction should not be applied. This is an important consideration when reporting negative impact information where the reporter might be inclined to represent situations in a more positive light.

TIPS AND TRICKS

Whenever possible, use internationally accepted approaches and methodologies applicable for your industry sector(s) and, if possible, have your use of them verified.

Faithful representation:

In the context of ESG reporting, faithful representation refers to the quality of reported information that accurately reflects the underlying transactions, events, and conditions in a manner that is complete, neutral, and free from material error. This concept is essential in ESG reporting to ensure that the disclosed information provides a true and fair view of the company's performance.

- A *complete* depiction includes all material information, without omission, necessary for the reader to understand the impacts, risks, or opportunities associated with a reporting criterion. This ensures that stakeholders have a comprehensive understanding of how the organisation has adapted its strategy, risk management, and governance in response to that impact, risk, or opportunity, as well as the metrics identified to set targets and measure performance.
- A *neutral* depiction is balanced, without bias, in its selection or disclosure of information and free from any intentional or unintentional

misrepresentation. Neutral information should not be manipulated to favour a particular viewpoint or hide negative aspects of the company's sustainability performance, so be sure to check that there is no slant, weighting, emphasis, de-emphasis that could portray a topic more or less favourably than reality. It is also important that any aspirational information, for example, future targets and plans, should be portrayed in a neutral way. Exercising prudence is a key element of neutrality whereby caution should be used when making judgements under conditions of uncertainty. Additionally, caution should be used with any information that is netted to avoid potentially obscuring relevant data. This is especially true for issues like carbon-offsetting, which must be reported separately, or even not included in a report under some circumstances.

TIPS AND TRICKS

Often an ESG report is prepared for publication by the marketing department as part of the annual corporate report communication, but before signing off, make sure an ESG lawyer has reviewed the text. Portraying information in a neutral light is much harder than it may seem at first.

- It goes without saying that ESG information must be *free from material error*, but that does not mean that reported data must be perfectly precise in all respects. Organisations should demonstrate that they have implemented adequate processes and controls to ensure information accuracy and integrity, as well as avoid material errors or material misstatements. Estimates must be accompanied by clear explanations of precision limitations and uncertainties, which may depend on the nature of the information and/or related impacts, risks, or opportunities.

TIPS AND TRICKS

For estimations, it is wise to outline the main factors restricting greater accuracy and the efforts your organisation makes to identify verifiable sources.

Forecasts should be clearly identified and related to accepted industry projections.

Verifiability

Verifiability enhances the credibility and reliability of ESG reports by reassuring readers that information is complete, neutral, and accurate. Normally this requires an independent party, such as auditor, analyst, or regulator, to review the data and confirm its accuracy and reliability using available evidence, methodologies, and standards. It is not required that verifiers should reach a complete agreement, but rather a consensus to agree that a particular depiction is a faithful representation. Report verifiability generally requires:

- relating information to publicly available sources, most importantly, the organisation's financial statements as well as from other similar organisations or about the external environment;
- including information about inputs, assumptions, and methods of calculation used to produce estimates or approximations; and
- an auditor's statement or similar information that the ESG report has been reviewed and agreed by the administrative, management, and supervisory bodies or their proxies.

Understandability

ESG reports tend to be lengthy and detailed, so ensuring understandability is vital to readers. Being clear and concise enables any reasonable knowledgeable user to readily grasp and comprehend the information being communicated. Therefore, it is wise to avoid generic 'boilerplate' information, which is not specific to the undertaking, and unnecessary duplication of information, including information also provided in financial statements. It is sometimes tempting to lapse into sector jargon, avoid doing so, and check that you have used clear language and well-structured sentences and paragraphs. Furthermore, the CSRD requires ESG reports prepared under the ESRS to be machine readable, requiring XBRL tagging of over 1000 data points, should they be included in your report.

The completeness, clarity, and comparability of sustainability reports rely on information being presented as a coherent whole. Coherency requires that they shall be presented in a way that explains the context, the relationships between the topics and, most importantly, how topics relate to impacts, risks, and opportunities contained within the organisation's financial statements.

TIPS AND TRICKS

The principle of continuous improvement underpins ESG practices but often has somewhat static information, for example policies and strategy, from year to year. So, to enhance the clarity of your reporting, consider distinguishing between information that is relatively unchanged from period to period, and highlighting that which has changed.

Additionally, some impacts may warrant their own highlighting from period to period, for example, those over a longer period and require constant tracking, like GHG transitioning or biodiversity impacts.

Relevancy

Sustainability information is relevant when it may make a difference in the decisions of users under a double materiality approach – see Chapter 3: Single and Double Materiality Assessment. Relevancy is also considered even if readers do not take advantage of the information provided or are already aware of it from other sources. There are three types of relevancies that you need to consider when deciding to report an ESG topic:

- **Predictive value:** Sustainability information has predictive value if it can be used as an input to processes employed by users to predict future outcomes. It does not need to be a prediction in your report, but rather, it could be of use to someone else.
- **Confirmatory value:** Does the information provide feedback about previous reported outcomes or evaluations, whether it be confirming or not.
- **Material value:** Materiality is an entity-specific aspect of relevance based on the nature or magnitude, or both, of the items to which the information relates. This is a large issue and is covered in detail in Chapter 3: Single and Double Materiality Assessment.

Relevancy is critical as it opens the door to interpretation or misinterpretation by stakeholders, as well as being a driver to your competitive advantage. Here are some key points that highlight the importance of relevancy in sustainability reporting:

- **Stakeholder engagement**: Relevant sustainability reporting engages stakeholders by providing them with information that is important to their decision-making processes. Curating the content and format of sustainability reports to address the specific concerns and interests of stakeholders helps to build trust and foster meaningful dialogue between companies and their stakeholders.

- **Strategic decision-making**: The language we choose to use helps to define action, so relevant sustainability information is an enabler for companies to make informed strategic decisions that align with their sustainability goals and values. By reporting on material issues that have a significant impact on the company's operations and stakeholders, organisations can prioritise resources effectively and drive positive change.
- **Risk management**: Focusing on relevant sustainability metrics and disclosures helps companies identify and manage material risks related to environmental, social, and governance factors. It may even encourage support and buy-in from concerned stakeholders who may help to alleviate potential risks and costs. Additionally, by reporting on issues that are most likely to affect the company's long-term performance, organisations can proactively address risks and safeguard their reputation and financial stability.
- **Competitive advantage**: Companies that report relevant sustainability information gain a competitive advantage by differentiating themselves in the marketplace. By highlighting their sustainability initiatives, performance, and impact, organisations can attract socially responsible investors, customers, and partners who value transparency and responsible business practices.
- **Performance evaluation**: Relevant sustainability reporting allows companies to track and evaluate their performance against key sustainability indicators and targets. By reporting on meaningful metrics that reflect the company's impact on society and the environment, organisations can assess their progress, identify areas for improvement, and demonstrate their commitment to sustainable practices.

Meaningfulness

The interplay between qualitative and quantitative aspects of ESG reporting can obscure or distract from important content. Additionally, the focus on utility and purpose within ESG reporting often overshadows the deeper significance of meaningfulness and the values that bind together and drive organisations forward, as well as hide issues that should be at the core of a broader conversation around corporate sustainability.

ESG reporting serves as a framework for companies to disclose their performance. Like every framework, however, it needs 'flesh'; otherwise, it is just a bare bone. The flesh is context and depth; it is the stuff stakeholders, including investors, consumers, employees, and communities, are interested in. It is the flesh that allows stakeholders to decide if they are interested in what you do and then assess the long-term value and sustainability of their engagements with your company.

As the ESG landscape continues to evolve, the emphasis on utility and purpose has become predominant. Let us take utility to refer to the practical usefulness of ESG reporting, such as its role in risk management, cost reduction, and attracting investment. Purpose revolves around the broader intent of ESG reporting, including fostering corporate responsibility, mitigating environmental impact, and enhancing social welfare. While utility and purpose are undeniably critical, the concept of meaningfulness tends to be lost or overlooked.

Meaningfulness in ESG reporting encompasses the depth and authenticity of a company's commitment to sustainability and ethical conduct. It goes beyond mere compliance with regulations and industry standards, delving into the actual impact of the company's initiatives on the environment, society, and governance structures. Meaningfulness involves the integration of ESG principles into the core values of the organisation, aligning them with the company's mission and long-term strategy. It requires a genuine dedication to effecting positive change, rather than engaging in superficial or performative actions solely for the sake of compliance or public relations. Meaningfulness is often the real reason why your staff come to work, why your suppliers try to make you stronger, and why your customers continue to choose you rather than someone else.

One of the primary reasons meaningfulness is often overshadowed in ESG reporting is the prevalence of greenwashing and focus on superficial ESG initiatives by companies who see sustainability as merely a reporting obligation and marketing tool. This can be exacerbated by the reduction of ESG responsibilities to junior staff members, or allocation of ESG reporting tasks to administrative functions, both of which restrict insightful analysis and interpretation. It is understandable that most managers are focused on finessing their role in the name of operational efficiency and competitiveness, but this creates the perfect habitat for greenwashing and social-washing. Such organisations seem intent on diverting attention from deeper, more meaningful sustainability efforts that are necessary for resilient growth, creating positive impacts and innovation. Similarly, superficial ESG initiatives, which prioritise optics over substance, can lead to a focus on utility and purpose at the expense of meaningful impact management.

Moreover, the current emphasis on metrics and standardisation in ESG reporting can inadvertently detract from meaningfulness. While metrics are essential for quantifying and comparing ESG performance, they often fail to capture nuanced qualitative aspects of a company's sustainability efforts. As a result, companies may prioritise activities that yield favourable metrics, even if they lack meaningful impact, thus perpetuating the disconnect between utility and meaningfulness.

Another challenge is the short-term mindset that pervades corporate decision-making. In a landscape where quarterly results often take precedence, the long-term significance and meaningfulness of ESG efforts can be overlooked. Companies may opt for quick, high-visibility projects that yield immediate benefits, rather than investing in more comprehensive, albeit slower, initiatives that could yield deeper and more meaningful ESG impact over time.

Additionally, the pressure to meet stakeholder expectations and industry benchmarks can lead to a compliance-driven approach to ESG reporting. This approach, while serving the purpose of meeting minimum requirements, may inhibit companies from pursuing more meaningful and innovative sustainability strategies. As a result, the true essence of meaningfulness can become obscured by a focus on meeting external standards rather than internalising a commitment to holistic sustainability.

To address these challenges, it is crucial to integrate meaningfulness into the core of your ESG reporting practices. This can be achieved through several key strategies, and it begins with redefining your approach to governance – begin with the intent to find meaningfulness in what you do and, if you don't find it, be sure to create some. It is essential to have a commitment to transparency and authenticity throughout your stakeholder engagement process and subsequently in your ESG reporting, including sharing clear and comprehensive information about your company's sustainability successes, failures, and challenges. Honest and transparent reporting is essential for building trust with all stakeholders and demonstrating a genuine commitment to meaningful ESG practices.

Furthermore, integrating meaningfulness into ESG reporting requires a shift in mindset – from a focus on short-term gains to a long-term perspective. Companies should prioritise initiatives that may not yield immediate gratification but have the potential to generate substantial and lasting positive impact. This shift in mindset necessitates a re-evaluation of performance metrics to accommodate the qualitative aspects of meaningfulness, in addition to the quantitative measures of utility and purpose.

Finally, fostering a culture of meaningfulness within organisations should be seen by senior management and the Board as essential. This involves embedding ESG principles into the company's values, mission, strategy, and day-to-day operations.[9] Ultimately, we all need to get better at looking beyond utility and purpose to find the hidden meaningfulness that drives and motivates us. When sustainability becomes an intrinsic part of a company's identity, meaningful ESG efforts are more likely to take root and flourish, and that is better for us and the planet.

[9] Adoption of ESG practices inevitably leads to a re-think of your organisational vision and strategy, which is covered in detail in *Organizing for Sustainability: A Guide to Developing New Business Models*, by Jan Jonker, and Niels Faber, 2020. Published by Management Impact.

2

Stakeholder Engagement

In this chapter: Roles and responsibilities in the ESG reporting process, engaging with stakeholders, discovering stakeholder concerns. How to solve potential data gathering bottlenecks in your value chain.

Of all the areas of ESG reporting, stakeholder management is the hardest to get right and the easiest to get wrong as it involves extensive and in-depth human communication. What adds extra complications are the broad scope of 'stakeholders', who include all individuals, groups, communities, and organisations that could influence or be impacted by your organisation's operations. So, depending on your organisation, there are a multitude of internal and external stakeholders, from staff to investors, from customers to suppliers, from legislators to local communities – in fact, all those who have a vested interest in your company's workings, products, services, and decision-making processes, and that may include the general public at times. Universally but not uniformly, stakeholders play a role in the success and sustainability of your organisation by contributing resources, or relying on outputs, or are affected by its decisions and actions.

Stakeholder engagement is a critical step in producing a sustainability report as it is the foundation of your materiality assessment and has a bearing on:

- **Informed decision-making**: Stakeholder feedback is invaluable in shaping ESG strategies. It provides a broad spectrum of viewpoints, incorporating the insights of those directly affected by environmental, social, and governance policies. This inclusive approach ensures that ESG initiatives are not only visionary but also grounded in practicality, leading to sustainable and socially responsible business practices.
- **Risk management**: Regular dialogue with stakeholders uncovers potential ESG-related risks early. These risks might range from environmental challenges, like pollution and resource depletion, to social

issues, such as labour rights violations or community displacement. Proactively identifying and addressing these risks can prevent larger crises and financial repercussions, and ensure regulatory compliance.

- **Reputation management**: A company's ESG reputation is significantly influenced by its stakeholder relationships. Positive engagement helps build a brand image that resonates with the values of sustainability and social responsibility. This enhanced reputation can be a magnet for socially conscious investors, customers, and talent who prioritise ESG criteria in their decision-making.
- **Compliance and transparency**: Effective stakeholder engagement is crucial for meeting the increasingly stringent regulatory requirements related to ESG. Open and honest communication about ESG policies, practices, and outcomes not only ensures compliance but also builds trust among stakeholders. This transparency is essential for demonstrating a genuine commitment to ESG principles, avoiding accusations of greenwashing, and fostering long-term stakeholder relationships.

2.1 How to Complete a Stakeholder Assessment in Five Steps

There are five steps to completing a stakeholder assessment, and it should be undertaken in a conscientious and sensitive manner. You will be engaging with the folks who make your organisation work, but although their success largely also depends on your success, it is potentially easy to cause a conflict during the information gathering phase. Remember that this is the beginning of a much longer engagement of data management that you will need to implement down the track.

STEP 1: Benchmarking and Governance

The approach to stakeholder engagement varies from sector to sector depending on the scope and the type of respondents to be included. For example, a mining company will need a comprehensive community outreach program, whereas an IT company will focus on their value chain, especially downstream impacts of use and maintenance. So, it makes sense to begin with an assessment of what your peers are doing and identifying best practices. You may find gaps in their processes that you could address with your own approach, so the idea is not to copy others but to broadly outline your own goals and process to align with or surpass sector-best practice.

Stakeholder engagement requires structure and accountability to be meaningful. So as part of your benchmarking assessment, define the best practice for management of your process. Typical issues that need to be addressed are:

- **Committee formation**: You need to create specific committees, like a Stakeholder Engagement Committee or an ESG Oversight Board. These committees should be composed of members with diverse expertise relevant to stakeholder engagement and ESG matters.
- **Define roles and responsibilities**: Clearly outline the roles and responsibilities of each committee. This includes monitoring engagement activities, ensuring adherence to ESG principles, and evaluating the effectiveness of engagement strategies.
- **Accountability processes**: Transparency and accountability in engagement activities are essential and there needs to be clear due process that can be audited if required. This could also include regular Board of Directors reports and setting up feedback mechanisms for continuous improvement.
- **Integration with policies and organisational strategy**: Ensure that these structures are integrated into the broader corporate governance framework of your organisation and align stakeholder engagement with the organisation's overall strategic objectives.

None of the major sustainability reporting standards require that you complete a stakeholder engagement to the specification of a particular framework, but it may be common in your sector to do so. The most widely used social standard is the Social Accountability International (SAI) SA8000, which sets out the requirements for fair and ethical treatment of workers and covers areas such as child labour, forced labour, health and safety, and freedom of association. Another important standard is the United Nations Global Compact (UNGC), which provides a framework for companies to align their operations with 10 principles covering human rights, labour, environment, and anti-corruption.

STEP 2: Identifying Your Stakeholders

There are a lot of stakeholders, even for a small company; so first it is necessary to identify them and then prioritise their potential inputs. Begin with supplier and customer inventories, investor and finance lists, surrounding communities and interest groups, and finally a brain-storming session that possibly includes some of your closest external stakeholders. Identify what is important to your organisation, and then identify what you think is important to your stakeholders. The greater the overlap, the more important is their opinion. It is essential that you do not 'pigeon hole' them as you are not yet fully aware of their inputs; so make sure you are flexible in your approach to 'tiering' them, as in the following example and Graphic 9:

- *Tier 1 stakeholders* are often the first to come to mind as they wield significant influence on the organisation's operations. This group should be high-touch and high-priority, with one-on-one or small group meetings and frequent communication.
- *Tier 2 stakeholders* will be familiar names and organisations, and their opinions are important but do not carry as much weight. Their feedback is high priority and often best sought in small group sessions or regular surveys both online and offline.
- *Tier 3 stakeholders* have a general and genuine interest in your organisation and its products and services, and their influence is as a broad group rather than individual opinions. Online surveys and occasional discussions are appropriate to canvas their views, and be prepared for the occasional important input from this group.
- *Tier 4 stakeholders* typically have low to moderate influence and interest in your organisation and rely on intermittent updates via newsletters or your website, but if they have an opinion, it could be very important as they are only likely to voice serious negatives or positives.

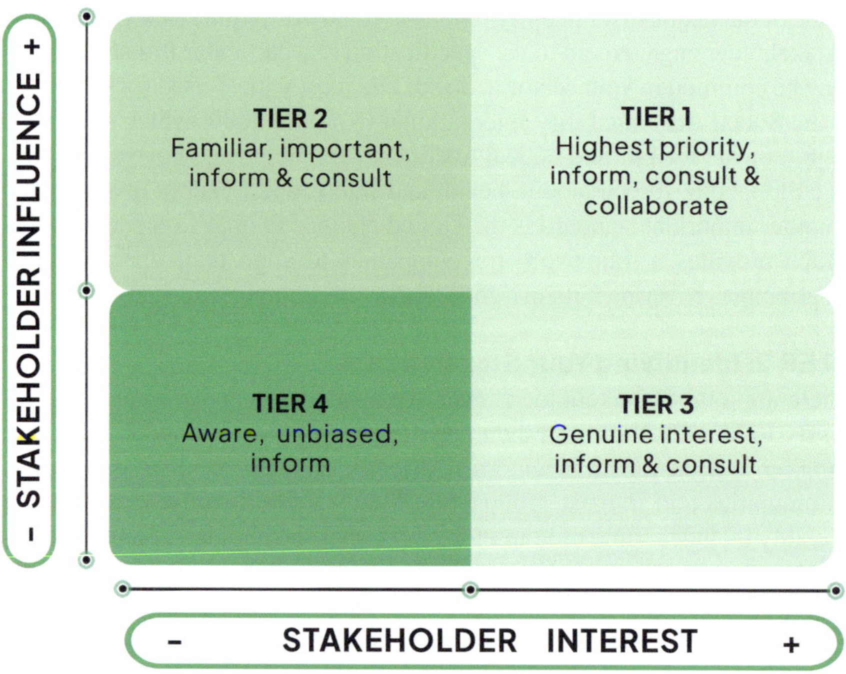

Graphic 9 Stakeholder identification. Source: Author.

Summary of key stakeholder categories:

Internal stakeholders:

- **Staff and management team**: In any organisation, your workforce and management are crucial for driving ESG initiatives. This group embodies the operational insights and strategic direction essential for embedding sustainable practices throughout your organisation. Acting as both implementers and strategists, they ensure that ESG values are deeply integrated into the company's culture and decision-making processes, making sustainability a lived reality.
- **Board of Directors**: The Board provides essential governance and oversight, ensuring that ESG initiatives are aligned with the company's overall mission and values. They hold the management accountable for ESG performance and guide long-term sustainability goals.

External stakeholders:

- **Customers**: Customers increasingly demand products and services that are environmentally friendly and socially responsible. Their purchasing choices can significantly influence a company's ESG strategy, prompting businesses to adopt sustainable practices to meet market demands.
- **Suppliers**: Suppliers play a pivotal role in the sustainability of value chains and mapping them and their supply network is an ongoing task. Organisations rely on them to uphold ESG standards, requiring sustainable sourcing and ethical labour practices. Collaborations with suppliers often lead to innovations in sustainable practices. In addition to your value chain partners, you also need to consider their related authorities, possibly including foreign regulators, supervisors, and central banks.
- **Investors**: Investors are focusing more on ESG criteria when making investment decisions. They assess a company's ESG performance as part of their risk management and long-term return strategies, influencing companies to prioritise ESG in their business models.
- **Regulatory bodies**: These organisations enforce compliance with environmental and social governance standards. Their regulations and policies shape the ESG landscape, pushing companies to adapt to evolving standards and practices.
- **Communities**: The communities where companies operate are directly affected by their actions. Community engagement and support are vital for maintaining a social license to operate. Their input is crucial in ensuring that company operations align with community needs and values.

- **NGOs**: Non-governmental organisations often serve as watchdogs for corporate ESG practices. They can influence company policies through advocacy, campaigns, and partnerships, especially in areas related to environmental protection and social justice.
- **Media**: The media plays a critical role in shaping public perception and awareness of a company's ESG efforts. Positive media coverage can enhance a company's reputation, while negative coverage can lead to public backlash.
- **Educational institutions**: These institutions contribute to ESG through research and development of sustainable technologies and practices. They are also instrumental in preparing future professionals and leaders with an understanding of ESG issues, thereby influencing the long-term trajectory of sustainable business practices.
- **Nature**: This may be considered as a silent stakeholder. In this case, ecological data and data on the conservation of species or biodiversity reports may be appropriate.

STEP 3: Engagement Plan and Implementation

Before you contact anyone, determine your goals for engagement; perhaps your approach needs to be process-centric, or specific to certain operational issues, or perhaps more wide-ranging. In other words, stakeholder engagement planning must focus on answering two questions – 'what to ask?' and 'how to ask it?'. Additionally, your engagement goals should be SMART:

- **Specific**: Clearly define what is to be achieved.
- **Measurable**: Ensure there are quantifiable indicators to track progress.
- **Achievable**: Set realistic and attainable objectives.
- **Relevant**: Align goals with the broader business strategy and stakeholder interests.
- **Time-bound**: Include a clear timeline for achieving these objectives.

When considering 'what to ask?' and 'how to ask it?', it is essential that you embrace all potential impact topics that might be included in your ESG report. This includes questions that relate to specific topics and issues covered in Chapters 5–15. Therefore, your Green Team must include people with direct and relevant knowledge of all internal and external (value chain) operations and processes that might need to be included in your ESG report. Also consider establishing scope-review sessions to refer back to line management and operational staff to check that you are capturing all relevant information.

Then consider your available resources, as often they are insufficient to contact all stakeholders and cover all subjects to the extent that you would wish. Make sure that you thoughtfully and considerately schedule engagements and follow-ups so that each stakeholder is aware of your expectations, especially if you are planning a long-term ongoing communication process. It is worth noting that some stakeholders will not reply or dedicate much time at first, especially those with whom your organisation has low rapport; so some 'bridge building' may be required to effectively engage.

Design programs that facilitate open and transparent communication, such that your company can foster constructive dialogue, build trust, and gather valuable insights from those included in the interviews. After each engagement, thoroughly review responses and possibly adapt approaches. Often a facilitator enhances interactions and gains deeper insights.

TIPS AND TRICKS

Stakeholder engagement is an opportunity to improve trust and introduce a collaborative approach to problem-solving, risk management, and to drive innovation. The impact of your engagement plan should be considered beyond your ESG reporting process; your first out-reach is simply the first step to creating a continuous improvement feedback loop.

STEP 4: Engagement Analysis

The purpose of stakeholder engagement is to assess the expectation of your stakeholders regarding your organisation's impacts from operations. Engagement analysis typically tabulates responses from stakeholder groups, with their expectations, the related medium or channel for communicating those issues, and the related impact management mechanism. Let us look at a simplified example:

Table 6 Example stakeholder engagement analysis table.

Stakeholder group	Stakeholder expectations	Communication channel	Management mechanism
Investors and owners	Sustainability reporting	Annual report with sustainability report	Sustainability reporting according to CSRD standards
	Best practices: corporate governance and compliance	Third-party analyses and audits	Communicating responsibilities
	Transparent and available information	The company's website	Availability of information
	Clear and consistent reporting	Shareholder general meetings	Preparing and highlighting policies
	Quantitative goals	Quarterly presentations	Setting up relevant and quantifiable goals
	Risk mitigation	Investor meetings and seminars	Surveying risks and opportunities
Employees, including unions and representatives	Safe workplaces	Cooperation with trade unions	Zero harm ambition
	Excellent reputation	Competence programmes and on-the-job training activities	Internal and external communication measures
	Responsible and attractive workplaces	Whistle-blowing procedure	Competitive conditions
	Management and employee participation	General meetings and departmental meetings	Competence mapping and evaluation
	Compliance with employee rights	Intranet	Employee surveys
Reporting Authorities	Compliance with laws and regulations	Dialogue meetings and conferences	Reporting on progress/sustainability reporting
	Health, safety, and environment, energy, and climate measures	Supervision, audits, and verifications	Clear goals and ambitions for ESG
	Comprehensive risk management	Environmental and climate reporting	Proactive dialogue with authorities
Suppliers	Defined quality requirements and service level	Surety of supply and service portal	Enquiries, with a description of requirements and scope
	Predictability and long-term perspective	Supplier meetings and online enquiries	Transparent description of selection criteria
	Integrity	Audits and verifications	Qualifying criteria: ESG, quality, and code of conduct
Customers	Enhanced quality and service levels	The company's website	Customer surveys
	New products and updates	Online innovation centre	Proactive dialogue with customers
	Zero packaging	The company's website and supplier meetings	Reporting on progress/sustainability reporting

Source: CDR (EU) 2023/2772 supplementing Directive 2013/34/EU and the author.

For each stakeholder group, which was identified in Steps 1 and 2, match their expectations or issues raised to an appropriate communication channel and a mechanism for addressing them. Some mechanisms or channels may be appropriate for multiple expectations. Once completed and all expectations have been tabulated, it is wise to share the analysis with C-suite or Board of Directors as there are likely to be policy implications. Additional details can be added; for example, the degree of impact or specific risks and opportunities that have been flagged.

Depending on the level of complexity of your analysis and after you have identified and finalised the above table, it may be worth mapping the interrelatedness of expectations as they are frequently interlinked. Mindmaps are a useful tool as they can plot the relative size/scale of influence and relationship channels.

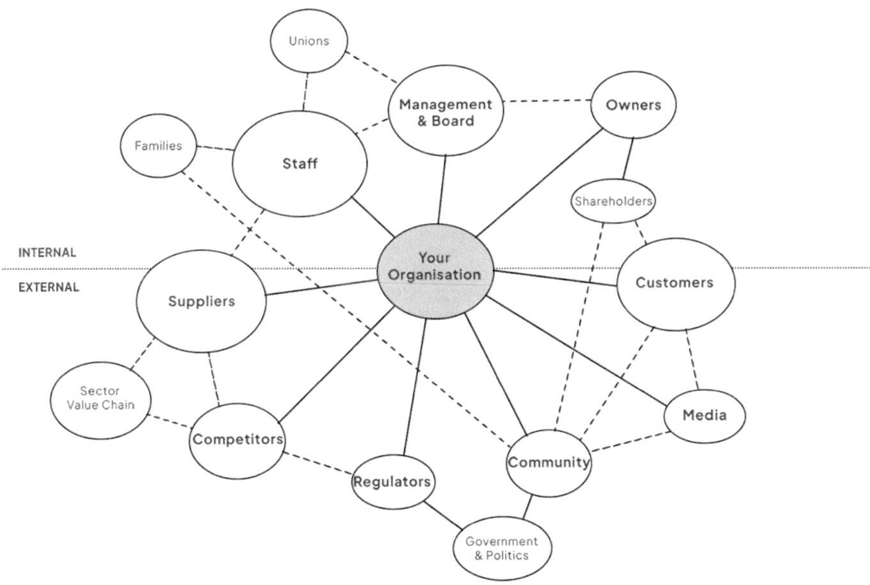

Graphic 10 Example stakeholder. Mindmap showing relative scale of influence and relationship channels. Source: Author.

In the above (Graphic 10), it is worth noting the relative number of channels available to Community, which, although relatively small in influence, connects with six stakeholder groups. So, it might be worthwhile further detailing influential community stakeholder groups and developing a dedicated engagement system, for example, that includes regular community impact assessments, inviting open questions at AGMs, or governance auditing.

TIPS AND TRICKS

During your analysis also categorise stakeholders regarding their attitude towards ESG standards and the degree to which they are likely to support your data-gathering requirements. If there is reticence or reluctance, it is wise to catch it now and begin building a rapport.

STEP 5: Sharing Information

A stakeholder engagement report should be published as part of your sustainability report relating to your materiality assessment (see Chapter 3). It is important that you invite feedback regarding the report to potentially identify stakeholders that were omitted or overlooked from the original process.

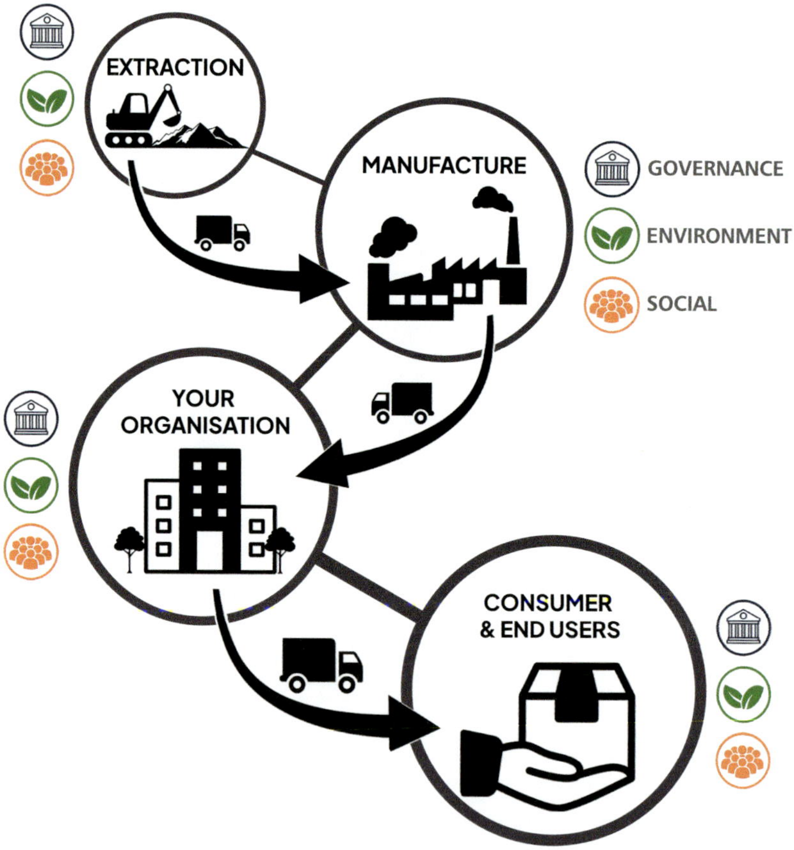

Graphic 11 Value chains are a critical part of ESG reporting. Source: Author.

2.2 Overcoming Value Chain Bottlenecks

Moving forward, your value chain is likely to pose a range of challenges to your ESG reporting system and stakeholder engagement process. Overcoming them is vital to ensuring accuracy, comparability and consistency, faithful representation, verifiability, understandability, relevancy, and meaningfulness. Typical reasons why your value chain partners are not as motivated or aligned to your ESG reporting goals are:

1. **Buy-in**: There is plenty of scepticism in the business world about the relevancy and benefits of ESG reporting despite scientific evidence and legislation. As mentioned at the beginning of your stakeholder engagement planning, you will find that some value chain partners will not actively get involved with your ESG reporting system despite your best efforts to convince them of the obligations you may face, the potential benefits to them, and the opportunity to mutually develop deeper trust and understanding. Paradoxically, if you are able to convince these companies of the importance of what you are asking them to do, they are likely to become advocates across their own value chains.

TIPS AND TRICKS

Should buy-in be a big problem for you, then it is wise to raise the issue with the senior management and possibly the Board of Directors. Perhaps pressure could be applied, or maybe it is time to find a new partner?

2. **Data collection**: Gathering ESG data from multiple partners with different reporting standards and systems is complex and time-consuming, and possibly too onerous for some companies to adapt to your requirements. There is no single approach to how companies gather the data that you will need for a multitude of indicators and that might even be the case within your own organisation.

TIPS AND TRICKS

It is important that as you consider stakeholder expectations, communication channels, and management mechanisms, you include metrics parameters at each stage; in other words, you try to find common ground or agree boundaries to the information that you require. For example, if you are calculating greenhouse gas emissions, then your energy usage figures should ideally be easily convertible, perhaps in kilowatt hours, mega-joules, or Btu. Or if you are reporting minimum working ages, then agree that at least two of these points are reported: relevant legislation in each jurisdiction, co-creation of a minimum age policy, and birth dates in employee records.

3. **Transparency**: Understandably, the idea of opening record keeping systems to an outsider can leave companies feeling nervous and uncooperative. Perhaps your relationship is new or concerns a copyrighted product or service that needs comprehensive protection, or maybe being transparent falls outside cultural norms, or there has not been clear direction from the senior management about which information to share. Whatever the reason, the lack of transparency could hinder the completeness of your reporting and needs to be addressed.

TIPS AND TRICKS

This is best achieved through a non-disclosure agreement that includes the following: (1) you are clear and concise about why you need specific information, (2) you explain precisely the format and options that you require, and (3) you have shared and confirmed with the right authority in your partner's organisation (and within your own) for the release and sharing of the information. This often involves the legal department, so take a top-down approach to connecting with the right authority.

4. **Verification**: Evidence-based is a fundamental principle of ESG reporting. Without an audit trail to a physical document or location, there is no way to verify information or claims of relevancy. Not only must you ask your value chain partners for information, you are going to ask them to provide evidence for what they are telling you, which can cause disruption and confusion as the actual evidence is often not held by the person organising the data.

TIPS AND TRICKS

Online document sharing systems, from Google Drive to OneHub to Zoho, are excellent solutions and can be formatted to reinforce your specific data requirements.

5. **Existence**: Quite simply, the data that you seek might not actually exist. A simple example is water usage for a company in a co-working space, which is a common reporting obstacle for IT staff and consultants who work remotely. The best solution that will be accepted by all ESG reporting standards is that you determine an appropriate estimation methodology based on industry averages, but without undue cost or effort. None of the reporting standards expect you to spend additional time and money creating specialised tools and methodologies to replace information that is not available. However, you will need to explain the detail of your estimation methodology and your plans to obtain better or more accurate information in the future.

6. **Too much work**: You can expect to hear this excuse, especially from SMEs, for some time to come. Within ESG standards, there is normally some flexibility for the first year or two in the event that not all necessary information regarding your upstream and downstream value chain is available. So, it is wise to anticipate that you will face difficulties with value chain reporting and therefore include the following in your ESG reporting from the outset:

   a) For information regarding policies, actions, and targets, you can limit upstream and downstream value chain information to that which is available in-house, such as data already available to you and/or publicly available.

   b) For details regarding metrics, you can state that you intend to exclude upstream and downstream value chain information, except for data-points concerning specific governance and organisational reporting requirements, which will be identified in Part 2.

   Additionally, in the CSRD/ESRS, there are some phase-in criteria that may mean your organisation can omit elements of the ESRS reporting, for example:

   c) *In the first year that you apply the standards*, organisations with less than 750 employees may omit Scope 3 GHG emissions data and 'own workforce' indicators.

   d) *In the two years that you apply the standards*, organisations with less than 750 employees may omit biodiversity, value chain workers, affected communities, and consumers and end-users indictors.

e) **All undertakings may omit the following information in the first year that you apply the standards**:

   i. anticipated financial effects related to non-climate environmental issues (pollution, water, biodiversity, and resource use); and
   ii. certain datapoints related to their own workforce (social protection, persons with disabilities, work-related ill-health, and work–life balance).

f) **Voluntary disclosures**: For example, biodiversity transition plans; certain indicators about 'non-employees' (self-employed persons or persons provided by workplace agencies or similar); and an explanation of why the undertaking may consider a particular sustainability topic not to be material.

g) **Flexibilities in specific disclosures**: For example, there are additional flexibilities in the disclosure requirements on the financial effects arising from sustainability risks and on engagement with stakeholders, and in the materiality assessment process methodology. Furthermore, the indicators regarding corruption and bribery, and regarding the protection of whistle-blowers have limited flexibility regarding what might be considered to have infringed on the right not to self-incriminate, although you should take legal advice before omitting information from these indicators.

3

Single and Double Materiality Assessment

In this chapter: Conducting materiality tests. Preparing materiality reports. Leveraging ESG reporting standards.

Materiality testing is the single most important process of all the activities in ESG reporting, as it defines what, why, and how you will be evaluating and reporting. Without a clearly explained materiality testing process, the reader will not be able to understand the context or relative importance of the information that you have chosen to include in your ESG report.

3.1 Contextualisation and Materiality

Contextualisation in ESG reporting refers to providing relevant information about a company's ESG performance within the broader context of its operations, industry, and stakeholder expectations. It involves presenting ESG data and metrics in a way that helps stakeholders understand the significance and implications of the reported information. Contextualisation can include comparisons to industry peers, historical data, targets, and relevant benchmarks to give a more comprehensive view of the company's ESG performance. You can think of contextualisation as a 'big canvas' approach to describing the organisation as it helps stakeholders make informed decisions and assessments about the company's sustainability practices.

Materiality takes a more focused approach to information, as it refers to the principle that companies should disclose information on sustainability issues that are considered significant to their organisation, financial performance, and stakeholders. Materiality helps organisations focus on disclosing ESG information that can impact decision-making processes of investors, customers, employees, regulators, and other stakeholders. There are a number of important aspects to materiality that we should clarify first:

- **Relevance**: Materiality ensures that the disclosed ESG information is relevant and significant in understanding the company's sustainability performance and its potential impact on financial results.
- **Impact**: Material ESG issues have the potential to influence the decision-making of stakeholders and can affect the company's ability to create long-term value.
- **Prioritisation**: Materiality helps companies prioritise which ESG issues to focus on, ensuring that the most critical topics relevant to the organisation and stakeholders are included.
- **Transparency**: By reporting material ESG information and highlighting specific impacts of concern, companies demonstrate transparency and accountability to their stakeholders, showing that they are addressing the most relevant sustainability challenges.

In summary, materiality ensures that organisations focus on reporting information on sustainability issues that are the most relevant, significant, and impactful to their operations and stakeholders. This principle helps companies enhance transparency, manage risks, make informed decisions, and build trust with their stakeholders.

It is worth noting that to facilitate both the process for reporting information and for stakeholders to receive comparable, consistent, and relevant information, standard-setters also develop sector-based standards. GRI have published some industry sector standards, including oil and gas, coal, mining, agriculture, aquaculture, and fishing with financial services and textiles and apparel due in 2024. EFRAG is due to start publishing sector-based standards in 2024 and 2025; however, the full extent of the range of sector standards has not yet been made public.

Both concepts are vital for enhancing the credibility and usefulness of ESG reporting by providing stakeholders with the right information to make informed decisions. Contextualisation helps in providing a richer understanding of the reported ESG data, and materiality guides companies in prioritising the most important ESG issues for reporting. Importantly, they both require companies to consider the broader context within which the organisation operates, whether it is the industry landscape, regulatory environment, or stakeholder expectations. However, there are two key differences that you need to keep in mind throughout this chapter.

- **Scope**: Materiality is more concerned with determining the significance of ESG factors on financial performance and stakeholder interests,

Table 7 Contextualisation vs. materiality.

Contextualisation	Materiality
Definition: Contextualisation in ESG reporting refers to the practice of providing relevant background information, explanations, and details necessary for understanding the reported data in its proper context.	**Definition**: Materiality in ESG reporting refers to the significance or relevance of an issue or impact in relation to a company's financial performance and its stakeholders.
Purpose: It helps stakeholders interpret the organisation's ESG information accurately, by presenting the data within the framework of the company's operations, industry standards, and broader socioeconomic and environmental contexts.	**Purpose**: Materiality helps organisations identify and prioritise ESG factors that have the most substantial impact on the company's ability to create value over the short, medium, and long term. It guides companies in determining which ESG issues to focus on in their reporting.
Example: Contextualisation can involve explaining the reasons behind significant changes in emission levels over a particular period, such as the implementation of new sustainability initiatives or the impact of external factors like regulatory changes.	**Example**: A materiality assessment concerning significant changes in emission levels might involve identifying and prioritising ESG issues based on their financial impact, stakeholder interests, industry norms, and the company's strategic priorities.

Source: Author.

while contextualisation aims to present data in a broader operational and societal context.

- **Focus**: Contextualisation emphasises building a comprehensive narrative around the reported data, while materiality focuses on identifying and prioritising the most critical ESG issues. In essence, contextualisation seeks to embrace all relevant information and explain the interrelatedness of impacts and actions, although it would be quite impractical to explain every single detail; so materiality is introduced to prioritise or omit details where appropriate.

Let us take a couple of simple examples. If a company reports a 20% decrease in carbon emissions, providing context would involve explaining whether this reduction aligns with industry standards, regulatory requirements, or the company's own sustainability goals. Materiality may consider this degree of reduction very significant and therefore include additional indicators to help explain the degree of impact, perhaps energy mix, or the amount and type of renewables, or a shift to vehicle fleet electrification.

Another example could be a company reporting a significant increase in diversity within its workforce. Contextualising this data would involve comparing it to industry averages or outlining the company's diversity initiatives and progress over time to give stakeholders a better understanding of the significance of the reported increase. Materiality may consider the diversity, equity, and inclusion (DE&I) issues to be highly material and therefore introduce indicators for diversity at various levels and functions, down from, say, Board level to all staff.

3.2 Severity and Irremediable Character

When discussing context and materiality, the seriousness of an impact, risk, or opportunity associated with an organisation's activities needs to be explained using the following factors:

1. **Scale**: How grave the negative impact is or how beneficial the positive impact is for people or the environment.
2. **Scope**: How widespread the negative or positive impacts are. In the case of environmental impacts, the scope may be understood as the extent of environmental damage or a geographical perimeter. In the case of impacts on people, the scope may be understood as the number of people adversely affected.
3. **Irremediable character**: Whether and to what extent the negative impacts could be remediated, i.e., restoring the environment or affected people to their prior state.

Any of the three characteristics (scale, scope, and irremediable character) can make a negative impact severe. In the case of a potential negative human rights impact, the severity of the impact always takes precedence over its likelihood.

3.3 Specific Metrics Identification

As we touched on in Chapter 2, during the process of identifying issues and impacts, you should also consider various options for measuring or expressing them either numerically or through KPIs or through text descriptions. The various elements of assessing an ESG indicator are called metrics, and for most it is easy to identify a suitable metric, while it can be a challenge for some.

Needless to say, ESG reporting involves a wide range of indicators/metrics that companies use to measure and disclose their sustainability performance. There is no pre-defined range of metrics as each organisation is different, and, of course, your range of metrics should relate superficially to your material impacts.

In each chapter of this manual, I suggest relevant metrics and indicators, but the list is by no means exhaustive. To reflect your organisation's uniqueness, you may well need to research additional metrics depending on your materiality assessment. Here are some examples of common metrics that you may have already come across related to ESG reporting:

Environmental metrics:

- Carbon footprint: Measures the greenhouse gas emissions produced directly or indirectly by a company's operations.
- Water usage: Tracks the amount of water consumed or withdrawn by a company in its operations.
- Waste generation: Quantifies the amount of waste produced by the company and how it is managed (recycled, reused, or disposed of).
- Renewable energy use: Indicates the proportion of energy derived from renewable sources in the company's energy mix.

Social metrics:

- Diversity and inclusion: Measures workforce diversity in terms of gender, ethnicity, age, and other factors.
- Employee turnover rate: Shows the percentage of employees who leave the company within a certain period.
- Health and safety incidents: Tracks the number of workplace accidents, injuries, or fatalities.
- Community engagement: Evaluates the company's relationships with local communities and the impact of its operations on them.

Governance metrics:

- Board diversity: Indicates the composition of the company's board in terms of gender, nationality, expertise, etc.
- Executive compensation ratio: Compares the CEO's compensation to the average employee's compensation.
- Ethics and compliance violations: Reports instances of ethical breaches, fraud, or legal violations within the company.
- Shareholder rights: Assesses the rights and protections granted to shareholders, such as voting rights and access to information.

These metrics help investors, stakeholders, and the public evaluate a company's overall sustainability performance and adherence to ESG principles. Companies often use a combination of these metrics, along with industry-specific indicators, to provide a comprehensive picture of their ESG efforts.

TIPS AND TRICKS

For a comprehensive range of ESRS support materials see the EFRAG website:https://www.efrag.org/en/projects/esrs-implementation-guidance-documents

3.4 Single and Double Materiality

Materiality testing introduces the identification of key issues that can impact your organisation's operations, financial performance, and stakeholder relationships, and then prioritise which information to disclose in your reports. There are four important reasons why consistency and thoroughness are vital when developing your materiality approach:

- **Enhanced credibility**: Materiality testing enhances the credibility of ESG reporting by demonstrating that the organisation has considered and disclosed information that is material to investors, customers, employees, regulators, and other stakeholders. It shows a commitment to transparency and accountability.
- **Risk management**: Identifying and reporting on material ESG issues through testing helps organisations manage risks effectively. By understanding the risks associated with these issues, organisations can develop strategies to mitigate them and enhance their long-term sustainability.
- **Strategic decision-making**: Materiality testing provides organisations with valuable insights to make informed strategic decisions. By understanding the ESG issues that are most important to stakeholders and the business itself, organisations can align sustainability efforts and their overall strategic goals.
- **Compliance and regulatory requirements**: Many jurisdictions around the world have regulations that require companies to report on material ESG issues. Materiality testing ensures that companies meet these reporting requirements and stay compliant with relevant laws and standards.

However, there is one factor above all others that will determine what and why you include specific issues and impacts in your ESG reporting – whether you assess impacts from a *single or double materiality* perspective. The concepts of single and double materiality are crucial for organisations to understand the type, complexity, relevancy, and detail of information required by ESG standards and frameworks.

Table 8 Single vs. double materiality.

<table>
<tr><th>Single materiality</th><th>Double materiality</th></tr>
<tr><td>Definition: Single materiality focuses on the impact of ESG factors on the financial performance of the organisation itself. It considers ESG issues that directly affect the company's financial bottom line by increasing costs, reducing revenues, or other factors.</td><td>Definition: Double materiality considers not only the impact of ESG factors on the organisation but also the impact of the organisation on the external environment, society, and economy. It recognises the interconnectedness between a company's activities and the broader external context.</td></tr>
<tr><td colspan="2">Both concepts aim to help companies prioritise and address the most relevant ESG issues for sustainable value creation and risk management.</td></tr>
<tr><td>Scope: Single materiality is inward-looking, focusing on the company's financial interests, often based on the past and without considering potentialities.</td><td>Scope: Double materiality is inward- and outward-looking, considering the broader societal and environmental impacts of the company's operations, and likelihood of potential adverse impacts.</td></tr>
<tr><td>Example: If a company identifies that climate change regulations directly impact its operational costs and revenue streams, it would consider climate-related risks as a single materiality issue.
A technology company may focus on single materiality by analysing how data privacy regulations could affect its revenue and costs.</td><td>Example: If a company recognises that its operations significantly impact the local community through pollution or resource depletion, it would consider these social and environmental impacts as part of double materiality.
In contrast, the same technology company may also consider its double materiality by assessing how its products and services impact issues like digital inclusion, environmental sustainability, and societal well-being beyond financial implications.</td></tr>
</table>

Source: Author.

3.5 Impact Materiality

Just to recap, in Chapter 1, you created your Green Team from various areas of your organisation and they were instrumental in developing and implementing a stakeholder engagement plan (Chapter 2). The outcome was a comprehensive list of stakeholders, their expectations, preferred communication channels, and the relevant internal management mechanisms. In fact, this is the first phase of Step 2 in preparing your ESG report and it needs to identify all associated impacts, risks, and opportunities for each material issue, otherwise called materiality testing.

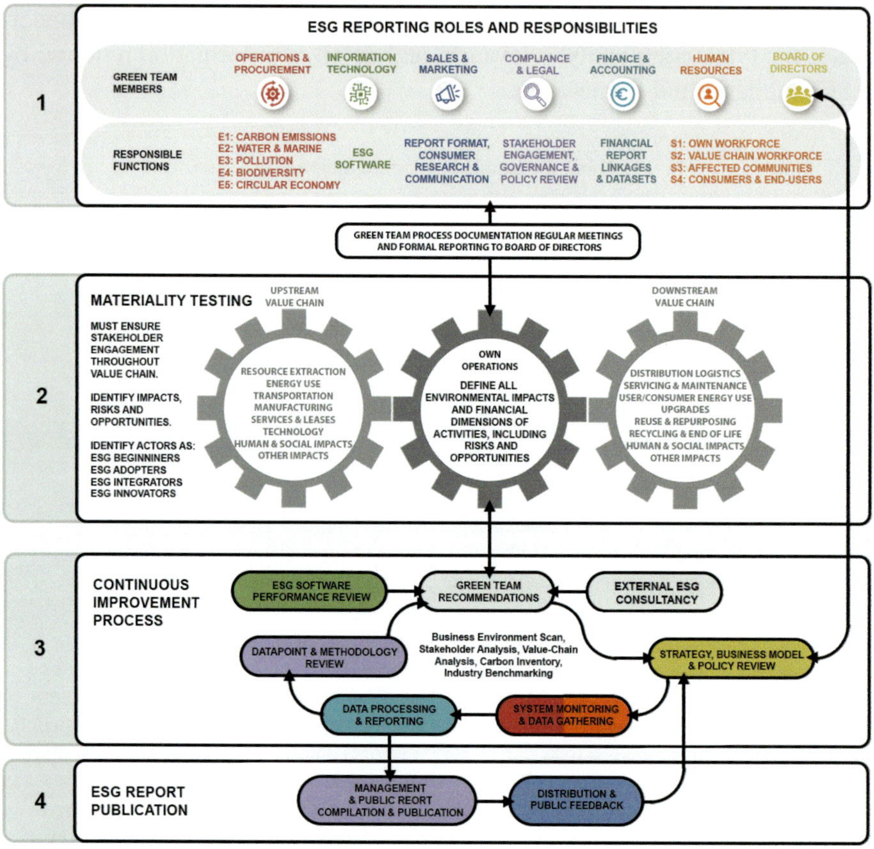

Graphic 12 ESG reporting systems roles and responsibilities.Source: Author.

Materiality testing begins with an analysis of the organisation's activities, business model, business relationships, and entire value chain (upstream and/or downstream) by mapping the geographic locations of the organisation's activities, products/services, and then mapping the organisation's relationships and upstream and/or downstream value chain, including type, nature, and extent of business relationships. You might need to revisit your stakeholder engagement data to complete the mapping process, but this should be an easy task. While revisiting, also review your timeline horizons (short-, medium-, and long-term) to be used in the materiality assessment process and ask the compliance and legal department to check that there are no unidentified legislations or regulations associated with each of the stakeholder expectations that have not already been flagged and incorporated into the scope of operations. In both cases, further stakeholder engagement could be required to validate the list of potentially material matters. Similarly, for impact materiality, the engagement could be focused on the scale aspect of severity and on time horizons. Ensure thorough documentation of this process.

It is now time to assess impact and financial materiality and determining what is to be reported (see Graphic 13 for a summary of Physical and Financial Materiality assessment steps). For each materiality point, you need to consider the following four steps:

1. understanding of the context in relation to its impacts including its activities, business relationships, sustainability context, and stakeholders;
2. identifying actual and potential impacts (both negative and positive), which may require relying on scientific and analytical research on sustainability matters;
3. assessing the materiality of its actual and potential impacts; and
4. deciding specific thresholds to determine which of the impacts will be covered in its sustainability statements – this is more complicated than it sounds!

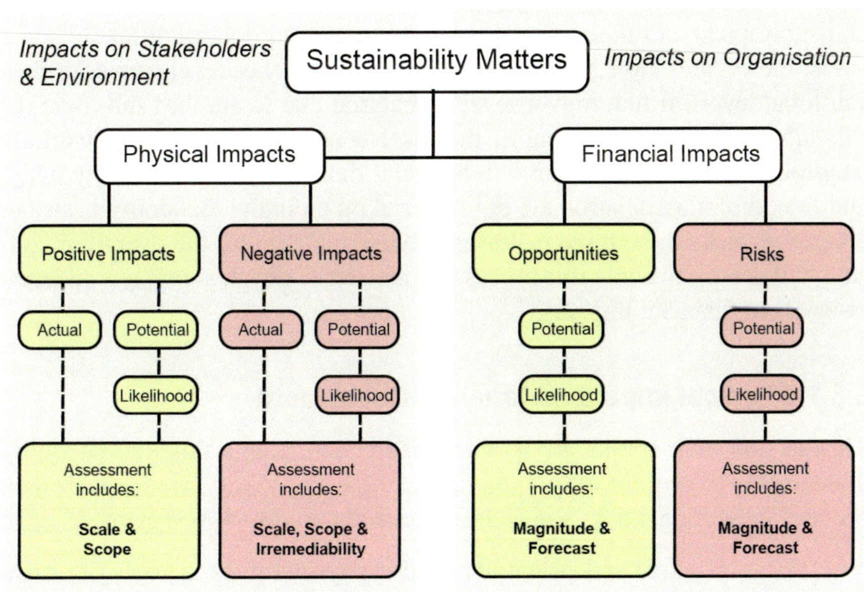

Graphic 13 Double materiality assessment. Source: Adapted from Lefebvre Sarrut's infographic in association with GRI (Global Reporting Initiative) and MEP Pascal Durand, CSRD rapporteur, A. Gilbert-D'Halluin, CSRD Essentials, 2024.

Steps 1 and 2 have resulted in our list that we are now going to use to assess actual and potential impacts and then determine thresholds. Now, for each material matter on the list, identify their related impacts, risks, and opportunities using both of these two methods:

A) **'Top-down' approach**: This is the identification of the list of potential material matters at topic/sub-topic/sub-sub-topic level followed by the assessment of the existence of the related potential material impacts (actual/potential, negative or positive), risks, or opportunities. Depending on the level of maturity of such assessments by the organisation, the identification could begin with the list in Table 9, or use your own list generated by an enterprise risk management system and then compared with the list below.

B) **'Bottom-up' approach**: Potential material matters are listed based on impacts, risks, and opportunities that are identified at a granular level. The organisation may then group these impacts, risks, and opportunities into topics or sub-topics, which become the list of potential material matters to be compared with your version of Table 9. I should note that this approach is less common than the top-down.

The outcome of this step is a defined list of potential material matters (topics and sub-topics) and their related impacts, risks, and opportunities. In some cases, the organisation may aggregate them at a higher level (topic level or sub-topic level), if that higher level is material due to another sub-topic (or sub-sub-topic) being material at the level below. An example, is working conditions in an organisation where social dialogue, collective bargaining, and freedom of association are not material on an individual entity basis but altogether make these three factors combined material for the overall organisation. It is wise to begin this process with material physical impacts and then progress to financial impacts.

3.5.1 Physical impact materiality assessment

This section only applies to organisations undertaking a double materiality assessment. If you are only applying a single materiality assessment, then proceed to Section 3.5.2, 'Financial Impact Materiality Assessment'.

- For each actual and potential negative physical impact, apply objective criteria, using appropriate quantitative and/or qualitative thresholds, to assess scale, scope, and irremediable character of the impact.

Table 9 Potential material matters at topic/sub-topic/sub-sub-topic level.

Topical ESRS	Topic	Sub-topic	Sub-sub-topics	EU Legal requirement
General information				
ESRS 2 BP-1	Business process	General basis for preparation of the sustainability statements: the undertaking shall disclose the general basis for preparation of its sustainability statements: provide an understanding of how the undertaking prepares its sustainability statements, including the scope of consolidation, the up and down stream value chain information.		
ESRS 2 BP-2		Disclosures in relation to specific circumstances: must contain: • Time horizon • Value chain estimation • Sources of estimation and outcome uncertainty • Changes in preparation or presentation of sustainability information • Reporting errors in prior periods • Disclosures stemming from local legislations or generally accepted sustainability reporting pronouncements • Incorporation by reference		
ESRS 2 GOV-1	Business governance	The role of the administrative, management, and supervisory bodies: the undertaking shall disclose the composition of the administrative, management, and supervisory bodies, their roles and responsibilities, and access to expertise and skills with regard to sustainability matters.		Yes
ESRS 2 GOV-2		Information provided to and sustainability matters addressed by the organisation's administrative, management, and supervisory bodies: how the administrative, management, and supervisory bodies are informed about sustainability matters and how these matters were addressed during the reporting period.		

(Continued)

Table 9 (*Continued*).

Topical ESRS	**Topic**	**Sub-topic**	**Sub-sub-topics**	**EU Legal requirement**
ESRS 2 GOV-3		Integration of sustainability-related performance in incentive schemes: information about the integration of its sustainability-related performance in incentive schemes.		
ESRS 2 GOV-4		Statement on sustainability due diligence: disclose a mapping of the information provided in its sustainability statements about the sustainability due diligence process(es).		Yes
ESRS 2 GOV-5		Risk management and internal controls over sustainability reporting: disclose the main features of its risk management and internal control system in relation to the sustainability reporting process(es).		
ESRS 2 SBM-1	Strategy, business model, and value chain	Market position, strategy, business model(s), and value chain: disclose its market position, the elements of its strategy that relate to or impact sustainability matters, its business model(s), and its value chain.		Yes
ESRS 2 SBM-2		Interests and views of stakeholders: disclose how the interests and views of its stakeholders are taken into account by the organisation's strategy and business model(s).		
ESRS 2 SBM-3		Material impacts, risks, and opportunities and their interaction with strategy and business model(s): disclose its material impacts, risks, and opportunities and how they interact with its strategy and business model(s).		
ESRS 2 IRO-1	Impacts, risks, and opportunities	Description of the processes to identify and assess material impacts, risks, and opportunities: disclose its processes to identify its impacts, risks, and opportunities and to assess which ones are material.		
ESRS 2 IRO-2		Impact, risk, and opportunity management: report on the disclosure requirements complied within its sustainability statements.		
ESRS 2 DC-P	Policies	Policies adopted to manage material sustainability matters: The undertaking shall apply the requirements for the content of disclosures in this provision when it discloses the policies it has in place with regard to each sustainability matter identified as material.		

ESRS 2 DC-A	Actions and resources	Actions and resources in relation to material sustainability matters: the undertaking shall apply the requirements for the content of disclosures in this provision when it describes the actions through which it manages each material sustainability matter including action plans and resources allocated and/or planned.		
ESRS 2 DC-M	Metrics	Metrics in relation to material sustainability matters: the undertaking shall apply the requirements for the content of disclosures in this provision when it discloses on the metrics it has in place with regard to each material sustainability matter.		
ESRS 2 DC-T	Target tracking	Tracking effectiveness of policies and actions through targets: the undertaking shall apply the requirements for the content of disclosures in this provision when it discloses information about the targets it has in place with regard to each material sustainability matter.		
Environmental				
ESRS E1	Climate change	• Climate change adaptation • Climate change transition plan • Climate change mitigation actions and targets • Energy consumption and mix • GHG Scopes 1, 2, and 3 and totals • GHG removals and mitigation projects • Internal carbon pricing		Yes
ESRS E2	Pollution	• Pollution of air • Pollution of water • Pollution of soil • Pollution of living organisms and food resources • Substances of concern • Substances of very high concern • Microplastics		Yes

(Continued)

Table 9 (*Continued*).

Topical ESRS	Topic	Sub-topic	Sub-sub-topics	EU Legal requirement
ESRS E3	Water and marine resources	Water and marine resources	• Water consumption • Water withdrawals • Water discharges • Water discharges in the oceans • Extraction and use of marine resources	Yes
ESRS E4	Biodiversity and ecosystems	Direct impact drivers of biodiversity loss	• Climate change • Land-use change, fresh-water-use change, and sea-use change • Direct exploitation • Invasive alien species • Pollution • Others	Yes
		Impacts on the state of species	Examples: • Species population size • Species global extinction risk	Yes
		Impacts on the extent and condition of ecosystems	Examples: • Land degradation • Desertification • Soil sealing	Yes
		Impacts and dependencies on ecosystem services		Yes
ESRS E5	Circular economy	• Resources inflows, including resource use • Resource outflows related to products and services • Waste		Yes

Social				
ESRS S1	Own workforce	Working conditions	• Secure employment, e.g., percentage of employees with temporary contracts, ratio of non-employees to employees, social protection • Working time, e.g., percentage of employees with part time or zero-hour contracts, employee satisfaction with working time • Adequate wages, e.g., EU, national, or local legal definitions of adequate wages, fair wages, and minimum wages • Social dialogue/existence of work councils/information, consultation, and participation rights of workers, e.g., extent of workplace, cross-border, and board-level representation through trade unions and/or work councils • Freedom of association/collective bargaining including the rate of workforce covered by collective agreements, e.g., percentage of own workforce covered by collective bargaining agreements, work stoppages • Work–life balance, e.g., family-related leave, flexible working hours, access to childcare • Health and safety, e.g., coverage by H&S system, rate of fatalities, non-fatal accidents, work-related ill health, work days lost	Yes

(*Continued*)

Table 9 (*Continued*).

Topical ESRS	Topic	Sub-topic	Sub-sub-topics	EU Legal requirement
		Equal treatment and opportunities for all	• Gender equality and equal pay for work of equal value, e.g., percentage of women in top management and workforce, male–female wage gap • Training and skills development, e.g., amount and distribution of training, percentage of employees with regular performance and development reviews • The employment and inclusion of persons with disabilities • Measures against violence and harassment in the workplace • Diversity, e.g., representation of women and/or ethnic groups or minorities in own workforce. Age distribution in own workforce. Percentage of persons with disabilities within the own workforce	Yes
		Other work-related rights associated with the type of operations and geographical area	• Child labour • Forced labour • Adequate housing • Privacy	Yes
ESRS S2	Value chain workers	Working conditions	• Secure employment • Working time • Adequate wages • Pay gap • Compensation equality • Social dialogue • Social protection	Yes

			• Freedom of association, the existence of works councils, and the information, consultation, and participation rights of workers • Collective bargaining, including rate of workers covered by collective agreements • Work–life balance • Health and safety • Incidents, complaints, and severe human rights impacts and incidents	
		Equal treatment and opportunities for all	• Gender equality and equal pay for work of equal value • Training and skills development • Employment and inclusion of persons with disabilities • Measures against violence and harassment in the workplace • Diversity	Yes
		Other work-related rights associated with the type of operations and geographical area	• Child labour • Forced labour • Adequate housing • Water and sanitation • Privacy	Yes
ESRS S3	Affected communities	Communities' economic, social, and cultural rights	• Adequate housing • Adequate food • Water and sanitation • Land-related impacts • Security-related impacts	Yes
		Communities' civil and political rights	• Freedom of expression • Freedom of assembly • Impacts on human rights defenders	Yes

(Continued)

Table 9 (*Continued*).

Topical ESRS	Topic	Sub-topic	Sub-sub-topics	EU Legal requirement
		Rights of indigenous peoples	• Free, prior, and informed consent • Self-determination Cultural rights	Yes
ESRS S4	Consumers and end-users	Information-related impacts for consumers and/or end-users	• Privacy • Freedom of expression • Access to (quality) information	Yes
		Personal safety of consumers and/or end-users	• Health and safety • Security of a person • Protection of children	Yes
		Social inclusion of consumers and/or end-users	• Non-discrimination • Access to products and services • Responsible marketing practices	Yes
Governance				
ESRS G1	Business conduct	• Corporate culture • Protection of whistle-blowers • Animal welfare • Political engagement and lobbying activities • Management of relationships with suppliers including payment practices		
		Corruption and bribery	• Prevention and detection including training • Confirmed incidents	

Source: CDR (EU) 2023/2772 supplementing Directive 2013/34/EU.

- For potential negative impacts, the organisation must also estimate the likelihood of the impact occurring and map it to the relevant time horizon.
- For actual positive impacts, the objective criteria need only be scale and scope, and the organisation must also estimate the likelihood of occurrence.

TIPS AND TRICKS

For negative impact objective criteria, it is often a good idea to use industry or sector benchmarks if available, and if not available, then the 90/10-rule or 80/20-rule or reasonable justification can be applied with clear justification.

During this stage of your assessment and whether it be for objective criteria or to assess materiality characteristics, it makes sense to again rely on stakeholder engagement. Key stakeholders can assess, validate, and ensure completeness of your final list of material impacts; in particular, for the scale and irremediable character determinations, as well as the estimation of likelihood occurrence.

Some examples of how impacts and dependencies are sources of material risks or opportunities:

- Increase in distribution transportation leading to higher fossil fuel usage.
- Dependency on natural resources, for example water, such that your organisation is likely to be affected by changes in the quality, availability, and pricing of the resource. During your stakeholder engagement, you will have begun mapping your supplier networks and their input flows, which should highlight potential hotspots or flag impacts that could be addressed through collaborative efforts.
- Pollution discharge into a community or biodiversity area that could become subject to stricter government regulation and/or the impact could trigger consequences of a reputational nature, thus increasing recruitment costs and/or reduced sales.
- Exposure of a value chain partner's workers to hazardous working conditions without adequate safety equipment.
- A supplier uses materials or products that in some part have been produced using child labour.

- Undeclared payments are made to secure a contract or influence an outcome of an agreement.
- Long-term pattern of failing to hire, support, and promote qualified women.
- Data leakage resulting from a lack of adequate confidentiality protection mechanisms.
- For a positive impact, an organisation in the energy production sector adopting measures that lower the cost of renewable energy for customers, thereby allowing more customers to switch to renewable energy, and thus contributing to mitigating climate change.

3.5.2 Financial impact materiality assessment

Similar to physical impacts, to assess the financial materiality of impacts, risks, and opportunities, appropriate quantitative and/or qualitative thresholds are required. However, rather than having to develop objective criteria, you can base your thresholds on the anticipated financial effects in terms of performance, financial situation, cash flows, access to, and cost of capital. Engagement throughout the business organisation, including operational units, investors, and other financial parties (e.g. banks) is appropriate to assess, validate, and ensure completeness of the list of material risk and opportunities. This is because the ultimate goal of financial materiality is to report on information that, if omitted, misstated, or obscured, could reasonably be expected to influence decisions that those stakeholders make on the basis of the organisation's ESG report.

It is important to note that financial materiality for your ESG report goes beyond the scope of the materiality used for financial statements. There are differences between financial materiality for financial statements and financial materiality for sustainability statements in a number of ways relating to timing and aggregation because ESG reports include disclosures of potential financial effects of material risks or opportunities that are not captured, or not yet fully captured by financial reporting at the reporting date, and that could be reasonably expected to result in financial effects for the organisation. From an ESG report user perspective, impacts, risks and opportunities are not limited to financial reporting definitions in the following ways:

So, how should the organisation set thresholds for financial materiality? Simply that once you have identified the impacts, risks, and opportunities, you need to determine which of them are material for reporting by using a combination of:

Table 10 ESG reporting vs. financial reporting.

Assets and liabilities	There may be potential adjustments to the carrying amount of assets and liabilities that may be of interest to ESG report users but may not meet the accounting criteria for recognition in financial statements at the reporting date.
Future realisation	ESG report users have interest in future assets, liabilities, income, and expenses, but these are only recognised at a later stage in financial reports when the accounting criteria are met.
Value generation	ESG report users are interested in factors of value creation that do not meet the financial accounting definition of assets and liabilities and/or the related recognition criteria but contribute to the generation of cash flows and more generally to the development of the organisation.
Entity aggregation	Information on the parent organisation and its subsidiaries are expanded in the ESG report to include information about material risks and opportunities arising from: i. the organisation's (parent and subsidiaries) relationships in the value chain; and ii. future events that may trigger anticipated sustainability-related risks and opportunities while financial risks are based upon past events.
Time horizons	Time horizons are longer in sustainability reporting as mid-term and long-term are considered for risks and opportunities and they are not constrained by the financial planning horizon or by the historical cost convention.
Natural and social resources	An ESG report must consider financial effects associated with dependencies on natural and social resources that do not meet (or do not meet yet) the criteria for accounting recognition.

Source: Author.

a) the likelihood of an occurrence resulting in outcomes that are below the 'more likely than not' threshold; and

b) the potential magnitude of financial effects determined on the basis of appropriate thresholds applied in your resilience and scenario analysis (see section 3.7).

For each material issue, consider its contribution of risks and opportunities to financial effects in the short, medium, and long term. This may result in the need to consider the cumulative effect of a material issue on revenues, costs, etc., over a long period. Similarly, a threshold for likelihood needs to consider the cumulative probability over a long period of time, to cover the long-term horizon.

You may refer to absolute monetary thresholds or to relative monetary thresholds, such as a percentage of the amount corresponding to a line item of your primary financial statements, revenues, costs, total assets, and net equity. Similar approaches are used to assess materiality of an item for the preparation of financial statements and may be a possible source of inspiration.

The organisation should also consider that a sustainability matter may be financially material, despite its financial effects not being (reliably) measurable at the reporting date. In this case, the thresholds will rely on qualitative factors and ranges of possible effects (high/medium/low). Also in this case, there is a similarity with financial reporting, where materiality is not confined to quantitative aspects, but an issue may be material due to its nature, i.e., reflecting a qualitative approach to materiality.

Stakeholder expectations and related issues that generate material physical impacts are also sources of material risks and opportunities in its financial materiality assessment. The organisation may assume that sustainability matters are financially material when they have been assessed to be material from a physical impact perspective, unless there is clear and demonstrable evidence that this is not true (see Graphic 13 for a summary comparison).

3.6 Identification of Information Sources and Alignment to Relevant SDGs

This stage is an essential step before aggregating your materiality test results and stakeholder feedback – identifying the information sources you have considered. Although not an essential exercise prescribed in ESG reporting standards, there is also merit at this stage in reviewing your list of ESG issues that are important to your organisation and your stakeholders by cross-referencing them with the Sustainable Development Goals (SDGs).

Table 11 Example of internal and external information sources.

Internal sources	External sources
Company business model, strategies, and policies	Stakeholder engagement and consultation
Financial and non-financial disclosures	Studies and surveys commissioned by your organisation
Enterprise risk management reports	Ongoing dialogue with suppliers, local communities, civil society, NGOs, business community, industry associations, governments, regulators, investors, and the media
Employee surveys	Sustainability topics identified by leaders and peers
Previous materiality assessments	Relevant sustainability standards and reporting frameworks have informed assessment: • Corporate Sustainability Reporting Standard CSRD/ESRS; • GRI Standards (list relevant sector standard); • International Financial Reporting Standards (IFRS) S1 and S2; • the recommendations of the Task Force on Climate-related Financial Disclosures (TCFD); • the UN Global Compact (UNGC).

Source: Author.

Now it is time to cross reference your list of ESG issues with the SDGs and your relevant standard. Note that Part 2 of this book reviews all prescribed CSRD/ESRS indicator metrics and datapoints, but Table 12 outlines a selection of common indicator metrics with the related standard and SDG.

3.7 Resilience and Scenario Analysis

Conducting an ESG resilience analysis is not stipulated in the CSRD/ESRS, but it does contribute to your governance reporting in a range of meaningful and relevant ways. However, a key materiality assessment process, scenario analysis, is required (in some form) by the CSRD/ESRS and will contribute to your impact, risk, and opportunity assessment system. Resilience and scenario analysis both involve assessing how well a company or organisation can respond to and recover from ESG-related impacts and challenges. Throughout your analysis process, you should consider involving an external expert, who could coordinate and manage the whole process for you. Following are the steps for conducting an ESG resilience analysis:

1. **Identify key ESG risks**: Your materiality assessment outcomes so far now need to be linked with possible environmental, social, and governance impacts. These could include climate change, supply chain disruptions, labour practices, diversity and inclusion, regulatory changes, etc.

2. **Assess current ESG practices**: Evaluate your organisation's current ESG policies, practices, and performance metrics. Identify those related to your material issues.

3. **Scenario analysis**: Conduct scenario analysis to understand how your organisation would fare under different ESG-related stressors. For example, how would the company respond to a severe weather event, a data breach, or a product recall? It is important that you define the nature of each scenario, the most common being:

    - Climate change scenario: Assess the impact of extreme weather event, increases in average temperatures, higher sea levels, sporadic rainfall, seasonality changes on operations, and supply chains.

    - Social unrest scenario: Evaluate how social unrest in a key market or on your workforce could affect employee safety and production timing.

Table 12 Some common material topics related to the CSRD/ESRS and SDGs.

Material issue	**CSRD/ESRS reference**	**Impact on SDG**	**Topic boundary**
Greenhouse gas emissions Other emissions	ESRS 2 IRO-1 – Description of the processes to identify and assess material climate-related impacts, risks, and opportunities E1-3 – Actions and resources in relation to climate change policies E1-4 – Targets related to climate change mitigation and adaptation E1-5 – Energy consumption and mix	SDG 13 – Climate action Emission reduction ambitions in line with the United Nations (UN) Framework Convention on Climate Change (FCCC) Paris Agreement	Own operations and entire value chain
Energy efficiency	E1-3 – Actions and resources in relation to climate change policies E1-5 – Energy consumption and mix	SDG 7 – Affordable and clean energy	Own operations and entire value chain
Shift to renewable energy	E1-1 – Disclosure of transition plan for climate change mitigation E1-2 & MDR-P – Policies in place to manage its material impacts, risks, and opportunities related to climate change mitigation and adaptation E1-3 & ESRS 2 MDR-A – Actions and resources related to climate change mitigation and adaptation	SDG 13 – Climate action Emission reduction ambitions in line with the United Nations (UN) Framework Convention on Climate Change (FCCC) Paris Agreement	Own operations and entire value chain
Water and effluents	E3-1 & ESRS 2 MDR-P – Policies to manage its material impacts, risks, and opportunities related to water and marine resources E2-4 – Pollution of air, water, and soil E3-2 – Actions and resources related to water and marine resources	SDG 6 – Clean water and sanitation SDG 14 – Life below water	Own operations
Zero waste target	ESRS 2 IRO-1 – Description of the processes to identify and assess material resource use and circular economy-related impacts, risks, and opportunities E5-1 – Policies related to resource use and circular economy E5-2 – Actions and resources related to resource use and circular economy E5-3 – Targets related to resource use and circular economy	SDG 12 – Responsible consumption and production Making the most of resources while reducing waste and minimising the impact on the environment	Own operations

Diversity, equal opportunity, and inclusion	S2-1 – Policies related to value chain workers S2-5 – Targets related to managing material negative impacts, advancing positive impacts, and managing material risks and opportunities	SDG 5 – Gender equality SDG 10 – Reduced Inequalities	Own operations and entire value chain
Access to remedy	S1-3 – Processes to remediate negative impacts and channels for own workers to raise concerns G1-1– Corporate culture and business conduct policies	SDG 8 – Decent work and economic growth	Own operations
Local employment	S3-4 – Taking action on material impacts on affected communities, and approaches to mitigating material risks and pursuing material opportunities related to affected communities, and effectiveness of those actions	SDG 8 – Decent work and economic growth	Own operations
Value chain OH&S monitoring	S1-14 – Health and safety indicators	SDG 3 – Good health and well-being	Own operations and entire value chain
Responsible marketing	S4-2 – Processes for engaging with consumers and end-users about impacts S4-4 – Taking action on material impacts on consumers and end-users, and approaches to mitigating material risks and pursuing material opportunities related to consumers and end-users, and effectiveness of those actions	SDG 16 – Peace, justice, and strong institutions	Own operations
ESG board expertise	GOV-1 – The role of the administrative, management, and supervisory bodies	Goal 16 – Promote just, peaceful, and inclusive societies: good governance	Own operations

Source: CDR (EU) 2023/2772 supplementing Directive 2013/34/EU.

- Governance crisis scenario: Analyse the implications of a boardroom scandal or greenwashing allegations on investor confidence and reputation.

In each scenario, quantify the potential financial and non-financial impacts on the organisation. This could include costs related to business disruptions, reputational damage, regulatory fines, etc.

4. **Stress and sensitivity testing**: Stress-test your organisation's ESG resilience by combining scenarios and simulating 'extreme' scenarios to assess the impact on operations, reputation, value chain partners, and environmental and financial performance. Conduct sensitivity analysis to understand how small to large changes in assumptions or inputs could impact the outcomes of the scenarios.
5. **Engage stakeholders**: Gather insights from your stakeholders about your stress tests, especially assumptions that you have made about potential impacts on them.
6. **Research benchmarks**: Compare your scenarios and stress testing against industry peers to identify areas of strength and improvement.
7. **Develop and assess resilience strategies**: Based on the analysis, develop ESG resilience strategies to mitigate risks and capitalise on opportunities. This could involve enhancing supply chain transparency, setting emissions reduction targets, improving diversity and inclusion practices, etc. In each case, evaluate your organisation's current resilience to each scenario by considering existing policies, procedures, and risk management practices.
8. **Mitigation strategies**: Develop mitigation strategies to address vulnerabilities identified in each scenario, for example:

   - Climate resilience strategy: Implementing renewable energy sources to reduce carbon emissions and dependence on fossil fuels.
   - Supply chain diversification: Diversify suppliers to reduce the risk of disruptions due to social or political unrest.
   - Governance improvement plan: Strengthening board oversight and transparency to prevent governance crises.

9. **Integrate into decision-making**: Integrate the findings from the ESG resilience and scenario analysis into your strategic decision-making processes to enhance resilience and long-term value creation.

10. **Monitor and review process**: Continuously monitor and review the effectiveness of ESG resilience strategies and adjust them as needed based on changing circumstances. It should not be necessary to repeat the entire process every year, but you should certainly flag high risk factors and report on them each year. Perhaps consider reviewing this process every 2–3 years depending on the nature of your organisation.

3.8 Aggregating and Reporting Materiality Assessments

The objective of this final step is to consolidate the results of the previous steps and obtain a comprehensive list of material impacts, risks, and opportunities, which will be the basis for the preparation of your ESG report. The analysis performed at material topic/sub-topic or sub-sub-topic level is to be converted into material impacts, risks, and opportunities and should include:

a) application of the defined thresholds to the results of the physical and financial impact materiality assessments;

b) identification of assessment criteria, including suitable metrics, for each material impact;

c) aggregation of the results of impact and financial materiality assessment with an identification of those material impacts that will lead to material risks and opportunities, ensuring the most appropriate level of granularity;

d) validation of the aggregated double materiality results with management (to assess and validate the completeness of the list of material impacts, risks, and opportunities).

The aggregating process can be complimented by a review process, as outlined in Graphic 14 below, that determines (1) what type of information you will be disclosing, and (2) what are your likely next steps for each materiality issue. It makes sense to begin assigning materiality issues as not disclosable where appropriate.

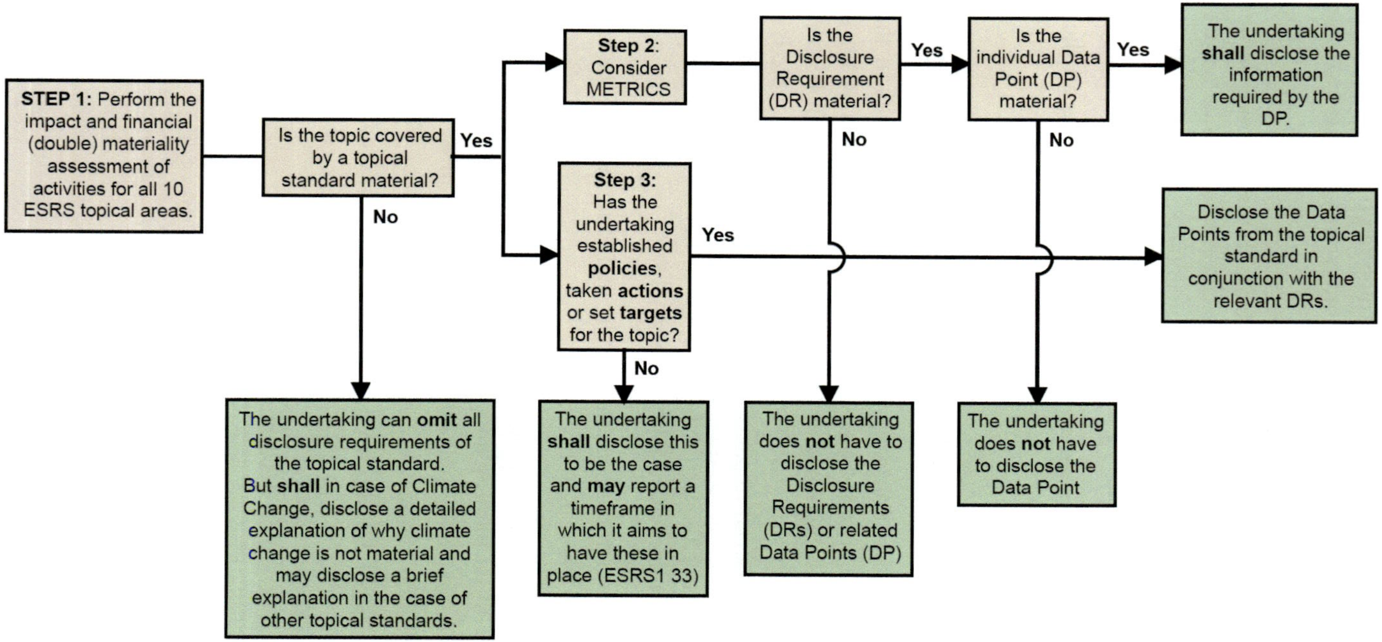

Graphic 14 Flowchart of determining inclusion of material information. Source: CDR (EU) 2023/2772 supplementing Directive 2013/34/EU.

Presentation of material impacts plays an important role in explaining relevancy, scope, scale, and irremediability. A materiality chart (see Table 13) is useful to explain the range of issues and impacts raised through your stakeholder engagement process and subsequent assessment.

Table 13 Hypothetical double materiality assessment table.

Potential material impact	**Severity assessment**			**Conclusion on materiality**
	Scale	**Scope**	**Irremediability**	
Impact #1				No
Impact #2				Yes
Impact #3				Yes
…				
Impact #*n*				Yes
Colour code				
Low		Medium		High

Source: CDR (EU) 2023/2772 supplementing Directive 2013/34/EU.

In terms of scale, this represents how grave is the impact for negative impacts and how beneficial the impact is or could be for positive impacts. Aspects that affect the scale can be whether the impact leads to non-compliance with laws or regulations; for example, if a negative impact leads to a violation of human rights. In this context, a reasonable objective measure of impacts to be used in the categorisation of the various impacts may not be available (i.e. a quantitative measure). Hence, the relevance of using the principle of relative severity among the various impacts to categorise them. The scale is also linked to the irremediable character. In the case of positive impacts, the irredeemable character is not a factor to consider.

In terms of scope, the measurement may vary depending on the sustainability matter in question. In some cases, the number of individuals affected could be the appropriate basis (for example, per cent of employees) or in other cases the number of countries affected by such matter may be relevant (for example, 40% of the suppliers are in countries where such negative impact arises). The scope factor could be assessed on a more quantitative basis than the other factors and hence translating this value into a soft quantitative scale such as the 1–5 mentioned above could be more direct, for example:

Table 14 Quantitative scaling for assessment of stakeholder expectations.

Significance to stakeholders	Impact type	Level of concern	Management mechanism outcome
Strategic (high)	Known to directly cause extensive or severe damage or harm	Widely acknowledged as a major issue or unacceptable situation	An opportunity, with the potential to generate lasting improvement and development
Significant	Known possibility to directly cause and frequent contributor to extensive or severe damage or harm	Widely acknowledged as an important issue or of ongoing concern	An opportunity, with the potential to generate immediate improvement or support for wider initiatives
Important (moderate)	Known contributor to declining state of ecosystems or socioeconomic conditions	Perceived as important in several locations or by some groups	Opportunities for localised improvement or support for wider initiatives
Informative	Weak ongoing contribution to declining ecosystems or socioeconomic conditions	Local concern that the issue is important	An opportunity for ongoing support for localised initiatives
Low	Known to have measurable but limited effects	Isolated or indirect criticism	Can provide minor but measurable improvement

Source: Author.

Table 15 Quantitative scaling for assessment of organisational issues.

Significance to organisation	Societal license to operate	Media, public, or political impact	Organisational strategy impact
Strategic (high)	Global, regional, or irreversible loss in societal license to operate	Strong criticism or anger expressed at an international level	Fundamental business change, including new opportunities
Significant	Substantial fines and extensive loss in societal license to operate	Intermittent strong public criticism at a regional or national level	Amendment required to strategy, policies, and management
Important (moderate)	Substantive enforcement, fines, or regulatory action	Prolonged coverage or local criticism (e.g. national)	Alteration to stated intentions or commitments
Informative	Minor enforcement, fines, or regulatory action	Ongoing localised coverage and frequent complaints	Public clarification of stated intentions and commitments
Low	Minor non-compliance without penalties	Short-term coverage or local complaints	Modifications to positions or systems

Source: Author.

Graphical representations of materiality assessments are popular as they can include additional quantitative factors, for example, risk of occurrence, cost of reparation, adaptation investment required or identifying issues as material topics or monitored topics:

- *Material topics* are those that have been classified as strategic or significant to either the organisation or stakeholders.
- *Monitored topics* are the remaining topics that are important for you to be a responsible business and may include informative or low significance.

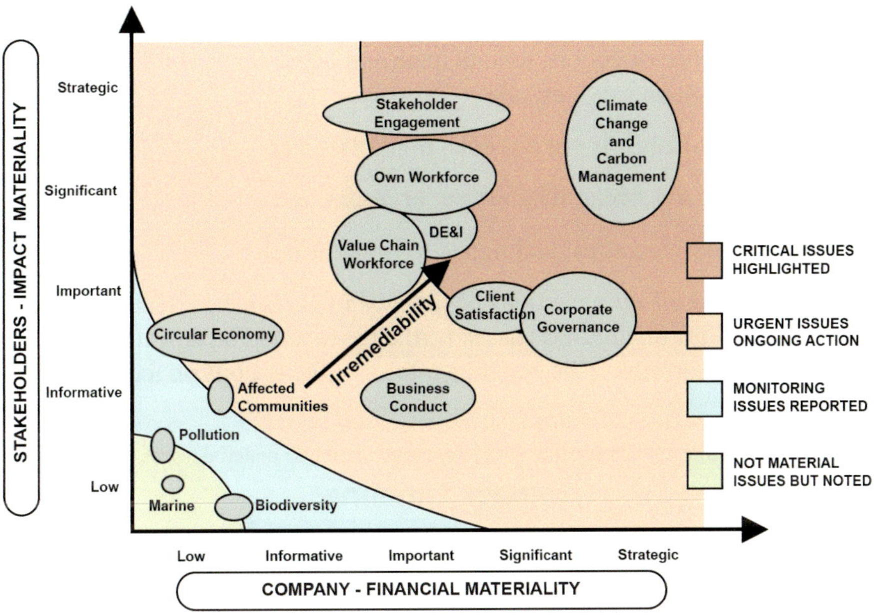

Graphic 15 Example of an energy company's double materiality assessment including irremediability and stakeholder concern. Source: Author.

In the above Graphic 15, each bubble requires comprehensive financial and impact disclosures regarding risks and opportunities, as well as associated metrics. Size of bubble indicates degree of stakeholder concern relating to topic materiality. Irremediability is indicated to show range and extent of which negative impacts could be easily remediated. Going a step further, a more detailed option may be to use two separate graphs for (1) impact materiality and (2) financial materiality and present material matters according to impacts, risks, and opportunities (IRO) timeline (on the x-axis) and magnitude of IRO (y-axis), thus incorporating time horizons to the visualisation.

Now that you have a firm grasp of the various material impacts, it is necessary to classify key elements, which are those that:

1. account for more than 10% of the organisation's revenue; and/or

2. are connected with material actual impacts or material potential negative impacts of the organisation.

When completing your materiality assessment, there may be actions that your organisation has put in place to avoid, minimise, mitigate, or rehabilitate impacts. It is important that you do include these, irrespective of the actions implemented, along with an explanation of the actions taken (according to the mitigation hierarchy) and the effects on the impact in terms of severity and likelihood. As defined by the CSBI and represented in Table 16,[1] the mitigation hierarchy is:

a) the sequence of actions to anticipate and avoid impacts on biodiversity and ecosystem services; and

b) where avoidance is not possible, minimise; and

c) when impacts occur, rehabilitate, or restore; and

d) where significant residual impacts remain offset.

Restoration and offsetting actions begin once impacts have already occurred. In these cases, the organisation has nevertheless to report on the impacts before those actions, describing the concrete effects of the actions on the impacts.

Table 16 Mitigation hierarchy table relating to material issues.

Impact Timeline →

(a)	POTENTIAL IMPACT				
(b)	Avoidance	Minimisation	Restoration	Significant Residual Impact	
(c)	Avoidance	Minimisation	Restoration	Offsets	Possible Additional Benefits

Source: CDR (EU) 2023/2772 supplementing Directive 2013/34/EU.

[1] Source: A cross-sector guide for implementing the Mitigation Hierarchy – Cross Sector Biodiversity Initiative (CSBI, 2013, pg. 12)

TIPS AND TRICKS

To make your data collection process easier, while you are finalising your materiality report, take time to identify the following sources of information for each material issue:

- Define the scope of the issue (what is included and not included).
- Determine a collection timeframe that is relevant to the issue.
- Define data owners and data sources.
- Educate data owners on required information needs.
- Select appropriate data gathering solutions.
- Define a verifiability process.
- Establish a proper quality check and control process.

Other examples of possible formats of visualisation, each highlighting certain aspects of the double materiality assessment:

a) A graph that portrays 'Likelihood of Impact, Risk or Opportunity' on the *x*-axis and the 'Magnitude of IRO' on the *y*-axis. This graph may also visualise a threshold that will position material matters above a certain cut-off point (easily displayed by a line, or by the upper corner of the graph).

b) Tables are also a way to present the many disclosures required for each material matters. Pictograms can be used in the different columns of the table to help visualise the information (e.g. where the impacts/risk is concentrated in the value chain, whether the matter is financially material, material from the impact perspective, or both).

3.9 Materiality Report and Action Plan

Your materiality report needs to outline the assumptions, process, analysis methodology, findings, and conclusions of your materiality assessment. Information needs to be presented in formats that make it understandable and meaningful to the reader, as well as outline future procedures in an action plan.

Report formats for potential impacts should combine the likelihood and severity (scale, scope, and irremediability) dimensions of the impacts. However, in the case of human rights impacts, severity takes precedence over likelihood when identifying material matters. The assessment of the three severity factors for a negative impact on violations of human rights could be as follows:

a) scale: extent of infringement of access to basic life necessities or freedoms (e.g., education, livelihood, etc.);

b) scope: number of people impacted, percentage of identifiable groups of people impacted; and

c) irremediable character: the extent to which the impact can be remediated (e.g., through compensation or restitution; whether the people affected can be restored to their exercise of the right in question).

In terms of likelihood, the likelihood of a potential negative impact refers to the probability of the impact happening. The likelihood of an impact can be measured or determined qualitatively or quantitatively, depending on the available information. It could be described using general terms (e.g., unlikely, highly likely) or mathematically using probability (e.g., 10 in 100, 10%) or a frequency over a given time period (e.g., once every 10 years).

A similar approach to the current impacts could be applied whereby the threshold for reporting material impacts is defined as the red area in the graph (as below).

Table 17 For human rights impacts, severity takes precedence over likelihood when identifying material matters.

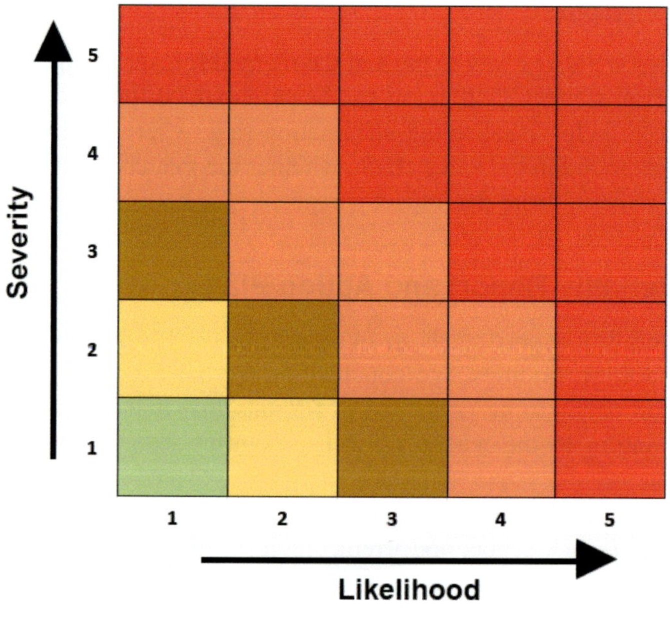

Source: Author.

Acquisitions should not be a reason for not reporting a material impact. For example, a mining company A is a large organisation with a material pollution matter regarding land next to a neighbouring community. In the following year, organisation A is acquired by group B; however, the scale of the land pollution does not change. Despite group B having several subsidiaries with other, different material environmental matters, the categorisation of scale or irremediable character would not change, although scope will.

Additionally, lack of action or inaction regarding a materiality issue, like not implementing a policy, action or target, should also be reported. For example, a hotel has identified training and development on health and safety as a material impact. However, it does not have a formal training policy or actions related to this matter. In this case, health and safety training is a material matter, which is disclosed by the organisation together with the fact that there are no policies or actions at the end of the reporting year, but the organisation may disclose that they are working on drafting and implementing such policies and actions in the future.

a) Your materiality assessment report needs to include a description of the process that you followed and present outputs based on:

b) Description of the processes to identify and assess material impacts, risks, and opportunities – see ESRS2 IRO-1 in Chapter 5.

c) Material impacts, risks, and opportunities and their interaction with strategy and business model – see ESRS 2 SBM-3 in Chapter 5.

The conclusion of your materiality assessment must produce a comprehensive action plan for your ongoing monitoring and reporting process. In your public materiality report, you must include an action plan overview. It is not necessary to detail each and every action in the report; however, key actions relating to significant material matters should be outlined using the SMART approach, such that they are Specific, Measurable, Achievable, Relevant, and Time-bound. Where appropriate, actions should also include responsible authority, designated roles, capital or financial requirements and involved stakeholders.

3.10 Materiality Assessment – How to Leverage Other Sources?[2]

The complex and constantly evolving world of ESG standards and frameworks is simplified to some extent by the recognition of the various standard-setters

[2] Draft EFRAG IG 1: Materiality assessment implementation guidance, 2 February 2024.

that interoperability could offer report streamlining. As of June 2024, there have been considerable coordination efforts between EFRAG (governing the ESRS) and the ISSB (governing the IFRS) and GRI (which will be governed by the ISSB in due course). The outcome is a series of statements of interoperability, a term define by the Oxford English dictionary defines as:

1. the ability of computer systems or software to exchange and make use of information;

   'interoperability between devices made by different manufacturers'

2. the ability of military equipment or groups to operate in conjunction with each other;

   'staff believe interoperability between forces is crucial to effectiveness'

So, although there is no formal definition with reference to sustainability standards, we may take interoperability to mean that there is some degree of commonality or possibility of exchange of information between standards and frameworks. There is some uncertainty about precisely how interoperability can be exploited by reporting companies who have invested time and energy in one standard or framework but now have to abide by a new regulation or reporting system. Only time and the continued efforts of EFRAG will shed sufficient light on how interoperability will manifest in reporting. For now, here are some general points about leveraging each.

Leveraging the GRI standards:[3]

The GRI framework is interoperable with the CSRD/ESRS and includes double materiality assessments. The impact dimension is the same for GRI and ESRS even if the scope of ESG matters under GRI may not exactly be the same with that of ESRS. Therefore, an assessment performed under GRI constitutes a good basis for the assessment of impacts under ESRS. Several synergies can be observed for those organisations that are currently reporting under the GRI Universal Standards 2021 framework in terms of:

1. Impact materiality assessment process: ESRS 2 IRO-1 and IRO-2 are aligned in substance with GRI Universal Standards and the GRI materiality assessment can be the starting point for the ESRS double materiality assessment, with appropriate integration of financial materiality.

[3] [draft] GRI-ESRS Interoperability Index: https://www.globalreporting.org/media/z2vmxbks/gri-standards-and-esrs-draft-interoperability-index_20231130-final.pdf

2. Universe of potential impacts identified using the GRI standards: They are compatible in principle with the list of matters in ESRS and as a result with the ESRS architecture of topics. In addition, GRI materiality assessment can inform the process of identifying your organisation-specific impacts.
3. Role of due diligence and stakeholder engagement: They are central in both GRI and ESRS frameworks in informing the impact materiality assessment.

Leveraging the ISSB/IFRS standards:[4]

In the intentions and expectations of EFRAG, the criteria for financial materiality and materiality of information in ESRS and the corresponding materiality approach in IFRS S1 are, in principle, aligned.

The financial materiality assessment in ESRS 1 corresponds to the identification of information that is considered material for primary users of general-purpose financial reports in making decisions relating to providing resources to your organisation. The definition of financial materiality is aligned aligned in both the standards.

Reflecting the equivalence of the scope of financial materiality in the two frameworks, an organisation that applies ESRS is expected to be able to comply with the identification of the risks and opportunities to be disclosed under IFRS, using the outcome of its ESRS assessment of financial materiality. In other terms, the same financial materiality assessment process can support the identification of the risks and opportunities for both IFRS and ESRS purposes.

The general criteria for assessing materiality to support the identification of relevant indicators/data points to be reported is expected, in most cases, to rely on decision-usefulness, which is also the criteria used in IFRS S1 to identify the information to be reported (as material). While in ESRS the assessment of what is decision-useful considers both investors and other stakeholders, in IFRS this is limited to the needs of investors. However, for the financial materiality component of ESRS, the two perspectives are broadly to be equivalent.

[4] ESRS–ISSB Standards: https://www.ifrs.org/content/dam/ifrs/supporting-implementation/issb-standards/esrs-issb-standards-interoperability-guidance.pdf

Leveraging international instruments of due diligence:
As defined in the international instruments of the UN Guiding Principles on Business and Human Rights and the OECD Guidelines for Multinational Enterprises, the materiality assessment of an organisation's impacts, risks, and opportunities is informed by the outcome of its due diligence process. Furthermore, on 23 February 2022, the European Commission adopted a proposal for a Directive on corporate sustainability due diligence (aka CSDDD[5]) and on 24 May 2024 the Council of the European Union approved the political agreement, thereby completing the adoption process. There is now a two-year period for member states to transpose the Directive into national law. The aim of the CSDDD is to foster sustainable and responsible corporate behaviour in companies' operations and across their global value chains.

The new rules will ensure that companies in scope identify and address adverse human rights and environmental impacts of their own operations, products, or services through its business relationships inside and outside Europe. This can be particularly useful when analysing the organisation's value chain and identifying the impacts originated there. Additionally, due diligence instruments provide criteria to prioritise actions, based on the severity and likelihood of the impacts previously identified.

In that sense, the due diligence process can help an organisation both (a) to identify and assess its potential and actual negative impacts, as well as (b) to assess their materiality for reporting purposes based on the criteria of severity and likelihood. The identification of material impacts also supports the identification of material sustainability risks and opportunities, which are often a consequence of such impacts. Through this process, the organisation can also identify affected stakeholders, which should be engaged when assessing the materiality of the impacts, risks, and opportunities.

The CSDDD was 'watered down' from the original proposal to only apply to very large organisations, but we should expect that applicability thresholds will reduce over time:

- **Large EU limited liability companies and partnerships**: +/– 6000 companies – >1000 employees and >EUR 450 million turnover (net) worldwide.
- **Large non–EU companies**: +/– 900 companies – > EUR 450 million turnover (net) in EU.

[5] For further information see: https://commission.europa.eu/business-economy-euro/doing-business-eu/corporate-sustainability-due-diligence_en

The Directive contains provisions to facilitate compliance and limit the burden on companies, both in scope and in the value chain. Micro companies and SMEs are not covered by the proposed rules. However, the Directive provides supporting and protective measures for SMEs, which could be indirectly affected as business partners in value chains.

4

Data Collection and Analysis

In this chapter: Data gathering and overcoming analysis challenges. Reporting assurance. Leveraging technology for data management.

Data collection is probably the biggest headache for organisations preparing an ESG report, and unfortunately there are no major shortcuts. Gathering relevant data, and in a timely manner, can quickly become a resource-intensive, time-consuming, and frustrating process. Added to your internal data gathering is the need to harness inputs from operational departments and stakeholders, and undertake research into appropriate indicators related to various topics. Phew! Where to begin?

Actually, your data collecting process began with your stakeholder engagement, when you were able to discuss suitable metrics. Then after your materiality assessment, you should have a comprehensive list of material and monitored topics, each requiring data points of specific information (these can be text or numerical bits of data) for reporting the associated indicator and metric. So, you have a basic list of what you need to know. Now we have to go find it.

In reality, organisations typically begin with internal data sources and available tools, and then migrate to specialised tools and APIs, which will be covered in more detail in section 4.3, ESG Software. Table 18 outlines the major pros and cons of each approach, but the major bonus with external data management tools is easy access to sector guidance and benchmarking, which can be extremely difficult to source. However, such information is not required at the beginning of your data gathering process, so I recommend that you also begin with internal systems.

The first and easiest locations to source data are from your financial reports and accounting system, as well as your human resources department.

Table 18 Comparison of traditional and specialised data tools.

Traditional data sources and tools (Excel, Google drive, internal systems, etc.)		**Specialised ESG data tools (Online tools/platforms and specialised APIs)**	
Pros	**Cons**	**Pros**	**Cons**
✓ Low initial cost ✓ Easy to implement ✓ Ideal for small data volumes ✓ Suits privacy-concerned external sources	✗ Time-consuming to aggregate data ✗ Human error rates due to manual data management ✗ No missing or data gap analysis ✗ No benchmarking tools	✓ Sector-specific guidance and benchmarks ✓ Visualisation tools, e.g., dashboards for real-time progress monitoring ✓ Regulation or methodology update notifications ✓ Long-term cost savings ✓ Low errors	✗ Higher setup costs ✗ Expensive for external sources ✗ Problematic system integration across platforms

Source: Author.

Legal and compliance are second on the list, and then you are off to operational departments. In fact, the majority of internal information should be easy to find on the condition you have standardised systems and procedures, but often, this is not the case. As you collect data, you will find gaps, anomalies, and contradictions; this is 'normal' and should be expected, and then it is time to get the Green Team involved.

One of the most important activities for the Green Team is to problem-solve data collection, for example, offices with different billing periods, or deciding a methodology to proportion flat rate figures, or find missing information lost in an office move, or finding the right departmental contact since a restructuring. In fact, Green Team should be involved with every aspect of the data management cycle, (see Graphic 16) as they have the authority and connections for the job.

4.1 Information Relating to Financial Statements

The ESRS expects cross-referencing of ESG data with financial statements wherever appropriate, including general disclosures that concern financial impacts, risks, and opportunities, and extends to all monetary amounts or other quantitative data points that are considered material and are presented in financial statements.

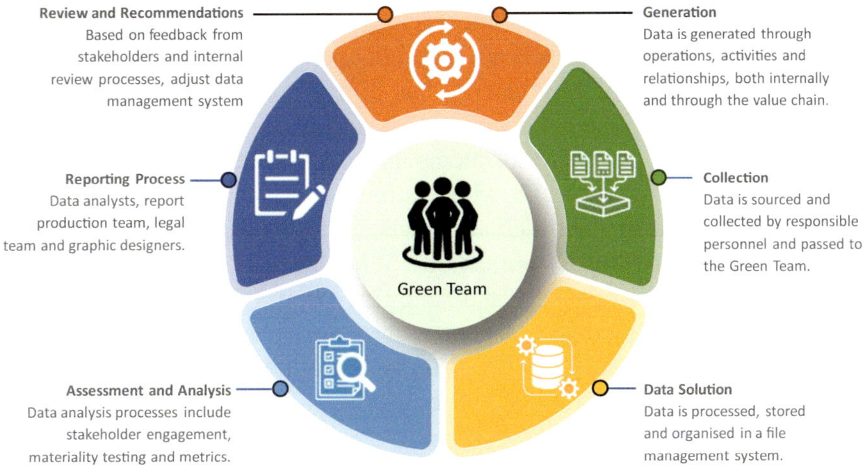

Graphic 16 Green Team roles in data management cycle. Source: Author.

As you review and extract information from your financial statements, check that all information describing cross-references is clear and concise, and includes a reference to the relevant paragraph in the financial statements. Consistency needs to be at the level of a single data point and include a reference to the relevant line item/paragraph of a footnote of the financial statements.

It is important to consider context when reviewing your financial statements. When describing the relationships between different pieces of information you need to develop a connecting narrative that links governance, strategy, and risk management to related metrics and targets. For example, your sustainability strategy may include adaption to fully renewable energy supplies, which is likely to have an effect on your financial statements or financial plans, or on metrics and targets used to measure progress against performance. Similarly, you might need to explain how the use of natural resources and changes within your supply chain could amplify, change, or reduce material impacts, risks, and opportunities. Accordingly, you may need to link this information to the potential or actual effect(s) on production costs, your organisation's strategic response to mitigate such impacts or risks and subsequent related investment in new assets. This information may also need to be linked to information in the financial statements and to specific metrics and targets.

Sometimes direct cross-referencing can be problematic, perhaps because direct reconciliation is not possible or the nature of qualitative information deemed sustainably material has different characteristics to related information in the financial reports. This is especially true when comparing metrics that in the financial statements are historical but forward looking in the sustainability report. Or when estimating the recoverable amount of assets, the amount of liabilities, or provisions in financial statements associated with material impact projections. When there is no direct or indirect link, you need to explain why and state (based on a threshold for material information) the consistency of data, assumptions used, and how the qualitative information included in the sustainability statements corresponds with the data, assumptions, and qualitative information included in the financial statements.

4.2 Common Problems with Data

Timing issues are common when gathering data, but there are more that you will come across. Let us review the most common:

- **Lack of standardisation**: There is a lack of standardised metrics, methodologies, and reporting frameworks for many ESG topics, making it difficult to compare data. So where possible, use recognised international approaches like the Greenhouse Gas Protocol or SBTi for calculating Scope 1, 2, and 3 emissions. In each of the following chapters, I will discuss potential methodologies and approaches for each topic.
- **Data quality and accuracy**: Ensuring the accuracy and reliability of ESG data is continually challenging. Fragmented information is commonplace and accuracy questionable, both of which can lead to potential greenwashing or allegations of misrepresentation.
- **Data availability**: Some ESG data, particularly on social factors, may not be readily available or easily quantifiable. Fortunately, the CSRD/ESRS does recognise that there are significant constraints to sourcing some data and that it will take time for appropriate tools and methodologies to become available. However, for material topics, you do need to make a serious effort to identify possible data sources and for missing data, and if necessary explain in full why you are unable to do so.

- **Scope and materiality changes**: Acquisitions, divestments, strategy shifts, and similar major changes to your organisation and in your value chain can be very disruptive to data gathering as well as your materiality assessments. Time periods can also pose a major complicating issue as some indicators/topics may require several years, including:
    - identification and analysis of value chain impacts;
    - calculation of scope 3 GHG emissions;
    - analysis of long-term sustainability data to support decision-making and management of underlying impacts, risks, and opportunities; development and approval of climate transition plan, adaptive technologies, or new investments.
- **Regulatory environment**: The regulatory landscape for ESG reporting is rapidly evolving, with varying requirements across jurisdictions. This lack of consistency can create compliance challenges for companies operating globally.
- **Complexity and interconnectedness of social impacts**: Social impact issues are often interconnected, making it challenging to isolate specific factors and consistent data sources. For example, community members change, opinions shift, and seemingly unrelated external factors can suddenly become relevant, further complicating analysis.
- **Estimations and duplications**: You will not be able to collect every single piece of data that you need to complete every indicator metric, so at some point you will make estimations, which can fragment data. An added problem is the potential for duplications (for example, board management statement, non-financial statement of the board, financial report, corporate report, ESG report, marketing communication, internal communication, internal audit reporting, internal monitoring of strategy implementation) or omissions in data, and both pose a major risk to organisations. Without clear methodologies, reporting organisations rely on their own estimations, which can be vague or misinterpreted. In turn, this can result in errors in your dataset leading to a cascading inaccuracy through your reporting. So, when you are collecting data, be explicit about the required metric detail including scope definition(s) and data collection methods. Another data challenge involves overlapping reporting boundaries within your value chain, many of which

are based on practices and processes beyond your control. You may need to host some value chain training events to ensure data collection consistency.

A major concern is in identifying hotspots – areas of significant environmental or social impact – which can be inconsistent through the use of duplicated, omitted, estimated, or redundant data that may obscure the identification of critical areas requiring attention or improvement. Verifiability is always critical, even more so when describing the true extent of significant impacts based on information sourced from multiple organisations who are using different methodologies or scopes.

Errors in ESG reports can erode stakeholder trust, even from the value chain partners who have helped to collect the data. When stakeholders encounter inconsistencies or redundancies in reported data, they may question the transparency and reliability of your sustainability disclosures. Impaired trust can have far-reaching consequences, affecting relationships with investors, customers, employees, and regulatory bodies, and undermining your organisation's credibility in its commitment to sustainability. Avoid multiple data reviews with value chain partners who will probably have to go through the whole gathering process again as they do not have the same system as you. Repetition of effort increases administrative burden and frustration, the prime causes of data fatigue, which then erodes energy levels for implementing transition plans and finding positive impact solutions.

Enhancing collaboration and coordination among your value chain is vital in minimising duplications and errors in your reporting. Work with your value chain partners to establish common reporting methodologies, share data, and align sustainability priorities. Collaborative initiatives, industry alliances, and partnerships facilitate a unified approach to ESG reporting, reducing redundancies and ensuring a more cohesive assessment of materiality and coordination in data gathering efforts.

This is all really good stuff, but it takes time, is often given a low priority compared to day-to-day internal operational issues, and does not fully address the problems as transparency is limited between organisations. Organisations should proactively communicate their commitment to minimising redundancies, explain reporting methodologies, and seek stakeholder input on ways to improve reporting efficiency. Transparent communication fosters trust and aligns stakeholders with the organisation's sustainability goals.

Leveraging technology through software for data integration throughout your value chain provides a framework for collaborations, agreements on standards, and drives data accuracy and integrity. Dispersed, integrated software can streamline processes and reduce duplications. Implementing dispersed integrated systems allows seamless data exchange, and reporting across the value chain enhances efficiency. Technologies such as machine learning, data-sharing platforms, and AI-based solutions can promote real-time data accessibility, accuracy, and consistency in ESG reporting.

TIPS AND TRICKS

- Identify multiple internal and external focal points using established communication channels for data collection.
- Prepare explanations of required data and identify data collection gaps.
- Analyse data sources and potential for automating data gathering and aggregation.
- Specify verifiability requirements with a single source of truth for data elements.
- Identify and implement with your partners specialised in data management tools and software to support the data collection process and management.
- Record everything! Prepare explanatory notes on data collection methodology and any significant trends to include in the report.

4.3 ESG Software

To improve the governance of ESG data collection and reporting, and to prepare for a transition to reasonable assurance, you are almost certainly going to need the support of third-party software and service providers. Additionally, the need for real-time monitoring will grow as firms look to wardoff claims of greenwashing and become more proactive in ESG risk management. ESG software can enhance ESG performance, improve transparency, and meet the growing demands for sustainable and responsible business practices by:

1. **Data collection and management**: Enables efficient data collection, aggregation, and management of information. It can help you gather data from various sources, track key performance indicators (KPIs), and streamline the process of reporting and disclosure.
2. **Performance monitoring and reporting**: Allows you to monitor your organisation's sustainability performance and track progress towards

specific goals and targets. It helps generate comprehensive reports and visualisations that showcase your ESG performance to stakeholders, investors, and regulatory bodies.

3. **Compliance and risk management**: Helps you stay compliant with relevant regulations and standards related to sustainability and ESG reporting. It provides tools for assessing and managing sustainability risks, identifying areas for improvement, and implementing appropriate mitigation strategies.

4. **Stakeholder engagement and transparency**: Facilitates stakeholder engagement by providing a centralised platform to communicate your organisation's ESG initiatives and progress. It enhances transparency by sharing relevant sustainability information with investors, customers, employees, and other stakeholders who are increasingly interested in a company's ESG practices.

5. **Benchmarking and comparison**: Benchmarking capabilities that allow you to compare your organisation's ESG performance against industry peers or established sustainability indices. This helps identify areas of strength and areas where improvements can be made, fostering a culture of continuous improvement.

6. **Decision-making support**: Generates insights to support strategic decision-making. It helps identify opportunities for sustainable innovation, assess the impact of potential initiatives, and align ESG considerations with broader business strategies.

7. **XBRL tagging**: The CSRD/ESRS requires ESG reports to be in both human and machine readable formats using XBRL. Using the appropriate software, the closer your ESG report can follow the ESRS statement structure, and the easier XBRL tagging becomes (see Chapter 16, Report Formation).

4.4 Assurance

The CSRD/ESRS does not prescribe the types of documentation to be presented as evidence, nor has the verification process been confirmed to date (as at October 2024), although all standards do require at least a limited assurance process.

Table 19 Limited vs. reasonable assurance.

Limited assurance	**Reasonable assurance**
Both limited and reasonable assurance involve an external party who, based on sufficient appropriate evidence, can assess that the risk of material misstatement has been reduced to an acceptably low level in the context of the engagement.	
Limited assurance allows for the assurance of subject matter, which is less well-defined and for which the control environment is less mature and robust.	Reasonable assurance is a more comprehensive process than a limited assurance engagement.
The assurance provider performs different or fewer tests than those required for reasonable assurance or uses smaller sample sizes for the tests performed. The auditor focuses on understanding the process used to compile the reported information and also on inquiry, observation, and analytical procedures (e.g., observing the data at an aggregated level).	The assurance provider must conduct extensive procedures as described in the standard. This includes identifying risks and assessing the possibility that any matters may be unfairly represented; testing the operating effectiveness of the organisation's internal controls on which the auditor intends to rely; and substantive procedures.
Based on the procedures performed and evidence obtained, the auditor expresses a conclusion in a negative manner, such as, 'Based on the procedures performed and evidence obtained, nothing has come to our attention that causes us to believe that the subject matter information has not complied, in all material respects, with XYZ law or applicable criteria'.	The auditor's conclusions are stated in a positive manner, such as, 'In our opinion, the [appropriate party's] statement that the key performance indicators are presented in accordance with XYZ criteria is, in all material respects, fairly stated'. As in an audit of financial statements, this is the highest level of assurance but is still not absolute.

Source: Author.

The CSRD/ESRS modifies the Accounting Directive, which requires a certain level of documentation to be produced for internal purposes, such as to inform those in charge of the governance over the process of sustainability reporting, in order to be able to prepare the disclosure and to allow assurance providers to perform their work. Additionally, the CSRD/ESRS does require a detailed explanation of the monitoring and assessment process to assess impacts, risks, and opportunities as well as potential financial impacts. Finally, the Accounting Directive requires an assurance provider to 'express an opinion based on a limited assurance engagement as regards the compliance of the sustainability reporting adopted pursuant to Article 29b or Article 29c, i.e. the sustainability reporting prepared using ESRS including … the process carried out by the organisation to identify the information reported pursuant to those sustainability reporting standards, ...',[1] so your assurer will need verification documentation.

[1] EU Accounting Directive 2013/34/EU

The main goal of the CSRD/ESRS is to ensure consistency across EU reporting (see more in Chapter 16, Report Formation), so it establishes a new framework for EU-wide assurance in both human- and machine-readable formats. It also sets guidelines for reporting in accordance with the proposed IFRS framework. External assurance can be provided by statutory auditors and audit firms who are already experienced in auditing financial statements, as well as sustainability experts who have expertise in the reporting standard. In the sustainability report audit, you will need to provide detailed explanations of how financial data and sustainability performance are interconnected.

Repeating the materiality assessment process may not be necessary each reporting period (yearly) as practices may vary. Some organisations may prefer to update their materiality analysis (for example, process, criteria, research on sustainability matters, multi-stakeholder engagement, etc.) on a two- or three-year cycle, while others may prefer performing targeted updates on an annual basis (for example, list of impacts or risks and opportunities, while not repeating the initial materiality steps). However, it should be noted that materiality assessment is a dynamic process that will be subject to the inherent evolution of the organisation and needs to be updated on an ongoing basis.

TIPS AND TRICKS

The EU has repeatedly stated the desire to see ESG report formulation and ratings kept separate in every way possible, so ensure that any organisation you engage with has no direct or indirect relationship with a ratings organisation – or you both might be judged to have a conflict of interest.

PART 2

ESG Reporting Disclosures

ESG reporting structures differ depending on their core user group needs but basically include similar disclosures, metrics, and methodologies; for example, all accept the SBTi approach to climate change reporting. In summary, we can say:

- The GRI focuses on sector-based sustainability reporting, requiring organisations to disclose economic, environmental, and social impacts through a structured set of indicators and metrics that reflect investor and user needs within specific sectors. Over time, their aim is to develop standards for 40 sectors.
- The IFRS primarily focuses on combining financial reporting with sustainability disclosures, requiring organisations to disclose financial information such as balance sheets, income statements, and cash flow statements following specific accounting standards and that relate to sustainability matters.
- The ESRS integrates elements of both GRI and IFRS, requiring organisations to disclose information related to environmental, social, and governance factors alongside traditional financial disclosures to provide a comprehensive view of the organisation's performance and impact.

As the CSRD/ESRS includes all elements of both the GRI and IFRS standards, I will use it as a guide through ten 'topical standards' with seven cross-cutting themes for the following chapters. In each chapter, I will review linkages between standards where there is a common approach and/or indicator and metric.

The CSRD/ESRS prescribes that each topical standard must include disclosures regarding policies and actions as well as metrics and targets:

- Governance: business conduct and business process, including general disclosures.
- Environment: climate change, pollution, water and marine, biodiversity, and circular economy.
- Social: own workforce, workers' value chain, communities, and consumers/end-users.

Cross-cutting standards are sector agnostic general requirements that are mandatory for all organisations regardless of the results of your materiality assessment – see Chapter 3. In the CSRD/ESRS, the ESRS 1 prescribes general requirements and principles applicable to all reporting topics, but does not include any specific disclosure requirements. Most of the ESRS 1 general principles have already been covered in previous chapters as they include approaches to quantitative and qualitative information, severity, etc. The cross-cutting disclosure requirements are provided in ESRS 2, which will be discussed in Chapter 5.

It is important to note that the ESRS permits organisations to omit all disclosure requirements in a topical standard if it assesses that the topic in question is not material. In that case, it may disclose a brief explanation of the conclusions of the materiality assessment for that topic but shall disclose a detailed explanation in the case of ESRS E1 climate change (Chapter 6). Similarly, the ESRS recognises that due to the size, capacity, resources, and skills of the organisation disclosure requirements in relation to action plans, targets, policies, scenario analysis, and transition plans may not be complete at the time of reporting and therefore proportionate in nature. Detailed explanation may be required for material matters.

A disclosure on any topic should contain structured information that covers:

1. Industry-specific issues and benchmarks by appropriate contextualisation and explanation of industry norms. This includes localised context based on geographical or regional challenges.
2. Defining internal and external linkages and boundaries helps to explain the scope of topics under your organisation's control (direct operations) and which are external (suppliers, distributors, etc.).
3. Completeness by encompassing all relevant information to provide a holistic view of impacts, risks, and opportunities, which may

include how you have ensured consistency in data collection and methodologies.

4. Stakeholder engagement information about how you have considered the perspectives of various stakeholders, such as investors, customers, employees, and communities.
5. Applying consistent materiality testing when identifying the most significant ESG issues for your organisation and its stakeholders.

In the following chapters, we will focus on statutory reporting indicators and metrics defined in the CSRD/ESRS as well as identify linkages with ISSB and GRI standards, the draft CSRD standard for listed small- and medium-sized enterprises (LSMEs) and the voluntary CSRD standard for unlisted SMEs (VSMEs).

It is important to understand the inter-relationship between disclosures in ESRS 2, including the minimum general disclosures (MDR), which is reflected in the machine readable XBRL structure discussed in Chapter 16. From your materiality assessment, topics, sub-topics, and sub-sub-topics are identified, for which impacts, risks, and opportunities (IROs) are then identified. For each IRO, there needs to be an associated policy and metric(s) that define actions and subsequent targets and tracking mechanisms, see Graphic 17 for interrelationships. Details for each are discussed in the following disclosure requirements.

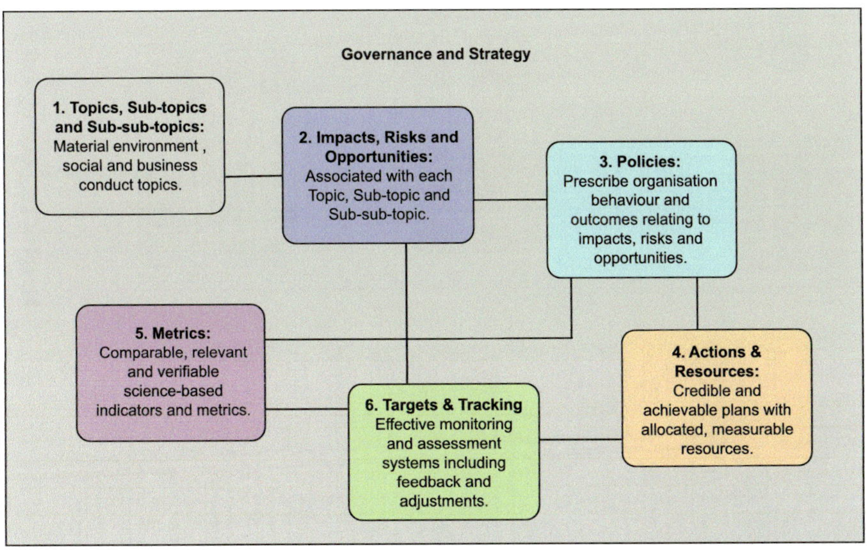

Graphic 17 Relationship between topics, policies, actions, metrics, and targets. Source: Author.

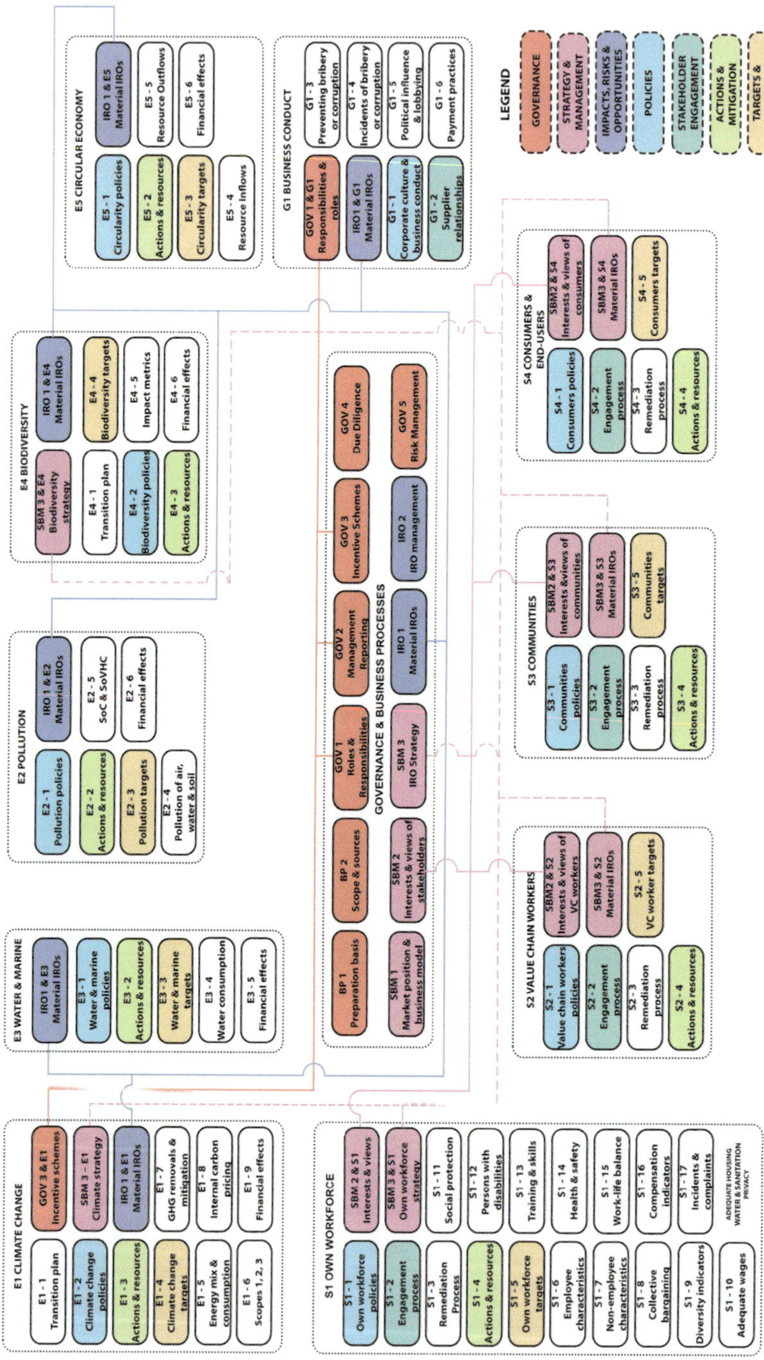

Graphic 18 Overview of the CSRD/ESRS reporting structure. Source: Author.

5

Minimum and General Disclosures

In this chapter: Minimum disclosure requirements. General disclosures concerning the role of administration, management, and supervisory bodies in governance, strategy, business process, impacts, risks, and opportunities. Due diligence and due process leading to policies, actions, and resource allocations.

ESG reports begin with disclosures about the nature of the organisation and its overall approach to sustainability and report preparation, which are called general disclosures. Throughout the CSRD/ESRS, indicators and metrics are referred to as data points (DPs), which are not pre-defined but are subject to your materiality testing. In other words, if your materiality testing deems an issue as not material, then the associated DPs do not need to be reported, except for some climate change and general organisational details, which are always mandatory and covered in the minimum disclosure requirements (MDR). In the CSRD/ESRS, the MDR includes your organisation's approach to policies, actions, metrics, and targets, and requires you to explain why you might not yet have developed and implemented policies or outcome-orientated measurable targets for some topics or data points.

General disclosures are defined in the ESRS 2 and are a dive into your business practices, governance, strategies and business model(s), and key information concerning impacts, risks, and opportunities that are covered in more detail in later chapters. The following CSRD/ESRS disclosures are always mandatory irrespective of the outcomes of the company's materiality assessment:

- all information covered in ESRS 2 general disclosures (called cross-cutting measures);
- all information covered in ESRS E1 climate change;

- information on policies, actions, and targets as outlined in ESRS S1 Own Workforce if the company has 250 or more employees (S1-1 up to S1-9); and
- specific data points across other topical standards, the SFDR, Pillar 3, Benchmark Regulation and EU Climate Law that are presented in Appendix C to ESRS 2 (identified in tables as Appendix C column).

The first section of the CSRD/ESRS is ESRS 1 General Application Requirements, which do not include any specific data points to report on, but does prescribe the characteristic of processes and disclosures, which has largely been addressed in preceding chapters.

ESRS 2 & MDR
BASIS FOR ESG REPORT PREPARATION
BP-1: General basis for preparation of the sustainability statements
BP-2: Disclosures in relation to specific circumstances

GOVERNANCE	STRATEGY	IMPACTS, RISKS, & OPPORTUNITIES	METRICS & TARGETS
• GOV1: *The role* of the administrative, management, and supervisory bodies includes *board gender diversity* • GOV2: Information *provided to* and sustainability matters *addressed by* the organisation's administrative, management, and supervisory bodies • GOV3: Integration of sustainability strategies and performance in *incentive schemes* • GOV4: Statement on sustainability *due diligence* • GOV5: *Risk management and internal controls* over sustainability reporting	• SBM1: *Market positioning*, strategy, business model(s), and value chains • SBM2: Interests and views of *stakeholders* • SBM3: Material impacts, risks, and opportunities and their *interaction* with strategy and business model(s) • INCLUDES CROSS-CUTTING ISSUE: • Intellectual property, know-how, or results of innovation • INCLUDES Declaration if company is active in: • Fossil fuel sector, chemicals production, controversial weapons, activities negatively affecting biodiversity and operations, geographies, or commodities at significant risk of incidents of forced, compulsory, or child labour.	MATERIALITY ASSESSMENT DISCLOSURE • IRO1: Description of *processes to identify and assess* material impacts, risks, and opportunities • IRO2: Disclosure requirements statement includes *tabulated index of disclosures* INCLUDES CROSS-CUTTING POLICES AND ACTIONS • DC-P: Policies adopted to manage material sustainability matters • DC-A: Actions and resources in relation to material sustainability matters	CROSS-CUTTING RELATING TO MATERIALITY ASSESSMENT DC-M: *Metrics* in relation to material sustainability matters DC-T: Tracking effectiveness of *policies and actions* through targets (sustainability strategy goals, ESG performance – covering all material topics)

5.1 Minimum Disclosure Requirements (MDR)

The MDR includes your organisation's approach to policies, actions, metrics, and targets, and requires you to explain why you might not (yet) have developed and implemented policies or outcome-orientated measurable targets for some topics or data points.

The MDR mandates disclosures of some data points (metrics and indicators) if they have been assessed as material (and if not assessed as material, then an explanation of why they should not be stated) according to your materiality assessment for: E1-2; E1-3; E1-4; E2-1; E2-2; E2-3; E3-1; E3-2; E3-3; E4-2; E4-3; E4-4; E5-1; E5-2; E5-3; S1-1; S1-4; S1-5; S2-1; S2-4; S2-5; S3-1; S3-4; S3-5; S4-1; S4-4; S4-5; G1-1 (policies); G1-4 (actions) – each section will be flagged for reference.

5.1.1 MDR-P – Minimum disclosure: Policies adopted to manage material sustainability matters

The organisation is required to disclose information about the policies you have adopted to manage and monitor material sustainability matters. Your policy must, at a minimum, include corporate culture, business conduct, and climate change, as well as any topics that your materiality assessment has identified as significant impacts, risks, or opportunities.

Your disclosure must include the following information:

(a) a description of the key contents of the policy, including its general objectives and which material impacts, risks, or opportunities the policy relates to and the process for monitoring;

(b) a description of the scope of the policy, or of its exclusions, in terms of activities, upstream, and/or downstream value chain, geographies, and, if relevant, affected stakeholder groups;

(c) the most senior level in your organisation that is accountable for the implementation of the policy;

(d) a reference, if relevant, to third-party standards or initiatives the organisation commits to respect through the implementation of the policy;

(e) if relevant, a description of the consideration given to the interests of key stakeholders in setting the policy; and

(f) if relevant, whether and how the organisation makes the policy available to potentially affected stakeholders, and stakeholders who need to help implement it.

NOTE: If a single policy covers both an environmental matter and a social matter, the organisation may report on the policy in the environmental section of its sustainability statement. In this case, it should include in the social section a cross-reference to the environmental section where the policy is reported. Equally a policy may be reported in the social section with a cross-reference to it in the environmental section.

5.1.2 MDR-A – Minimum disclosure: Actions and resources in relation to material sustainability matters

Where the implementation of a policy requires actions (see Section 3.9, 'Materiality Report and Action Plan'), to achieve its objectives, as well as when actions are implemented without a specific policy, your organisation needs to disclose the following information:

a) the list of key actions taken in the reporting year and planned for the future, their expected outcomes, and, where relevant, how their implementation contributes to the achievement of policy objectives and targets;

b) the scope of the key actions (i.e., coverage in terms of activities, upstream, and/or downstream value chain, geographies, and, where applicable, affected stakeholder groups);

c) the time horizons under which your organisation intends to complete each key action;

d) if applicable, key actions taken (along with results) to provide for and cooperate in or support the provision of remedy for those harmed by actual material impacts;

e) if applicable, quantitative and qualitative information regarding the progress of actions or action plans disclosed in prior periods; and

f) where the implementation of an action plan requires significant operational expenditures (OpEx) and/or capital expenditures (CapEx), the organisation shall:

   i) in the form of a table, describe the type of current and future financial and other resources (broken down between capital expenditure and operating expenditures) allocated to the action plan, including, if applicable, the relevant terms of sustainable finance instruments, such as green bonds, social bonds, and green loans, the environmental or social objectives, and whether the ability to implement the

Table 20 MDR-P: Policies adopted to manage material sustainability matters.

ESRS Topic	ESRS data type	ESRS paragraph	Description	Data type	Tips and tricks	Related CDR	VSMEs	ISSB IFRS	GRI
ESRS 2	MDR-P	65 a	Description of key contents of your ESG/sustainability policy	Narrative	Either a single or multiple policy document(s)			IFRS S2.25(a)(vi)	
ESRS 2	MDR-P	65 b	Description of scope of policy or of its exclusions	Narrative	Policy(s) coverage, including value chain, geographic locations, and possibly stakeholders				GRI 2-23, GRI 3-3
ESRS 2	MDR-P	65 c	Description of the most senior level in the organisation that is accountable for the implementation of policy	Narrative	Brief biography of responsible Board member and/or bios of the Green Team				GRI 2-23, GRI 2-24
ESRS 2	MDR-P	65 d	Disclosure of third-party standards or initiatives that are respected through the implementation of policy	Narrative	International, regional, and national standards including ISO and industry codes of practice				
ESRS 2	MDR-P	65 e	Description of consideration given to interests of key stakeholders in setting the policy	Narrative	Policy feedback from key stakeholders during engagement and materiality assessment				
ESRS 2	MDR-P	65 f	Explanation of how the policy is made available to potentially affected stakeholders and stakeholders who need to help implement it	Narrative	Make the policy public on your website with a dedicated page and feedback mechanism				GRI 2-23

Source: CDR (EU) 2023/2772 supplementing Directive 2013/34/EU, *IFRS S1 General Requirements for Disclosure of Sustainability-related Financial Information, IFRS S2 Climate-related Disclosures, GRI 1, 2, 3: 11, 12, 13, 14, 306, 101, 201, 202, 203, 204, 205, 206, 207,301, 302, 303, 304, 305, 306, 308,401, 402, 403, 404, 405, 406, 407, 408, 409, 410, 411, 413, 414, 415, 416, 417, 418*, and the author.

Table 21 MDR-A: Actions and resources in relation to material sustainability matters.

ESRS Topic	ESRS data type	ESRS paragraph	Description	Data type	Tips and tricks	Related CDR	VSMEs	ISSB IFRS	GRI
ESRS 2	MDR-A	68 a	Disclosure of key action	Narrative	Action plan summary of those relating to significant material matters			IFRS S2.14(a)(v)	
ESRS 2	MDR-A	68 b	Description of scope of key action	Narrative	Describe action coverage, including value chain, geographies, and possibly affected stakeholders			IFRS S2.14(a)(v), IFRS S2.14(a)(ii), IFRS S2.14(a)(iii)	GRI 3-3
ESRS 2	MDR-A	68 c	Time horizon under which key action(s) to be completed	Semi-narrative	Planned timeline, with deadline(s)			IFRS S2.14(a)(v)	
ESRS 2	MDR-A	68 c	Year when key action is to be completed	GYear					
ESRS 2	MDR-A	68 d	Description of key action(s) taken, and results, to provide for and cooperate in, or support provision of remedy for those harmed by actual material impacts	Narrative	When harm has been caused by material impacts, describe how they have been remedied			IFRS S2.14(a)(v)	
ESRS 2	MDR-A	68 e	Disclosure of quantitative and qualitative information regarding the progress of actions or action plans disclosed in prior periods	Narrative	Only where there are relevant ongoing issues relating to a current material matter			IFRS S2.14(a)(v), IFRS S2.14(c)	

(Continued)

Table 21 (*Continued*).

ESRS Topic	ESRS data type	ESRS paragraph	Description	Data type	Tips and tricks	Related CDR	VSMEs	ISSB IFRS	GRI
ESRS 2	MDR-A	69 a	Disclosure of the type of current and future financial and other resources allocated to the action plan	Narrative	Only for actions requiring significant capital or operational resources, tabulate requirements, source of funds, funding preconditions (if any), and amounts			IFRS S2.14(a)(iv), IFRS S2.14(a)(v), IFRS S2.14(b)	
ESRS 2	MDR-A	69 b	Current financial resources allocated to the action plan (CapEx)	Monetary				IFRS S2.14(a)(iv)	
ESRS 2	MDR-A	69 b	Current financial resources allocated to the action plan (OpEx)	Monetary				IFRS S2.14(a)(iv), IFRS S2.14(a)(v), IFRS S2.14(b)	
ESRS 2	MDR-A	69 c	Future financial resources allocated to the action plan (CapEx)	Monetary				IFRS S2.14(a)(v), IFRS S2.14(b)	
ESRS 2	MDR-A	69 c	Future financial resources allocated to the action plan (OpEx)	Monetary				IFRS S2.14(a)(v), IFRS S2.14(b)	

Source: CDR (EU) 2023/2772 supplementing Directive 2013/34/EU, *IFRS S1 General Requirements for Disclosure of Sustainability-related Financial Information, IFRS S2 Climate-related Disclosures, GRI 1, 2, 3: 11, 12, 13, 14, 306, 101, 201, 202, 203, 204, 205, 206, 207,301, 302, 303, 304, 305, 306, 308,401, 402, 403, 404, 405, 406, 407, 408, 409, 410, 411, 413, 414, 415, 416, 417, 418*, and the author.

actions or action plan depends on specific preconditions, e.g., granting of financial support or public policy and market developments;

ii) provide the amount of current financial resources and explain how they relate to the most relevant amounts presented in the financial statements; and

iii) provide the amount of future financial resources.

5.1.3 MDR-M – Minimum disclosure: Metrics in relation to material sustainability matters

For each significant material matter that you have identified in your assessment, you need to disclose the relevant metric(s) that you use to evaluate performance and effectiveness in relation to impact(s), risk(s), or opportunity(s). Metrics can be those defined in the relevant following chapter or that your organisation has identified, whether taken from other sources or developed by the organisation itself. Each metric requires the following information:

a) label and define the metric using meaningful, clear, and precise names and descriptions, including when a currency is specified, the unit of measure is the same as in the financial statements;

b) disclosure of the methodologies and significant assumptions behind the metric, including the limitations of the methodologies used; and

c) disclosure of whether the measurement of the metric is validated by an external body other than the assurance provider and, if so, which body.

5.1.4 MDR-T – Minimum disclosure: Tracking effectiveness of policies and actions through targets

As we have seen in previous chapters, the main purpose of ESG reporting is to report continuous progress of your actions in reducing or mitigating negative impacts over time. In disclosing information about tracking the progress of your actions, you should begin with a recap of your materiality report and whether and how stakeholders have been involved in target setting for each material sustainability matter. When targets address the prevention or mitigation of social impacts, quote the related SDG or international convention that relates to effects on human rights, welfare, or positive outcomes for affected stakeholders.

Your action plan is the main tool for tracking measurable time-bound outcome-oriented targets and your overall progress towards the adopted

Table 22 5.1.3 MDR-M: Metrics in relation to material sustainability matters.

ESRS topic	ESRS data type	ESRS paragraph	Description	Data type	Tips and tricks	Related CDR	VSMEs	ISSB IFRS	GRI
ESRS 2	MDR-M	75	Description of metrics used to evaluate performance and effectiveness, in relation to material impact, risk, or opportunity	Narrative	Ensure currencies and labels are the same as in your financial statements				
ESRS 2	MDR-M	77 a	Disclosure of methodologies and significant assumptions behind the metric	Narrative	Including scope, recommend a tabulated summary of material matters, related metrics, and methodology references			IFRS S2.29(a)(iii)(1–3)	GRI 3-3
ESRS 2	MDR-M	77 b	Type of the external body other than the assurance provider that provides validation	Narrative	Choose internationally recognised frameworks and/or methodologies that align with them. If using software check that it is compliant and has verification from the relevant framework, like SBTi, ISO, etc.			IFRS S1.50(c)	

Source: CDR (EU) 2023/2772 supplementing Directive 2013/34/EU, *IFRS S1 General Requirements for Disclosure of Sustainability-related Financial Information, IFRS S2 Climate-related Disclosures, GRI 1, 2, 3: 11, 12, 13, 14, 306, 101, 201, 202, 203, 204, 205, 206, 207,301, 302, 303, 304, 305, 306, 308,401, 402, 403, 404, 405, 406, 407, 408, 409, 410, 411, 413, 414, 415, 416, 417, 418*, and the author.

targets over time. You need to set a target for each material impact, risk, and opportunity, and include the relevant metrics to track it. Target tracking in the CSRD/ESRS prioritises absolute rather than relative measurements, for example, the number of tonnes of greenhouse gases emitted rather than a percentage reduction, although relative measures can be used alongside absolute values.

The best way to present your measurable time-bound outcome-oriented targets and progress towards the related policy's objectives is in a comprehensive table (including baseline and target value, milestones, and achieved performance over the prior periods) and defined in terms of expected results for people, the environment, or your organisation regarding material impacts, risks, and opportunities. For each target, the disclosure shall include the following information:

a) a description of the relationship of the target to the policy objectives as a percentage or decimal;

b) the defined target level to be achieved, including, where applicable, whether the target is absolute or relative and in which unit it is measured;

c) the scope of the target, including the organisation's activities and/or its upstream and/or downstream value chain where applicable and geographical boundaries;

d) the baseline value and base year from which progress is measured;

e) the period to which the target applies and, if applicable, any milestones or interim targets;

f) the methodologies and significant assumptions used to define targets, including, where applicable, the selected scenario, data sources, alignment with national, EU or international policy goals, and how the targets consider the wider context of sustainable development and/or local situation in which impacts take place;

g) whether your organisation's targets related to environmental matters are based on conclusive scientific evidence;

h) whether and how stakeholders have been involved in target setting for each material sustainability matter;

i) any changes in targets and corresponding metrics or underlying measurement methodologies, significant assumptions, limitations, sources, and processes to collect data adopted within the defined time horizon.

Table 23 MDR-T: Tracking effectiveness of policies and actions through targets.

ESRS topic	ESRS data type	ESRS paragraph	Description	Data type	Tips and tricks	Related CDR	VSMEs	ISSB IFRS	GRI
ESRS 2	MDR-T	80 a	Relationship with policy objectives	Decimal	For each metric, identify the progress of measurable, time-bound outcome-orientated targets towards specific policy objective(s)			IFRS S2.33(b)	
ESRS 2	MDR-T	80 b	Measurable target	Decimal or integer	Defined target in absolute or relative terms with the unit of measurement.			IFRS S2.B67, IFRS S2.33(g)**	
ESRS 2	MDR-T	80 b	Nature of target	Semi-narrative				IFRS S2.B67, IFRS S2.33(g)**	
ESRS 2	MDR-T	80 c	Description of scope of target	Narrative	Target coverage, including value chain, geographies, and possibly affected stakeholders			IFRS S2.33(c)	GRI 3-3
ESRS 2	MDR-T	80 d	Baseline value	Integer	Baseline value and base year from which the progress is measured			IFRS S2.33(e)	
ESRS 2	MDR-T	80 d	Baseline year	Integer				IFRS S2.33(e)	
ESRS 2	MDR-T	80 e	Period to which target applies	Semi-narrative	Define the period to which the target applies and, if applicable, any achieved or missed milestones or interim targets			IFRS S2.33(d), IFRS S2.33(f)	
ESRS 2	MDR-T	80 e	Indication of milestones or interim targets	Narrative				IFRS S2.33(d), IFRS S2.33(f)	

ESRS 2	MDR-T	80 f	Description of methodologies and significant assumptions used to define the target	Narrative	References to your selected methodologies should include data sources, alignment with external policy goals (e.g., EU, UN), scenario type (e.g. limiting climate change to 1.5 C) and related contexts			IFRS S2.33(h)	GRI 303-1
ESRS 2	MDR-T	80 g	Target related to environmental matters is based on conclusive scientific evidence	Semi-narrative					
ESRS 2	MDR-T	80 h	Disclosure of how stakeholders have been involved in target setting	Narrative	Reference to your materiality report and stakeholder engagement process				
ESRS 2	MDR-T	80 i	Description of any changes in target and corresponding metrics or underlying measurement methodologies, significant assumptions, limitations, sources, and adopted processes to collect data	Narrative	If relevant, link to General Disclosure ESRS 2 BP-2: Disclosures in relation to specific circumstances For most companies, this is not relevant to their first CSRD/ ESRS report			IFRS S2.29(a) (iii)(1–3), IFRS S2.34(d)	
ESRS 2	MDR-T	80 j	Description of performance against disclosed target	Narrative	Explain whether progress is in line with initial target and related circumstances			IFRS S2.34(b), IFRS S2.34©, IFRS S2.35	

** Additional relevant information can be found in Section 4. IFRS S2 (climate) to ESRS: information that an entity starting with ISSB Standards needs to know when also applying ESRS to enable compliance with both sets of standards.

Source: CDR (EU) 2023/2772 supplementing Directive 2013/34/EU, *IFRS S1 General Requirements for Disclosure of Sustainability-related Financial Information, IFRS S2 Climate-related Disclosures, GRI 1, 2, 3: 11, 12, 13, 14, 306, 101, 201, 202, 203, 204, 205, 206, 207,301, 302, 303, 304, 305, 306, 308,401, 402, 403, 404, 405, 406, 407, 408, 409, 410, 411, 413, 414, 415, 416, 417, 418*, and the author.

This includes an explanation of the rationale for those changes and their effect on comparability (see BP-2 in General Disclosures, in relation to specific circumstances); and

j) the performance against its disclosed targets, including information on how the target is monitored and reviewed and the metrics used, whether the progress is in line with what had been initially planned, and an analysis of trends or significant changes in the performance of your organisation towards achieving the target.

5.1.5 Minimum disclosure: If targets have not been set or are missing

In the case that the organisation has not set measurable time-bound outcome-oriented targets, you need to disclose whether and how your organisation tracks the effectiveness of its actions to address material impacts, risks, and opportunities and measures the progress in achieving its policy objectives. The detail required for each target that has not been set or is missing should include:

a) you may disclose whether such targets will be set in the future and the timeframe for setting them, or the reasons why your organisation does not plan to set such targets;

b) you must disclose whether it nevertheless tracks the effectiveness against a baseline of your policies and actions in relation to the material sustainability-related impact, risk, and opportunity, and if so:

   i) any processes through which you do so;

   ii) the defined level of ambition to be achieved and any qualitative or quantitative indicators you use to evaluate progress, including the base period from which progress is measured. For example, your organisation may assess an increase of wages by a certain percentage for those below a fair wage, or may assess the quality of its relationships with local communities by reference to the proportion of issues raised by communities that were resolved to their satisfaction. The baseline and the assessment of the progress shall be related to the impacts, risks, and opportunities that underpin the materiality of the matter addressed by the policy.

Table 24 Indicators and metrics related to missing or incomplete minimum disclosures.

ESRS topic	ESRS data type	ESRS paragraph	Description	Data type	Tips and tricks	Related CDR	VSMEs	ISSB IFRS S1 & S2	GRI
ESRS 2	MDR-P	62	Disclosure of reasons for not having adopted policies	Narrative	Even if you are lacking a policy or direct action, check to see if there is some other form of target tracking of material impacts, possibly related to another indicator or metric. Timeframes should be within the first two years of the first reporting period.				
ESRS 2	MDR-P	62	Disclosure of the timeframe in which your organisation aims to adopt policies	Narrative			V		
ESRS 2	MDR-A	62	Disclosure of reasons for not having adopted actions	Narrative					
ESRS 2	MDR-A	62	Disclosure of the timeframe in which your organisation aims to adopt actions	Narrative			V		
ESRS 2	MDR-T	81 a	Disclosure of the timeframe for the setting of measurable outcome-oriented targets	Narrative	Priority data points that need targets: E1-2; E1-3; E1-4; E2-1; E2-2; E2-3; E3-1; E3-2; E3-3; E4-2; E4-3; E4-4; E5-1; E5-2; E5-3; S1-1; S1-4; S1-5; S2-1; S2-4; S2-5; S3-1; S3-4; S3-5; S4-1; S4-4; S4-5; G1-1		V		
ESRS 2	MDR-T	81 a	Description of reasons why there are no plans to set measurable outcome-oriented targets	Narrative					
ESRS 2	MDR-T	81 b	Effectiveness of policies and actions is tracked in relation to material sustainability-related impact, risk, and opportunity	Semi-narrative	Cross-reference with above MDR-P and A, paragraph 62 (above)				

(Continued)

Table 24 (*Continued*).

ESRS topic	ESRS data type	ESRS paragraph	Description	Data type	Tips and tricks	Related CDR	VSMEs	ISSB IFRS S1 & S2	GRI
ESRS 2	MDR-T	81 b i	Description of processes through which effectiveness of policies and actions are tracked in relation to material sustainability-related impact, risk, and opportunity	Narrative	Ensure materiality assessment process can embrace missing or incomplete policies, actions, and targets				
ESRS 2	MDR-T	81 b ii	Description of defined level of ambition to be achieved and of any qualitative or quantitative indicators used to evaluate progress	Narrative	Identify key desired outcomes and relevant metrics				
ESRS 2	MDR-T	81 b ii 80 d	Base year from which progress is measured	GYear	Should be the same as your baseline year for all other indicators and metrics				

Source: CDR (EU) 2023/2772 supplementing Directive 2013/34/EU, *IFRS S1 General Requirements for Disclosure of Sustainability-related Financial Information, IFRS S2 Climate-related Disclosures, GRI 1, 2, 3: 11, 12, 13, 14, 306, 101, 201, 202, 203, 204, 205, 206, 207,301, 302, 303, 304, 305, 306, 308,401, 402, 403, 404, 405, 406, 407, 408, 409, 410, 411, 413, 414, 415, 416, 417, 418*, and the author.

5.2 General Disclosures

General disclosures are defined in the ESRS 2 and are a dive into your business practices (BP), governance (GOV), strategies and business model(s) (SBMs), and key information concerning impacts, risks, and opportunities (IROs). It is important to not confuse general disclosures with business conduct disclosures, which are covered in a separate reporting chapter; see Chapter 15.

Some general instructions for disclosures:

- In all the following general disclosures, where it states, 'shall disclose' specific data points then they must be disclosed irrespective of the outcome of your materiality assessment.
- Additionally, in BP-2, all disclosures relating to paragraph 17 must be disclosed.
- However, for organisations or groups not exceeding on their balance sheet dates, the average number of 750 employees during the financial year (on a consolidated basis where applicable) may omit the information required by ESRS E4, ESRS S1, ESRS S2, ESRS S3, or ESRS S4 unless required by EU Laws or Regulations (CDRs).
- Information regarding revenue by ESRS sectors (SBM1, paragraph 40b, 40c) and anticipated financial effects of material risks and opportunities on financial position, financial performance and cash flows over short, medium, and long term (SBM3, paragraph 48e) may be phased in; see Annex 2: List of Phased-in Disclosure Requirements.

5.2.1 BP-1: General basis for preparation of the sustainability statement

The first Business Process disclosures are contextual, whereby the purpose is to provide an understanding of how your organisation prepares its ESG report, including the scope of consolidation (specifically, is it the same as financial statements? And are there any specific exemptions or omissions?), information about key impacts, risks, and opportunities associated with upstream and downstream value chain and, where relevant, whether the organisation has used any of the options for omitting of information.

Table 25 Indicators and metrics related to Business Process BP-1 – General Disclosures.

ESRS topic	ESRS data type	ESRS paragraph	Description	Data type	Tips and tricks	Related CDR	VSMEs	ISSB IFRS S1 & S2	GRI
ESRS 2	BP-1	3	Disclosure of general basis for the preparation of sustainability statement	Narrative	Including organisational details, reporting period, and contact point from ESRS 1. Contextual information concerning preparation of your organisation's ESG report. Note that information may be omitted if: a) It is a secret, commercially or otherwise, and your organisation has taken reasonable steps to keep it a secret. b) It is specific information relating to intellectual property, know-how, or results of innovation. However, omission is not permitted to impair the overall relevance and meaning of the disclosure.				GRI2-1 GRI2-3
ESRS 2	BP-1	5 a	Basis for the preparation of sustainability statement on a consolidated or individual basis	Semi-narrative					GRI 2-2
ESRS 2	BP-1	5 b i	Scope of consolidation of consolidated sustainability statement is the same as for financial statements	Narrative					GRI 2-2
ESRS 2	BP-1	5 b ii	Indication of which subsidiary entities included in the consolidation are exempted from annual or consolidated sustainability reporting	Narrative					
ESRS 2	BP-1	5 c	Disclosure of extent to which sustainability statement covers upstream and downstream value chain	Narrative					GRI3-1
ESRS 2	BP-1	5 d	Option to omit specific piece of information corresponding to intellectual property, know-how, or results of innovation has been used	Semi-narrative					
ESRS 2	BP-1	5 e	Option allowed by member states to omit disclosure of impending developments or matters in course of negotiation has been used	Semi-narrative					

Source: CDR (EU) 2023/2772 supplementing Directive 2013/34/EU, *IFRS S1 General Requirements for Disclosure of Sustainability-related Financial Information, IFRS S2 Climate-related Disclosures, GRI 1, 2, 3: 11, 12, 13, 14, 306, 101, 201, 202, 203, 204, 205, 206, 207,301, 302, 303, 304, 305, 306, 308,401, 402, 403, 404, 405, 406, 407, 408, 409, 410, 411, 413, 414, 415, 416, 417, 418*, and the author.

5.2.2 BP-2: Disclosures in relation to specific circumstances

The second set of Business Process disclosures contains disclosures that may not apply to first time reports as they concern timing or presentation of information changes; however, you may need to include estimations regarding the value chain, sources of uncertainty, details concerning any corrections to previous period reporting, and information regarding whether you have relied on European Standardisation System (ISO/IEC or CEN/CENELEC) standards. Additionally, paragraph 17 contains a list of disclosures and data points relating to the outcome of your materiality assessment and if they are fully reported or being phased-in. Some more specific details of the required metrics are:

- *Deviation from time horizons* as defined in ESRS 1 as (a) for the short-term time horizon: the period adopted by the organisation as the reporting period in its financial statements; (b) for the medium-term time horizon: from the end of the short-term reporting period per (a) above to five years; and (c) for the long-term time horizon: more than five years.
- *Value chain estimations* when metrics include upstream and/or downstream value chain data estimated using indirect sources, such as sector-average data or other proxies, the organisation shall: (a) identify the metrics; (b) describe the basis for preparation; (c) describe the resulting level of accuracy; and (d) where applicable, describe the planned actions to improve the accuracy in the future.
- **Sources of estimation and outcome uncertainty**: Identify and disclose assumptions, approximations, and judgements that relate to quantitative metrics and monetary amounts that are subject to a high level of measurement uncertainty. By, for example, describing the dependence of the amount on the outcome of a future event, on a measurement technique or on the availability and quality of data from your organisation's upstream and/or downstream value chain; and when disclosing forward-looking information, the organisation may indicate that it considers such information to be uncertain.
- **Changes in preparation or presentation of sustainability information**: (a) Explain the changes and the reasons for them, including why the replaced metric provides more useful information; (b) disclose revised comparative figures, unless it is impracticable to do so. When it is impracticable to adjust comparative information for one or more prior periods, the organisation shall disclose that fact; and (c) disclose

the difference between the figure disclosed in the preceding period and the revised comparative figure.

- **Reporting errors in prior periods**: (a) The nature of the prior period material error; (b) to the extent practicable, the correction for each prior period included in the sustainability statement; and (c) if correction of the error is not practicable, the circumstances that led to the existence of that condition.
- **Phase-in for ESRS E4 Biodiversity, ESRS S1 Own Workforce, ESRS S2 Workers' Value Chain, ESRS S3 Affected Communities or ESRS S4 Consumers and End-Users:** If an organisation, on its balance sheet date, has less than an average of 750 employees during the financial year and decides to omit the information required by ESRS E4 Biodiversity, ESRS S1 Own Workforce, ESRS S2 Workers' Value Chain, ESRS S3 Affected Communities, or ESRS S4 Consumers and End-Users, it shall nevertheless disclose whether the sustainability topics have been assessed to be material. In addition, if one or more of these topics has been assessed to be material, the organisation shall, for each material topic:

  a) disclose the list of matters that are assessed to be material and briefly describe how the organisation's business model and strategy takes account of the impacts of the organisation related to those matters;

  b) briefly describe any time-bound targets it has set related to the matters in question, the progress it has made towards achieving those targets, and whether its targets related to biodiversity and ecosystems are based on conclusive scientific evidence;

  c) briefly describe its policies in relation to the matters in question;

  d) briefly describe actions it has taken to identify, monitor, prevent, mitigate, remediate, or bring an end to actual or potential adverse impacts related to the matters in question, and the result of such actions; and

  e) disclose metrics relevant to the matters in question.

Table 26 Indicators and metrics related to Business Process BP-2 – General Disclosures.

ESRS topic	ESRS data type	ESRS paragraph	Description	Data type	Tips and tricks	Related CDR	VSMEs	ISSB IFRS S1 & S2	GRI
ESRS 2	BP-2	6	Disclosures in relation to specific circumstances that may affect disclosures and must be reported alongside the disclosures to which they refer	Narrative	For first time reports, items of these disclosures will not apply. However, if you have reported ESG data in the past, then you will have to identify any changes.				
ESRS 2	BP-2	9	Medium- or long-term time horizons defined by ESRS 1 have been deviated from	Semi-narrative					
ESRS 2	BP-2	9 a	Disclosure of definitions of medium- or long-term time horizons	Narrative				IFRS S2.10(d)	
ESRS 2	BP-2	9 b	Disclosure of reasons for applying different definitions of time horizons	Narrative				IFRS S2.10(d)	
ESRS 2	BP-2	10	Metrics include value chain data estimated using indirect sources	Semi-narrative	Value chain estimations should be referenced with plans to improve accuracy in the future. For specific significant uncertainties, explain your assumptions to justify the lack of accuracy and plans to improve in the future; this may involve strategy or business model considerations, which should be cross-referenced (SBM 2 and SBM 3).				
ESRS 2	BP-2	10 a	Disclosure of metrics that include value chain data estimated using indirect sources	Narrative					
ESRS 2	BP-2	10 b	Description of basis for preparation of metrics that include value chain data estimated using indirect sources	Narrative					

(Continued)

Table 26 (*Continued*).

ESRS topic	ESRS data type	ESRS paragraph	Description	Data type	Tips and tricks	Related CDR	VSMEs	ISSB IFRS S1 & S2	GRI
ESRS 2	BP-2	10 c	Description of resulting level of accuracy of metrics that include value chain data estimated using indirect sources	Narrative					
ESRS 2	BP-2	10 d	Description of planned actions to improve accuracy in future of metrics that include value chain data estimated using indirect sources	Narrative					
ESRS 2	BP-2	11 a	Disclosure of quantitative metrics and monetary amounts disclosed that are subject to high level of measurement uncertainty	Narrative					
ESRS 2	BP-2	11 b i	Disclosure of sources of measurement uncertainty	Narrative					
ESRS 2	BP-2	11 b ii 12	Disclosure of assumptions, approximations, and judgements made in measurement	Narrative					
ESRS 2	BP-2	13 a	Explanation of changes in preparation and presentation of sustainability information and reasons for them	Narrative	For first time reports, items of these disclosures will not apply. However, if you have reported ESG data in the past, then you will have to identify any changes.				GRI 2-4
ESRS 2	BP-2	13 b	Disclosure of revised comparative figures	Narrative					GRI 2-4

ESRS 2	BP-2	13 c	Disclosure of difference between figures disclosed in preceding period and revised comparative figures	Narrative					GRI 2-4
ESRS 2	BP-2	14 a	Disclosure of the nature of prior period material errors	Narrative					GRI 2-4
ESRS 2	BP-2	14 b	Disclosure of corrections for prior periods included in sustainability statement	Narrative					GRI 2-4
ESRS 2	BP-2	14 c	Disclosure of why correction of prior period errors is not practical	Narrative					
ESRS 2	BP-2	15	Disclosure of other legislation or generally accepted sustainability reporting standards and frameworks based on which information has been included in sustainability statement	Narrative	This can include international, regional, and national standards including ISO and industry codes of practice.				
ESRS 2	BP-2	15	Disclosure of reference to paragraphs of standard or framework applied	Narrative					
ESRS 2	BP-2	AR 2	Which, if any, European standards approved by European Standardisation System (ISO/IEC or CEN/CENELEC standards) have been relied on	Semi-narrative			V		

(Continued)

Table 26 (*Continued*).

ESRS topic	ESRS data type	ESRS paragraph	Description	Data type	Tips and tricks	Related CDR	VSMEs	ISSB IFRS S1 & S2	GRI
ESRS 2	BP-2	AR 2	Disclosure of extent to which data and processes that are used for sustainability reporting purposes have been verified by external assurance provider and found to conform to corresponding ISO/IEC or CEN/CENELEC standard	Narrative			V		
ESRS 2	BP-2	16	List of DRs or DPs mandated by a disclosure requirement	Narrative	Comprehensive table of material matters and related disclosures; this can be in the form of a table or index.				
ESRS 2	BP-2	17	Topic has been assessed to be material	Semi-narrative	For organisations with less than 750 employees, it is possible to phase-in some material disclosures; however, they do require explanation of impacts, risks, and opportunities as well as mitigation measures, associated actions, and targets to the best extent possible. See Annex 2: List of phased-in disclosure requirements.				
ESRS 2	BP-2	17 a	List of sustainability matters assessed to be material (phase-in)	Semi-narrative					
ESRS 2	BP-2	17 a	Disclosure of how business model and strategy take account of impacts related to sustainability matters assessed to be material (phase-in)	Narrative					

ESRS 2	BP-2	17 b	Description of any time-bound targets set related to sustainability matters assessed to be material (phase-in) and progress made towards achieving those targets	Narrative					
ESRS 2	BP-2	17 c	Description of policies related to sustainability matters assessed to be material (phase-in)	Narrative					
ESRS 2	BP-2	17 d	Description of actions taken to identify, monitor, prevent, mitigate, remediate, or bring end to actual or potential adverse impacts related to sustainability matters assessed to be material (phase-in) and result of such actions	Narrative					
ESRS 2	BP-2	17 e	Disclosure of metrics related to sustainability matters assessed to be material (phase-in)	Narrative					

Source: CDR (EU) 2023/2772 supplementing Directive 2013/34/EU, *IFRS S1 General Requirements for Disclosure of Sustainability-related Financial Information, IFRS S2 Climate-related Disclosures, GRI 1, 2, 3: 11, 12, 13, 14, 306, 101, 201, 202, 203, 204, 205, 206, 207,301, 302, 303, 304, 305, 306, 308,401, 402, 403, 404, 405, 406, 407, 408, 409, 410, 411, 413, 414, 415, 416, 417, 418*, and the author.

5.2.3 GOV-1: The role of the administrative, management, and supervisory bodies

GOV-1 disclosures involve the composition and diversity of management bodies, their roles and responsibilities for your organisation's ability to manage impacts, risks, and opportunities, and specific supervisory bodies concerned with business conduct and climate change. Included in the disclosures should be a comprehensive explanation of management's expertise and skills in sustainability matters or access to such expertise and skills. Additional information required:

a) The number of executive and non-executive members.

b) Representation of employees and other workers.

c) Experience relevant to the sectors, products, and geographic locations of the organisation.

d) Percentage by gender and other aspects of diversity that the organisation considers. The board's gender diversity shall be calculated as an average ratio of female to male board members, **see S1-9 Number and percentage of women board members**.

e) The percentage of independent board members. For organisations with a unitary board, this corresponds to the percentage of independent non-executive board members. For organisations with a dual board, it corresponds to the percentage of independent members of the supervisory body.

f) The identity of the administrative, management, and supervisory bodies (such as a board committee or similar) or individual(s) within a body responsible for oversight of impacts, risks, and opportunities, especially concerning Climate Change and Business Conduct.

g) How each body's or individual's responsibilities for impacts, risks, and opportunities are reflected in your organisation's terms of reference, board mandates, and other related policies.

h) A description of management's role in the governance processes, controls, and procedures used to monitor, manage, and oversee impacts, risks, and opportunities, including:

- whether that role is delegated to a specific management-level position or committee and how oversight is exercised over that position or committee;

- information about the reporting lines to the administrative, management, and supervisory bodies;
- whether dedicated controls and procedures are applied to the management of impacts, risks, and opportunities and, if so, how they are integrated with other internal functions.

i) How the administrative, management, and supervisory bodies and senior executive management oversee the setting of targets related to material impacts, risks, and opportunities, and how they monitor progress towards them.

j) The sustainability-related expertise that the bodies, as a whole, either directly possess or can leverage, for example, through access to experts or training.

k) How those skills and expertise relate to the organisation's material impacts, risks, and opportunities.

5.2.4 GOV-2: Management reporting of sustainability

GOV-2 contains disclosures on information provided to and sustainability matters addressed by your organisation's administrative, management, and supervisory bodies, including the Green Team and Board level:

a) whether by whom and how frequently the administrative, management, and a) supervisory bodies, including their relevant committees, are informed about material impacts, risks, and opportunities (see **5.2.11 IRO–1 – Identify and assess material impacts, risks, and opportunities**), the implementation of due diligence, and the results and effectiveness of policies, actions, metrics, and targets adopted to address them;

b) how the administrative, management, and supervisory bodies consider impacts, risks, and opportunities when overseeing your organisation's strategy, its decisions on major transactions, and its risk management process, including whether they have considered trade-offs associated with those impacts, risks, and opportunities;

c) a list of the material impacts, risks, and opportunities addressed by the administrative, management, and supervisory bodies, or their relevant committees during the reporting period.

Table 27 Indicators and metrics related to GOV-1 – Management composition, roles and responsibilities disclosures.

ESRS topic	ESRS data type	ESRS paragraph	Description	Data type	Tips and tricks	Related CDR	VSMEs	ISSB/IFRS S1 & S2	GRI
ESRS 2	GOV-1	21	Information about composition, diversity, roles, and responsibilities of members of administrative, management, and supervisory bodies	Narrative	Senior management organogram and brief biometrics/ introduction.		V		GRI 2-9
ESRS 2	GOV-1	21 a	Number of executive members	Integer	Match financial statements				GRI 2-9
ESRS 2	GOV-1	21 a	Number of non-executive members	Integer					GRI 2-9
ESRS 2	GOV-1	21 b	Information about representation of employees and other workers in administrative, management, and supervisory bodies	Narrative	At Board and senior management level				GRI 2-9
ESRS 2	GOV-1	21 c	Information about member's experience relevant to sectors, products, and geographic locations	Narrative	This should include an ESG focus			IFRS S1.27(a)(ii), IFRS S2.6(a)(ii)	GRI 2-9
ESRS 2	GOV-1	21 d	Percentage and diversity ratio of members of Board, administrative, management, and supervisory bodies by gender and other aspects of DE&I	Per cent	See S1-9 number and percentage of women board members.	SFDR	V		GRI 2-9, GRI 405-1

ESRS 2	GOV-1	21 e	Percentage of independent board members	Per cent		SFDR			GRI 2-9
ESRS 2	GOV-1	22 a	Information about identity of administrative, management, and supervisory bodies or individual(s) within body responsible for oversight of impacts, risks, and opportunities	Narrative	These criteria cover your organisational capacity to identify, assess, and report impacts, risks, and opportunities and subsequently implement actions and targets. Disclosures should feature your green team, and how these capacities link with other topics, especially to E1: Climate Change, Own Workforce and G1: Business Conduct		V	IFRS S2.6(a)	GRI 2-9
ESRS 2	GOV-1	22 b	Disclosure of how each body's responsibilities for impacts, risks, and opportunities are reflected in the organisation's terms of reference, board mandates, and other related policies	Narrative			V	IFRS S2.6(a)(i)	
ESRS 2	GOV-1	22 c	Description of management's role in governance processes, controls, and procedures used to monitor, manage, and oversee impacts, risks, and opportunities	Narrative				IFRS S2.6(b)	GRI 2-12
ESRS 2	GOV-1	22 c i	Description of management's role in assessing and managing impacts, risks, and opportunities, including whether that role is delegated to a specific management-level position or committee and how oversight is exercised over that position or committee	Narrative				IFRS S2.6(b)(i)	GRI 2-12

(Continued)

Table 27 (*Continued*).

ESRS topic	ESRS data type	ESRS paragraph	Description	Data type	Tips and tricks	Related CDR	VSMEs	ISSB/IFRS S1 & S2	GRI
ESRS 2	GOV-1	22 c ii	Information about the reporting lines to the administrative, management, and supervisory bodies, whether dedicated controls and procedures are applied to the management of impacts, risks, and opportunities	Narrative				IFRS S1.27(b)(i)–(ii), IFRS S2.6(b)(i)–(ii)	GRI 2-12
ESRS 2	GOV-1	22 c iii	If applicable, disclosure of how dedicated controls and procedures are integrated with other internal functions	Narrative				IFRS S2.6(b)(ii)	
ESRS 2	GOV-1	22 d	Disclosure of how administrative, management, and supervisory bodies and senior executive management oversee setting of targets related to material impacts, risks, and opportunities and how progress towards them is monitored	Narrative				IFRS S2.6(a)(v)	

ESRS 2	GOV-1	23	Disclosure of how administrative, management, and supervisory bodies determine whether appropriate skills and expertise are available or will be developed to oversee sustainability matters	Narrative				IFRS S2.6(a)(ii)	GRI 2-9, GRI 2-17
ESRS 2	GOV-1	23 a	Information about sustainability-related expertise that bodies either directly possess or can leverage	Narrative				IFRS S1.27(a)(ii), IFRS S2.6(a)(ii)	GRI 2-9, GRI 2-17
ESRS 2	GOV-1	23 b	Disclosure of how sustainability-related skills and expertise relate to material impacts, risks, and opportunities	Narrative				IFRS S1.27(a)(ii), IFRS S2.6(a)(ii)	GRI 2-9, GRI 2-17

Source: CDR (EU) 2023/2772 supplementing Directive 2013/34/EU, *IFRS S1 General Requirements for Disclosure of Sustainability-related Financial Information, IFRS S2 Climate-related Disclosures, GRI 1, 2, 3: 11, 12, 13, 14, 306, 101, 201, 202, 203, 204, 205, 206, 207,301, 302, 303, 304, 305, 306, 308,401, 402, 403, 404, 405, 406, 407, 408, 409, 410, 411, 413, 414, 415, 416, 417, 418*, and the author.

Table 28 Indicators and metrics related to GOV-2 – Management reporting of sustainability.

ESRS Topic	ESRS data type	ESRS paragraph	Description	Data type	Tips and tricks	Related CDR	VSMEs	ISSB IFRS S1 & S2	GRI
ESRS 2	GOV-2	26 a	Disclosure of whether, by whom, and how frequently administrative, management, and supervisory bodies are informed about material impacts, risks, and opportunities, implementation of due diligence, and results and effectiveness of policies, actions, metrics, and targets adopted to address them	Narrative	Related to GOV-1, disclose relevant SOPs, internal reporting channels linked to roles and responsibilities, as well as list of records of impact occurrences, associated risks, and opportunities within the reporting period.			IFRS S2.6(a)(iii), IFRS S1.27(a)(iii) IFRS S2.6(a)(iii)	GRI 2-12, GRI 2-13, GRI 2-16
ESRS 2	GOV-2	26 b	Disclosure of how administrative, management, and supervisory bodies consider impacts, risks, and opportunities when overseeing strategy, decisions on major transactions, and risk management process	Narrative				IFRS S2.6(a)(iv)	GRI 2-12, GRI 2-24
ESRS 2	GOV-2	26 c	Disclosure of list of material impacts, risks, and opportunities addressed by administrative, management, and supervisory bodies or their relevant committees	Narrative				IFRS S1.26–27, IFRS S2.5–6	
ESRS 2	GOV-2	AR 6	Disclosure of how governance bodies ensure that appropriate mechanism for performance monitoring is in place	Narrative			V		

Source: CDR (EU) 2023/2772 supplementing Directive 2013/34/EU, *IFRS S1 General Requirements for Disclosure of Sustainability-related Financial Information, IFRS S2 Climate-related Disclosures, GRI 1, 2, 3: 11, 12, 13, 14, 306, 101, 201, 202, 203, 204, 205, 206, 207,301, 302, 303, 304, 305, 306, 308,401, 402, 403, 404, 405, 406, 407, 408, 409, 410, 411, 413, 414, 415, 416, 417, 418*, and the author.

5.2.5 GOV-3: Integration of sustainability-related performance in incentive schemes

The organisation shall disclose the following information about the incentive schemes and remuneration policies linked to sustainability matters for members of your organisation's administrative, management and supervisory bodies, where they exist:

a) a description of the key characteristics of the incentive schemes;

b) whether performance is being assessed against specific sustainability-related targets and/or impacts, and if so, which ones;

c) whether and how sustainability-related performance metrics are considered as performance benchmarks or included in remuneration policies;

d) the proportion of variable remuneration dependent on sustainability-related targets and/or impacts; and

e) the level in the organisation at which the terms of incentive schemes are approved and updated.

Table 29 Indicators and metrics related to GOV-3 – Integration of sustainability-related performance in incentive schemes.

ESRS topic	ESRS data type	ESRS paragraph	Description	Data type	Tips and tricks	Related CDR	VSMEs	ISSB IFRS S1 & S2	GRI
ESRS 2	GOV-3	29	Incentive schemes and remuneration policies linked to sustainability matters for members of administrative, management, and supervisory bodies exist	Semi-narrative	For listed organisations, this disclosure requirement should be consistent with the remuneration report prescribed in articles 9a and 9b of the Directive 2007/36/EC on the exercise of certain rights of shareholders in listed companies. Reference to this remuneration could be made in footnotes.				
ESRS 2	GOV-3	29 a	Description of key characteristics of incentive schemes and remuneration policies linked to sustainability matters for members of administrative, management, and supervisory bodies exist	Narrative					GRI 2-19
ESRS 2	GOV-3	29 b	Description of which, if any, specific sustainability-related targets and/or impacts used to assess performance of members of administrative, management, and supervisory bodies	Narrative				IFRS S1.21(b)	GRI 2-19
ESRS 2	GOV-3	29 c	Disclosure of whether and how sustainability-related performance metrics are considered as performance benchmarks or included in remuneration policies	Narrative				IFRS S1.21(b), IFRS S2.29(g)(i)	GRI 2-19
ESRS 2	GOV-3	29 d	Description of the key characteristics of the incentive schemes including the proportion of variable compensation dependent on such metrics	Per cent					
ESRS 2	GOV-3	29 e	Description of the responsibility level in your organisation that approves and updates the terms of incentive schemes	Narrative					GRI 2-20

Source: CDR (EU) 2023/2772 supplementing Directive 2013/34/EU, *IFRS S1 General Requirements for Disclosure of Sustainability-related Financial Information, IFRS S2 Climate-related Disclosures, GRI 1, 2, 3: 11, 12, 13, 14, 306, 101, 201, 202, 203, 204, 205, 206, 207,301, 302, 303, 304, 305, 306, 308,401, 402, 403, 404, 405, 406, 407, 408, 409, 410, 411, 413, 414, 415, 416, 417, 418*, and the author.

5.2.6 GOV-4: Statement on sustainability due diligence

As discussed in 1.6 Integrated Due Diligence Process, your organisation needs to provide a mapping table that explains how and where its application of the main aspects and steps of the due diligence process are reflected in your sustainability statement, to allow a depiction of the actual practices of the organisation with regard to due diligence. This disclosure requirement does not mandate any specific behavioural requirements with regard to due diligence actions and does not extend or modify the role of administrative, management, and supervisory bodies as mandated by other legislation or regulation. Additionally, the following table must be considered to include the main references in the international instruments (see Annex 1: List of data points in cross-cutting and topical standards that are required by EU law): of the UN Guiding Principles on Business and Human Rights and the OECD Guidelines for Multinational Enterprises to the core elements of the due diligence process:

Table 30 Table of GOV-4 – Due diligence elements.

Core elements of due diligence	**Paragraphs in the sustainability statement**
a) Embedding due diligence in governance, strategy, and business model	
b) Engaging with affected stakeholders in all key steps of the due diligence	
c) Identifying and assessing adverse impacts	
d) Taking actions to address those adverse impacts	
e) Tracking the effectiveness of these efforts and communicating	

Source: CDR (EU) 2023/2772 supplementing Directive 2013/34/EU.

Table 31 GOV-4 – Statement on sustainability due diligence.

ESRS Topic	ESRS Data type	ESRS paragraph	Description	Data type	Tips and tricks	Related CDR	VSMEs	ISSB IFRS S1 & S2	ISSB IFRS paragraph	GRI
ESRS 2	GOV-4	30; 32	Disclosure of a mapping that explains how and where your organisation applies the main aspects and steps of the sustainability due diligence process are reflected in its sustainability reporting, to allow a depiction of the actual practices of the undertaking with regard to due diligence	Narrative	Note that there is no mandate of any specific behavioural requirements with regard to sustainability due diligence actions and does not extend to or modify the role of administrative, management, and supervisory bodies as mandated by other legislation or regulation, which should be noted where appropriate.	SFDR				GRI 2-23

Source: CDR (EU) 2023/2772 supplementing Directive 2013/34/EU, *IFRS S1 General Requirements for Disclosure of Sustainability-related Financial Information, IFRS S2 Climate-related Disclosures, GRI 1, 2, 3: 11, 12, 13, 14, 306, 101, 201, 202, 203, 204, 205, 206, 207,301, 302, 303, 304, 305, 306, 308,401, 402, 403, 404 405, 406, 407, 408, 409, 410, 411, 413, 414, 415, 416, 417, 418*, and the author.

5.2.7 GOV-5: Risk management and internal control systems

The most important aspect of the GOV-5 disclosure is the identification of the main risks identified in the materiality assessment and a detailed description of the mitigation and management system, including:

a) the scope, main features, and components of the risk management and internal control processes and systems in relation to sustainability reporting;

b) the risk assessment approach followed, including the risk prioritisation methodology;

c) the main risks identified and their mitigation strategies including related controls;

d) a description of how the organisation integrates the findings of its risk assessment and internal controls as regards the sustainability reporting process into relevant internal functions and processes; and

e) a description of the periodic reporting of the findings referred to in point (d) to the administrative, management, and supervisory bodies.

5.2.8 SBM-1: Strategy, business model, and value chain

Your organisational strategy, business model, and value chain now become the focus of your ESG report as we relate them to material impacts. We begin by revisiting and mapping the key elements of our materiality assessment (see 3.8 **Aggregating and Reporting Materiality Assessments**), which impact general strategy relating to or affecting sustainability matters. To recap, key elements are:

1. account for more than 10% of the organisation's revenue; and/or

2. are connected with material actual impacts or material potential negative impacts of the organisation.

In mapping/tabulating significant activities, each needs to be cross-referenced to the ESRS disclosure reference code (for example, GOV-1) and contain a description of:

a) key activities, markets/customer groups, significant products and services, resources, distribution channels, and customer segments including changes in the reporting period (new/removed products and/or services);

b) key business relationships and their main characteristics, including relationships with customers and suppliers;

Table 32 Indicators and metrics related to GOV-5 – Risk management and internal control system.

ESRS topic	ESRS data type	ESRS paragraph	Description	Data type	Tips and tricks	Related CDR	VSMEs	ISSB IFRS S1 & S2	GRI
ESRS 2	GOV-5	36 a	Description of scope, main features, and components of risk management and internal control processes and systems in relation to sustainability reporting	Narrative	Recommend a standard operating procedure (SOP) for risk identification, monitoring, and reporting. This can be a separate system or related to OH&S systems already in place and cross-referenced to environmental risks and opportunities. Maintain and review a risk inventory with regular review. Incorporate relevant metrics into assessment; see 5.3 Risk and Opportunity Indicator Metrics.				GRI 2-14
ESRS 2	GOV-5	36 b	Description of risk assessment approach followed, including the risk prioritisation methodology	Narrative					GRI 2-14
ESRS 2	GOV-5	36 c	Description of main risks identified, actual, and potential, and their mitigation strategies including related controls	Narrative					GRI 2-14
ESRS 2	GOV-5	36 d	Description of how risk assessment findings and internal controls relating to sustainability reporting processes have been integrated into relevant internal functions and processes	Narrative					GRI 2-14
ESRS 2	GOV-5	36 e	Description of periodic reporting of findings of risk assessment and internal controls to administrative, management, and supervisory bodies	Narrative					GRI 2-14

Source: CDR (EU) 2023/2772 supplementing Directive 2013/34/EU, *IFRS S1 General Requirements for Disclosure of Sustainability-related Financial Information, IFRS S2 Climate-related Disclosures, GRI 1, 2, 3: 11, 12, 13, 14, 306, 101, 201, 202, 203, 204, 205, 206, 207,301, 302, 303, 304, 305, 306, 308,401, 402, 403, 404, 405, 406, 407, 408, 409, 410, 411, 413, 414, 415, 416, 417, 418*, and the author.

c) the cost structure and revenue of its business segments, in line with IFRS 8 Operating Segments disclosure requirements in the financial statement, where applicable;

d) the potential impacts, risks, and opportunities in its significant sector(s) and their possible relationship to its own business model or value chain;

e) headcount of employees by geographical areas;

f) where applicable and material, products, and services that are banned in certain markets;

g) a breakdown of total revenue, as included in its financial statements, by significant ESRS sectors. When the organisation provides segment reporting as required by IFRS 8 Operating Segments in its financial statements, this sector revenue information shall be, as far as possible, reconciled with IFRS 8 information;

h) a statement indicating, together with the related revenues, that the organisation is active in fossil fuels (including revenues from exploration, mining, extraction, production, processing, storage, refining or distribution, transportation, and trade), chemicals production, controversial weapons (such as anti-personnel mines, cluster munitions, chemical weapons, and biological weapons), and/or the cultivation and production of tobacco;

i) its sustainability-related goals in terms of significant groups of products and services, customer categories, geographical areas, and relationships with stakeholders;

j) the elements of the organisation's strategy that relate to or impact sustainability matters, including the main challenges ahead, critical solutions or projects to be put in place, when relevant for sustainability reporting.

With reference to the value chain, the organisation shall disclose a description of its business model and value chain integration and dependencies, including:

a) its inputs and its approach to gathering, developing, and securing those inputs;

b) its outputs and outcomes in terms of current and expected benefits for customers, investors, and other stakeholders; and

c) the main features of its upstream and downstream value chain and the organisation's position in its value chain, including a description of the main business actors (such as key suppliers, customers, distribution channels, and end-users) and their relationship to the organisation. When the organisation has multiple value chains, the disclosure shall cover the key value chains.

Table 33 Indicators and metrics related to SBM-1 – Strategy, business model, and value chain.

ESRS topic	ESRS data type	ESRS paragraph	Description	Data type	Tips and tricks	Related CDR	VSMEs	ISSB IFRS S1 & S2	GRI
ESRS 2	SBM-1	40 a i	Description of significant groups of products and/or services offered, including changes in the reporting period	Narrative	Cross-reference with impacts from materiality assessment		V		GRI 2-6
ESRS 2	SBM-1	40 a ii	Description of significant markets and/or customer groups served, including changes in the reporting period	Narrative			V		GRI 2-6
ESRS 2	SBM-1	40 a iii	Total number of employees (head count) by geographical area	Integer	Countries with >50 staff must be included				GRI 2-7
ESRS 2	SBM-1	40 a iii	Number of employees (head count)	Integer					
ESRS 2	SBM-1	40 a iv	If applicable and material, description of products and services that are banned in certain markets	Narrative					
ESRS 2	SBM-1	40 b	Breakdown of total revenue by ESRS sectors	Semi-narrative, monetary	Must be reconciled with IFRS 8 operating segments in financial statements		V		GRI 2-6
ESRS 2	SBM-1	40 c	List of additional significant ESRS sectors in which significant activities are developed or in which organisation is or may be connected to material impacts	Semi-narrative			V		GRI 2-6
ESRS 2	SBM-1	40 d i	Organisation is active in fossil fuel (coal, oil, and gas) sector	Semi-narrative	If not, simply state so	SFDR	V		

ESRS 2	SBM-1	40 d i	Revenue from taxonomy-aligned economic activities related to fossil gas (broken down coal, oil, and gas)	Monetary	If applicable, relate to E1 Climate Change 6.5, 6.6, and 6.7		V		
ESRS 2	SBM-1	40 d ii	Organisation is active in chemicals production	Semi-narrative	If not, simply state so	SFDR	V		
ESRS 2	SBM-1	40 d ii	Revenue from chemicals production	Monetary		SFDR	V		
ESRS 2	SBM-1	40 d iii	Organisation is active in controversial weapons	Semi-narrative	If not, simply state so	SFDR	V		
ESRS 2	SBM-1	40 d iii	Revenue from controversial weapons	Monetary		SFDR	V		
ESRS 2	SBM-1	40 d iv	Organisation is active in cultivation and production of tobacco	Semi-narrative	If not, simply state so	SFDR	V		
ESRS 2	SBM-1	40 d iv	Revenue from cultivation and production of tobacco	Monetary		SFDR	V		
ESRS 2	SBM-1	40 e	Description of sustainability-related goals in terms of significant groups of products and services, customer categories, geographical areas, and relationships with stakeholders	Narrative	Relate suppliers and value chain to policies where relevant				GRI 3-3
ESRS 2	SBM-1	40 f	Disclosure of assessment of current significant products and/or services, and significant markets and customer groups, in relation to sustainability-related goals	Narrative					
ESRS 2	SBM-1	40 g	A statement regarding the main challenges ahead, critical solutions or projects to be put in place, when relevant for sustainability reporting	Narrative			V		GRI 2-22

(Continued)

Table 33 (*Continued*).

ESRS topic	ESRS data type	ESRS paragraph	Description	Data type	Tips and tricks	Related CDR	VSMEs	ISSB IFRS S1 & S2	GRI
ESRS 2	SBM-1	41	List of ESRS sectors that are significant for your organisation with breakdown by revenue	Semi-narrative	Can be a table with revenue percentages				
ESRS 2	SBM-1	42	General description of business model and value chain	Narrative			V		
ESRS 2	SBM-1	42 a	With reference to business model and value chain, description of inputs and approach to gathering, developing, and securing inputs	Narrative	Such as key suppliers, customers distribution channels, and end-users, and their relationship to the organisation. When the organisation has multiple value chains, the disclosure shall cover only the key value chains				
ESRS 2	SBM-1	42 b	With reference to business model and value chain, description of outputs and outcomes in terms of current and expected benefits for customers, investors, and other stakeholders	Narrative					
ESRS 2	SBM-1	42 c	Description of main features of upstream and downstream value chain and organisation's position in value chain	Narrative					GRI 2-6

Source: CDR (EU) 2023/2772 supplementing Directive 2013/34/EU, *IFRS S1 General Requirements for Disclosure of Sustainability-related Financial Information, IFRS S2 Climate-related Disclosures, GRI 1, 2, 3: 11, 12, 13, 14, 306, 101, 201, 202, 203, 204, 205, 206, 207,301, 302, 303, 304, 305, 306, 308,401, 402, 403, 404, 405, 406, 407, 408, 409, 410, 411, 413, 414, 415, 416, 417, 418*, and the author.

5.2.9 SBM-2: Interests and views of stakeholders

SBM 2 is a summary of your stakeholder engagement analysis and what potential influence the process has on your organisational strategy, business model, and sustainability management. This is a straightforward summarising exercise in the disclosure of:

a) Your stakeholder engagement process, including a summary of key stakeholders, description of the process, and the subsequent organisational interpretation and understanding of the interests and views of its key stakeholders as they relate to the organisation's strategy and business model. Also include the extent that interests and views were analysed during the organisation's due diligence process and/or materiality assessment process (see **Chapter 2: Stakeholder Engagement** and **5.2.11 IRO–1 – Identify and assess material impacts, risks, and opportunities**).

b) Where applicable, describe changes to your organisation's strategy and/or business model, including:

   i) how your organisation has amended or expects to amend its strategy and/or business model to address the interests and views of its stakeholders;

   ii) any further steps that are being planned and in what timeline;

   iii) whether these steps are likely to modify the relationship with and views of stakeholders; and

c) whether and how the administrative, management, and supervisory bodies are informed about the views and interests of affected stakeholders with regard to your organisation's sustainability-related impacts.

Table 34 Indicators and metrics related to SBM-2 – Interests and views of stakeholders.

ESRS Topic	ESRS data type	ESRS paragraph	Description	Data type	Tips and tricks	Related CDR	VSMEs	ISSB IFRS S1 & S2	GRI
ESRS 2	SBM-2	45 a i	Description of how stakeholders' interests and views inform the undertaking's strategy and business model(s) and identify key stakeholders	Narrative	Summary of stakeholder engagement process and outcomes, Chapter 2		V		GRI 2-29
ESRS 2	SBM-2	45 a ii	Description of categories of stakeholders for which engagement occurs	Narrative			V		GRI 2-29
ESRS 2	SBM-2	45 a iii	Description of how stakeholder engagement is organised	Narrative					GRI 2-29
ESRS 2	SBM-2	45 a iv	Description of purpose of stakeholder engagement	Narrative					GRI 2-29
ESRS 2	SBM-2	45 a v	Description of how outcome of stakeholder engagement is taken into account	Narrative					
ESRS 2	SBM-2	45 b	Description of understanding of interests and views of key stakeholders as they relate to undertaking's strategy and business model	Narrative					
ESRS 2	SBM-2	45 c	Description of amendments to strategy and/or business model	Narrative	Not relevant for first reports				
ESRS 2	SBM-2	45 c i	Description of how strategy and/or business model have been amended or are expected to be amended to address interests and views of stakeholders	Narrative					
ESRS 2	SBM-2	45 c ii	Description of any further steps that are being planned and in what timeline	Narrative	Could be relevant for missing or omitted disclosures				
ESRS 2	SBM-2	45 c iii	Further steps that are being planned are likely to modify relationship with and views of stakeholders	Semi-narrative					
ESRS 2	SBM-2	45 d	Description of how administrative, management, and supervisory bodies are informed about views and interests of affected stakeholders with regard to sustainability-related impacts	Narrative	Report submission to management and publication channels				GRI 2-12

Source: CDR (EU) 2023/2772 supplementing Directive 2013/34/EU, *IFRS S1 General Requirements for Disclosure of Sustainability-related Financial Information, IFRS S2 Climate-related Disclosures, GRI 1, 2, 3: 11, 12, 13, 14, 306, 101, 201, 202, 203, 204, 205, 206, 207,301, 302, 303, 304, 305, 306, 308,401, 402, 403, 404, 405, 406, 407, 408, 409, 410, 411, 413, 414, 415, 416, 417, 418*, and author.

5.2.10 SBM-3: Material impacts, risks, and opportunities

SBM 3 is a summary of your materiality assessment, identification of risks and opportunities, the potential influence the process has on your organisational strategy, business model, and sustainability management. As you summarise your assessment, consider geographical areas, facilities or types of assets, inputs, outputs, and distribution channels when disclosing:

a) a brief description of material impacts, risks, and opportunities resulting from your materiality assessment (see **Chapter 3: Single and Double Materiality Assessment** and **5.2.11 IRO-1 – Identify and assess material impacts, risks, and opportunities**), including a description of where in its business model, its own operations and its upstream and downstream value chain these material impacts, risks, and opportunities are concentrated;

b) the current and anticipated effects of material impacts, risks, and opportunities on your organisation's business model, value chain, strategy and decision-making, and how it has responded or plans to respond to these effects, including any changes it has made or plans to make to its strategy or business model as part of its actions to address particular material impacts or risks, or to pursue particular material opportunities;

c) with reference to the organisation's material impacts:

   i. how the organisation's material negative and positive impacts affect (or, in the case of potential impacts, are likely to affect) people or the environment;

   ii. whether and how the impacts originate from or are connected to the organisation's strategy and business model;

   iii. the reasonably expected time horizons of the impacts; and

   iv. whether the organisation is involved with the material impacts through its activities or because of its business relationships, describing the nature of the activities or business relationships concerned;

d) the current financial effects of the organisation's material risks and opportunities on its financial position, financial performance, and cash flows and the material risks and opportunities for which there is a significant risk of a material adjustment within the next annual reporting period to the carrying amounts of assets and liabilities reported in the related financial statements;

e) the anticipated financial effects of the organisation's material risks and opportunities on its financial position, financial performance, and cash flows over the short, medium, and long term, including the reasonably expected time horizons for those effects. This shall include how the organisation expects its financial position, financial performance, and cash flows to change over the short, medium, and long term, given its strategy to manage risks and opportunities, taking into consideration:

   i. its investment and disposal plans (for example, capital expenditure, major acquisitions and divestments, joint ventures, business transformation, innovation, new business areas, and asset retirements), including plans the organisation is not contractually committed to; and

   ii. its planned sources of funding to implement its strategy;

f) information about the resilience (see **3.7 Resilience and Scenario Analysis**) of the organisation's strategy and business model regarding its capacity to address its material impacts and risks and to take advantage of its material opportunities, including critical assumptions and time horizons (including alignment with scenarios) about how the transition to a lower-carbon (see **6.1 Comprehensive Transition Plan for Climate Change Mitigation**) and resilient economy will affect its surrounding macroeconomic trends, energy consumption and mix, and technology deployment assumptions. The organisation shall explain which part of its own operations and upstream and downstream value chain as well as which material physical risks and transition risks may have been excluded from the analysis. The organisation shall disclose a qualitative and, when applicable, a quantitative analysis of the resilience, including how the analysis was conducted and the time horizons. When providing quantitative information, the organisation may disclose single amounts or ranges; and

g) changes to the material impacts, risks, and opportunities compared to the previous reporting period.

Table 35 Indicators and metrics related to SBM-3 – Material impacts, risks, and opportunities.

ESRS topic	ESRS data type	ESRS paragraph	Description	Data type	Tips and tricks	Related CDR	VSMEs	ISSB IFRS S1 & S2	GRI
ESRS 2	SBM-3	48 a	Description of material impacts, risks, and opportunities resulting from materiality assessment, including material negative or positive impacts that affect (or are expected to affect) people or the environment, the reasonably expected time horizons for those effects, whether the undertaking is involved with the material impacts through its activities or because of its business relationships, and whether they originate from or are connected to the undertaking's strategy and business model(s)	Narrative	Detail of your materiality assessment including all material impacts, risks, and opportunities identification, where they are concentrated, associated actions and targets, and cross-reference with IRO-1.		V	IFRS S2.10(a), IFRS S2.13(b)	GRI 3-2, GRI 201-2, GRI 303-1, GRI 306-1
ESRS 2	SBM-3	48 b	Disclosure of current and anticipated effects of material impacts, risks, and opportunities on business model, value chain, strategy, and decision-making, and how the organisation has responded through specific actions or plans to respond to these effects	Narrative			V	IFRS S2.13(a)	

(*Continued*)

Table 35 (*Continued*).

ESRS topic	ESRS data type	ESRS paragraph	Description	Data type	Tips and tricks	Related CDR	VSMEs	ISSB IFRS S1 & S2	GRI
ESRS 2	SBM-3	48 c i	Disclosure of how material negative and positive impacts affect (or are likely to affect) people or environment	Narrative	Cross-reference material impacts with SDGs				GRI 3-3, GRI 308-2, GRI 413-2, GRI 414-2
ESRS 2	SBM-3	48 c ii	Disclosure of how impacts originate from or are connected to strategy and business model	Narrative	Includes direct and indirect relationships				GRI 306-1, GRI 413-2
ESRS 2	SBM-3	48 c iii	Disclosure of reasonably expected time horizons of impacts	Narrative	Short, medium, or long term				GRI 413-2
ESRS 2	SBM-3	48 c iv	Description of nature of activities or business relationships through which undertaking is involved with material impacts	Narrative	Include linkages with value chain				GRI 3-3, GRI 306-1, GRI 308-2, GRI 413-2, GRI 414-2
ESRS 2	SBM-3	48 d	Disclosure of *current* financial effects of material risks, and opportunities on financial position, financial performance, and cash flows and material risks and opportunities for which there is significant risk of material adjustment within the next annual reporting period to carrying amounts of assets and liabilities reported in related financial statements	Narrative	Summary of resilience and sensitivity analysis with scenarios that influence strategy and business model and financial effects – see 5.3 Risk and Opportunity Indicator Metrics		V	IFRS S2.15(a), IFRS S2.16(a), IFRS S2.16(b)	GRI 2-27, GRI 201-2

ESRS 2	SBM-3	48 e	Disclosure of *anticipated* financial effects of material risks and opportunities on financial position, financial performance, and cash flows over short, medium, and long term including investment and disposal plans and funding source(s)	Narrative			V	IFRS S2.10(c), IFRS S2.15(b) , IFRS S2. 16©(i)-(ii), IFRS S2. 16(d)	GRI 201-2
ESRS 2	SBM-3	48 f	Information about resilience of strategy and business model regarding capacity to address material impacts and risks and to take advantage of material opportunities	Narrative					
ESRS 2	SBM-3	48 g	Disclosure of changes to material impacts, risks, and opportunities compared to the previous reporting period	Narrative	Not relevant for first report				GRI 3-2
ESRS 2	SBM-3	48 h	Disclosure of specification of impacts, risks, and opportunities that are covered by ESRS disclosure requirements as opposed to those covered by additional entity-specific disclosures	Narrative	Check relevancy for specific sector entities				

Source: CDR (EU) 2023/2772 supplementing Directive 2013/34/EU, *IFRS S1 General Requirements for Disclosure of Sustainability-related Financial Information, IFRS S2 Climate-related Disclosures, GRI 1, 2, 3: 11, 12, 13, 14, 306, 101, 201, 202, 203, 204, 205, 206, 207,301, 302, 303, 304, 305, 306, 308,401, 402, 403, 404, 405, 406, 407, 408, 409, 410, 411, 413, 414, 415, 416, 417, 418*, and the author.

5.2.11 IRO-1: Identify and assess material impacts, risks, and opportunities

IRO-1 is an overview of the materiality assessment analysis that you completed in **Chapter 3: Single and Double Materiality Assessment, financial reports and combined with analysis from each impact topic, especially E1: Climate Change**, which may mean returning to this section from time to time. So, in this section, you will summarise (with references) a description of the methodologies and assumptions that you applied when identifying, assessing, prioritising, and establishing a monitoring system for your organisation's potential and actual impacts on people and the environment. This information should be cross-referenced and enhanced by content from your organisation's due diligence process, such that you can explain whether and how the risk and opportunity identification process:

a) focuses on specific activities, business relationships, geographies, or other factors that give rise to heightened risk of adverse impacts, for example, exposure to climate change, industrial unrest, or political turbulence;

b) considers the impacts with which the organisation is involved through its own operations or as a result of its business relationships;

c) includes consultation with affected stakeholders to understand how they may be impacted and with external experts;

d) prioritises negative impacts based on their relative severity and likelihood, and, if applicable, positive impacts on their relative scale, scope, and likelihood, and determines which sustainability matters are material for reporting purposes;

e) an overview of the process used to identify, assess, prioritise, and monitor risks and opportunities that have or may have financial effects. The disclosure shall include:

   i. how your organisation has considered the connections of its impacts and dependencies with the risks and opportunities that may arise from those impacts and dependencies;

   ii. how your organisation assesses the likelihood, magnitude, and nature of effects of the identified risk and opportunities;

   iii. how your organisation prioritises sustainability-related risks relative to other types of risks, including its use of risk-assessment tools;

f) a description of the decision-making process at management and/or stakeholder level and the related internal/external control procedures and, where applicable, the extent to which and how the process to identify, assess, and manage opportunities is integrated into the organisation's overall management process;

g) the extent to which and how the process to identify, assess, and manage impacts and risks is integrated into the organisation's overall risk management process and used to evaluate the organisation's overall risk profile and risk management processes, which may involve a risk evaluation system;

h) the input parameters it uses, for example, data sources, the scope of operations covered (including extent of assets and liabilities included), and the detail used in assumptions; and

i) whether and how the process has changed compared to the prior reporting period, when the process was modified for the last time, and future revision dates of the materiality assessment.

When disclosing the information on the processes to identify and assess climate impacts, the undertaking shall explain how it has:

a) emission sources and, if applicable, drivers for other climate-related impacts (e.g., emissions of black carbon or tropospheric ozone or land-use change) in own operations and along the value chain; and

b) assessed its actual and potential impacts on climate change (i.e., its total GHG emissions).

When disclosing the information on the processes to identify and assess physical impacts, the organisation shall explain whether and how:

a) it has identified climate-related hazards (see table below) ovcr thc short, medium, and long term and screened whether its assets and business activities may be exposed to these hazards;

b) it has defined short-, medium- and long-term time horizons and how these definitions are linked to the expected lifetime of its assets, strategic planning horizons, and capital allocation plans;

c) it has assessed the extent to which its assets and business activities may be exposed and are sensitive to the identified climate-related hazards, taking into consideration the likelihood, magnitude, and duration of the hazards as well as the geospatial coordinates (such as Nomenclature of

Territorial Units of Statistics – NUTS for the EU territory) specific to the organisation's locations and supply chains; and

d) the identification of climate-related hazards and the assessment of exposure and sensitivity are informed by high emissions climate scenarios, which may, for example, be based on IPCC SSP5-8.5, relevant regional climate projections based on these emission scenarios, or NGFS (Network for Greening the Financial System) climate scenarios with high physical risk such as "Hot house world" or "Too little, too late". For general requirements regarding climate-related scenario analysis see 5.2.10 IRO-1: Identify and Assess Material Impacts, Risks and Opportunities.

Table 36 Classification of climate-related hazards.

	Temperature-related	**Wind-related**	**Water-related**	**Solid mass-related**
Chronic	Changing temperature (air, freshwater, marine water)	Changing wind patterns	Changing precipitation patterns and type (rain, hail, snow/ice)	Coastal erosion
	Heat stress		Precipitation or hydrological variability	Soil degradation
	Temperature variability		Ocean acidification	Soil erosion
	Permafrost thawing		Saline intrusion	Solifluction
	Pollinator abundance		Sea level rise	
			Water stress	
Acute	Heat wave	Cyclones, hurricanes, typhoons	Drought	Avalanche
	Cold wave/frost	Storms (including blizzards, dust, and sandstorms)	Heavy precipitation (rain, hail, snow/ice)	Landslide
	Wildfire	Tornado	Flood (coastal, fluvial, pluvial, ground water)	Subsidence

Source: CDR (EU) 2023/2772 supplementing Directive 2013/34/EU and TNFD *Conducting advanced scenario analysis*, 2023.

When disclosing the information on the processes to identify transition risks and opportunities, the organisation shall explain whether and how it has:

a) identified transition events (see the table with examples below) over the short, medium, and long term and screened whether its assets and business activities may be exposed to these events. In case of transition risks and opportunities, what is considered long term may cover more than 10 years and may be aligned with climate-related public policy goals;

Table 37 Examples of climate-related transition events based on TCFD classification.

Policy and legal	Technology	Market	Reputation
Increased pricing of GHG emissions	Substitution of existing products and services with lower emission options	Changing customer behaviour	Shifts in consumer preferences
Enhanced emissions-reporting obligations	Successful investment in new technologies	Uncertainty in market signals	Stigmatisation of sector
Mandates on and regulation of existing products and services	Costs of transition to lower emissions technology	Increased cost of new materials	Increased stakeholder concern
Mandates on and regulation of existing production processes			Negative stakeholder feedback
Exposure to litigation			

Source: CDR (EU) 2023/2772 supplementing Directive 2013/34/EU.

b) assessed the extent to which its assets and business activities may be exposed and are sensitive to the identified transition events, taking into consideration the likelihood, magnitude, and duration of the transition events.

c) informed the identification of transition events and the assessment of exposure by climate-related scenario analysis, considering at least a scenario consistent with the Paris Agreement and limiting climate change to 1.5 °C, for example, based on scenarios of the International Energy Agency (Net Zero Emissions by 2050, Sustainable Development Scenario, etc.), or NGFS (Network for Greening the Financial System) climate scenarios; and

d) identified assets and business activities that are incompatible with or need significant efforts to be compatible with a transition to a climate-neutral economy (for example, due to significant locked-in GHG emissions or incompatibility with the requirements for taxonomy-alignment under Commission Delegated Regulation (EU) 2021/2139).

The organisation shall explain how it has used climate-related scenario analysis that is commensurate to its circumstances to inform the identification and assessment of physical and transition risks and opportunities over the short, medium, and long term, including:

a) which scenarios were used, their sources and alignment with state-of-the-art science;

b) narratives, time horizons, and endpoints used with a discussion of why it believes the range of scenarios used covers its plausible risks and uncertainties;

c) the key forces and drivers taken into consideration in each scenario and why these are relevant to the undertaking, for example, policy assumptions, macroeconomic trends, energy usage and mix, and technology assumptions; and

d) key inputs and constraints of the scenarios, including their level of detail (e.g., whether the analysis of physical climate-related risks is based on geospatial coordinates specific to the organisation's locations or national- or regional-level broad data).

When conducting scenario analysis, the undertaking may consider the following guidance:

- TCFD Technical Supplement on "The Use of Scenario Analysis in Disclosure of Climate-Related Risks and Opportunities" (2017);
- TCFD "Guidance on Scenario Analysis for Non-Financial Companies" (2020);
- ISO 14091:2021 "Adaptation to climate change — Guidelines on vulnerability, impacts and risk assessment";
- any other recognised industry standards such as NGFS (Network for Greening the Financial System); and EU, national, regional, and local regulations.

The undertaking shall briefly explain how the climate scenarios used are compatible with the critical climate-related assumptions made in the financial statements.

Table 38 Indicators and metrics related to IRO-1 – Identify and assess material impacts, risks, and opportunities.

ESRS topic	ESRS data type	ESRS paragraph	Description	Data type	Tips and tricks	Related CDR	VSMEs	ISSB IFRS S1 & S2	GRI
ESRS 2	IRO-1	53 a	Description of methodologies and assumptions applied in process to identify impacts, risks, and opportunities	Narrative	Combined information from Chapter 3: Single and Double Materiality Assessment, financial reports and combined with methodologies from each ESRS chapter, especially E1 Climate Change				
ESRS 2	IRO-1	53 b	Description of process to identify, assess, prioritise, and monitor potential and actual impacts on people and environment, informed by due diligence process	Narrative			V		
ESRS 2	IRO-1	53 b i	Description of how process focuses on specific activities, business relationships, geographies, or other factors that give rise to heightened risk of adverse impacts	Narrative					GRI 3-1
ESRS 2	IRO-1	53 b ii	Description of how process considers impacts with which your organisation is involved through own operations or as a result of business relationships	Narrative	Stakeholder engagement – identification and prioritisation of sustainability-related matters, including severity, scope, and likelihood				GRI 3-1

(Continued)

Table 38 (*Continued*).

ESRS topic	ESRS data type	ESRS paragraph	Description	Data type	Tips and tricks	Related CDR	VSMEs	ISSB IFRS S1 & S2	GRI
ESRS 2	IRO-1	53 b iii	Description of how process includes consultation with affected stakeholders to understand how they may be impacted and with external experts	Narrative					GRI 3-1
ESRS 2	IRO-1	53 b iv	Description of how process prioritises negative impacts based on their relative severity and likelihood and positive impacts based on their relative scale, scope, and likelihood and determines which sustainability matters are material for reporting purposes	Narrative					GRI 3-1
ESRS 2	IRO-1	53 c	Description of process used to identify, assess, prioritise, and monitor risks and opportunities that have or may have financial effects	Narrative	Refer to your materiality assessment, specifically: 1. Double materiality assessment table 2. Quantitative scaling for assessment of stakeholder expectations			IFRS S2.25(a)(v), IFRS S2.25(b)	

ESRS 2	IRO-1	53 c i	Description of how connections of impacts and dependencies with risks and opportunities that may arise from those impacts and dependencies have been considered	Narrative	3. Irremediability and stakeholder concern 4. Mitigation hierarchy table relating to material issues				
ESRS 2	IRO-1	53 c ii	Description of how likelihood, magnitude, and nature of effects of identified risks and opportunities have been assessed	Narrative				IFRS S2.25(a)(iii)	
ESRS 2	IRO-1	53 c iii	Description of how sustainability-related risks relative to other types of risks have been prioritised	Narrative				IFRS S2.25(a)(iv)	
ESRS 2	IRO-1	53 d	Description of decision-making process and related internal control procedures	Narrative					GRI 2-14
ESRS 2	IRO-1	53 e	Description of extent to which and how the process to identify, assess, and manage impacts and risks is integrated into overall risk management process and used to evaluate overall risk profile and risk management processes	Narrative				IFRS S2.25(a)(iv), IFRS S2.25(c)	
ESRS 2	IRO-1	53 f	Description of extent to which and how the process to identify, assess, and manage opportunities is integrated into overall management process	Narrative	… and then link back to your management supervision and green team			IFRS S2.25(c)	

(Continued)

Table 38 (*Continued*).

ESRS topic	ESRS data type	ESRS paragraph	Description	Data type	Tips and tricks	Related CDR	VSMEs	ISSB IFRS S1 & S2	GRI
ESRS 2	IRO-1	53 g	Description of input parameters used in the process to identify, assess, and manage material impacts, risks, and opportunities	Narrative	Refer to impact boundaries defined in materiality assessment			IFRS S2.25(a)(i)	
ESRS 2	IRO-1	53 h	Description of how the process to identify, assess, and manage impacts, risks, and opportunities has changed compared to prior reporting period	Narrative				IFRS S2.25(a)(vi)	

Source: CDR (EU) 2023/2772 supplementing Directive 2013/34/EU, *IFRS S1 General Requirements for Disclosure of Sustainability-related Financial Information, IFRS S2 Climate-related Disclosures, GRI 1, 2, 3: 11, 12, 13, 14, 306, 101, 201, 202, 203, 204, 205, 206, 207,301, 302, 303, 304, 305, 306, 308,401, 402, 403, 404, 405, 406, 407, 408, 409, 410, 411, 413, 414, 415, 416, 417, 418*, and the author.

5.2.12 IRO-2: Disclosure requirements statement

It is necessary to tabulate or cross-reference your disclosures with EU legislation, which can include a content index or a similar list of your disclosure requirements that you have complied with. You should also include a table of all the data points that derive from other EU legislation (see **Appendix A**), indicating where they can be found in the sustainability statement and including those that the organisation has assessed as not material, in which case the organisation shall indicate "Not material" in the table.

If the organisation concludes that a topic other than climate change is not material and therefore omits all the disclosure requirements in the corresponding topical ESRS, it may provide a brief explanation of the conclusions of its materiality assessment for that topic.

The organisation shall provide an explanation of how it has determined the material information to be disclosed in relation to the impacts, risks, and opportunities that it has assessed to be material, including the use of thresholds and/or how it has implemented the disclosure criteria.

5.3 Risk and Opportunity Indicator Metrics

The following collections of indictors are commonly used when assessing material impact risks and opportunities and relate to minimum, general, and topical disclosures.

5.3.1 Total expenditures on research and development (R&D) (ESRS2 IRO-1)

Total expenditures on research and development (R&D) can be measured using two ratio indicators. The first indicator is the ratio of total R&D expenditures to total assets, and the second indicator is the ratio of total R&D expenditures to total revenue. The accounting treatment of R&D expenses may differ based on the applicable accounting standards (e.g., US GAAP or IFRS).

1. Accounting treatment:

   - Under US GAAP: All R&D costs are expensed as incurred and written off to the income statement.

   - Under IFRS: Research costs are expensed, while development expenditures are capitalised as intangible assets, amortised, and reported in the balance sheet.

Table 39 Indicators and metrics related to IRO-2 – Disclosure requirements in ESRS covered by the undertaking's sustainability statements.

ESRS topic	ESRS data type	ESRS paragraph	Description	Data type	Tips and tricks	Related CDR	VSMEs	ISSB IFRS S1 & S2	GRI
ESRS 2	IRO-2	56	Disclosure of list of data points and that derive from other EU legislation and information on their location in sustainability statement	Narrative	Must include a content index and description of material impact metrics including those that were considered not material and cross-referenced to CDRs - see Annex 1: List of data points in cross-cutting and topical standards that are required by EU law. Can also include SDGs. Note: If climate change is determined as not material, you must still provide a detailed explanation of the conclusions relating to current and future scenarios and potential impacts. For other not material topics, a brief explanation is acceptable.				
ESRS 2	IRO-2	56	Disclosure of list of ESRS disclosure requirements complied with preparing sustainability statement following outcome of materiality assessment, including topics that have been omitted as not material	Narrative					
ESRS 2	IRO-2	57	Explanation of negative materiality assessment for ESRS E1 Climate change	Narrative					
ESRS 2	IRO-2	58	Explanation of negative materiality assessment for ESRS E2 Pollution	Narrative			V		
ESRS 2	IRO-2	58	Explanation of negative materiality assessment for ESRS E3 Water and marine resources	Narrative			V		
ESRS 2	IRO-2	58	Explanation of negative materiality assessment for ESRS E4 Biodiversity and ecosystems	Narrative			V		
ESRS 2	IRO-2	58	Explanation of negative materiality assessment for ESRS E5 Circular economy	Narrative			V		

ESRS 2	IRO-2	58	Explanation of negative materiality assessment for ESRS S1 Own workforce	Narrative		V		
ESRS 2	IRO-2	58	Explanation of negative materiality assessment for ESRS S2 Value chain workers	Narrative		V		
ESRS 2	IRO-2	58	Explanation of negative materiality assessment for ESRS S3 Affected communities	Narrative		V		
ESRS 2	IRO-2	58	Explanation of negative materiality assessment for ESRS S4 Consumers and end-users	Narrative		V		
ESRS 2	IRO-2	58	Explanation of negative materiality assessment for ESRS G1 Business conduct	Narrative		V		
ESRS 2	IRO-2	59	Explanation of how material information to be disclosed in relation to material impacts, risks, and opportunities has been determined	Narrative				

Source: CDR (EU) 2023/2772 supplementing Directive 2013/34/EU, *IFRS S1 General Requirements for Disclosure of Sustainability-related Financial Information, IFRS S2 Climate-related Disclosures, GRI 1, 2, 3: 11, 12, 13, 14, 306, 101, 201, 202, 203, 204, 205, 206, 207,301, 302, 303, 304, 305, 306, 308,401, 402, 403, 404, 405, 406, 407, 408, 409, 410, 411, 413, 414, 415, 416, 417, 418*, and the author.

2. Calculation of indicators:

   Regardless of the accounting treatment, consider all R&D expenditures incurred during the reporting period.

   - Total R&D expenditures/total assets: Calculate the ratio by dividing the total R&D expenditures by the total assets of your organisation.
   - Total R&D expenditures/total revenue: Calculate the ratio by dividing the total R&D expenditures by the total revenue generated by your organisation.

Note: It is important to disclose which accounting approach (GAAP or IFRS) your organisation uses in its accounting and reporting of R&D expenses. This indicator provides insights into the proportion of R&D expenditures relative to your organisation's assets and revenue, regardless of the accounting treatment.

5.3.2 Damage ratio (E1-1, E1-9, E2-6, E3-5, E4-6)

Relevant SDGs: SDG 9: Industry, Innovation, and Infrastructure, SDG 11: Sustainable Cities and Communities, SDG 13: Climate Action.

The damage ratio is a metric used to assess the financial impact of a hazard on a company's operations and assets. It represents the ratio of the total damage caused by a hazard to the value of the assets at risk. The damage ratio provides insights into the potential financial losses and helps in evaluating the vulnerability of assets to specific hazards.

The damage ratio is determined by comparing the total damage caused by a hazard to the value of the assets at risk. The following steps are typically involved:

- **Assessment of total damage**: Using data from insurance claims, disaster databases, expert assessments, and financial records, evaluate and quantify the total damage caused by hazard events. This assessment can include direct physical damage, indirect losses, business interruption, and other financial impacts resulting from the hazard.
- **Valuation of assets at risk**: Determine the value of the assets that are exposed to the hazard based on historical data of events and their associated damages, as well as asset valuation information. This involves assessing the monetary worth of the assets, such as buildings, equipment, infrastructure, inventory, and other relevant components.
- **Calculation of damage ratio**: Divide the total damage by the value of the assets at risk to calculate the damage ratio. The formula for calculating the damage ratio is:

Damage Ratio = (Total Damage/Value of Assets at Risk)

By calculating the damage ratio, decision-makers can gain insights into the potential financial impacts of hazards on a company's operations and assets. This information can support risk management strategies, investment decisions, insurance assessments, and the development of resilient business practices.

5.3.3 Vulnerability index (E1-1, 1-9, E2-6, E3-5, E4-6)

Relevant SDGs: SDG 9: Industry, Innovation, and Infrastructure, SDG 11: Sustainable Cities and Communities, SDG 13: Climate Action

The vulnerability index is a measure used to assess the susceptibility of a company's assets and infrastructure to a specific hazard. It takes into account various factors such as the location of assets, their design characteristics, and the level of maintenance. The vulnerability index helps identify the assets and infrastructure that are most at risk from a particular hazard, enabling informed decision-making for risk reduction and resilience planning.

The vulnerability index is determined by evaluating various factors that contribute to the susceptibility of assets and infrastructure to a hazard. The following steps are typically involved:

- **Hazard identification**: Identify the specific hazard or hazards that need to be assessed using asset inventories, engineering assessments, hazard mapping, historical data on hazard impacts, and expert knowledge. This could include natural hazards such as floods, earthquakes, storms, or human-induced hazards such as technological failures or cyber-attacks.
- **Asset and infrastructure assessment**: Evaluate the characteristics of the company's assets and infrastructure that may contribute to their vulnerability to the identified hazard. This can include factors such as location, age, design standards, structural integrity, maintenance practices, and other relevant parameters.
- **Weighting and scoring**: Assign weights or scores to different vulnerability factors based on their relative importance and impact on the susceptibility of assets and infrastructure. This step involves subjective judgement and expert input to determine the significance of each factor.
- **Calculation of vulnerability index**: Aggregate the weighted scores or calculate an overall index value to represent the vulnerability of assets and infrastructure to the identified hazard. The specific formula for

calculating the vulnerability index may vary depending on the chosen approach and methodology, but is usually expressed as:

$$\text{Vulnerability Index} = (w1 * F1) + (w2 * F2) + \ldots + (wn * Fn)$$

where:

Vulnerability Index: The overall vulnerability index value.

$w1, w2, ..., wn$: Weights assigned to each vulnerability factor.

$F1, F2, ..., Fn$: Scores or ratings assigned to each vulnerability factor.

The weights assigned to vulnerability factors reflect their relative importance in contributing to overall vulnerability. Scores or ratings represent the level of vulnerability for each factor, which can be determined through expert assessment or quantitative analysis. The vulnerability index provides a quantitative measure of the susceptibility of assets and infrastructure to the identified hazard. By using the vulnerability index, decision-makers can prioritise resources and implement targeted measures to reduce risks and enhance the resilience of assets and infrastructure. This information supports effective risk management, emergency planning, and investment decisions.

5.3.4 Adaptive capacity (E1-1, E1-9, E2-6, E3-5, E4-6)

Relevant SDGs: SDG 9: Industry, Innovation, and Infrastructure, SDG 11: Sustainable Cities and Communities, SDG 13: Climate Action

The adaptive capacity index is a measure used to assess a company's ability to adapt to the impacts of climate change. It considers various factors, including financial resources, technical expertise, and governance structures, to determine the company's readiness to respond and recover from climate-related challenges. The index also encompasses broader measures such as local flood and storm surge defences, which play a crucial role in enhancing adaptive capacity.

The adaptive capacity index is determined by evaluating various factors that contribute to a company's ability to adapt to climate change impacts. The following steps are typically involved:

- **Identification of adaptive capacity factors**: Identify the key factors that influence a company's adaptive capacity with reference to financial reports, technical assessments, governance frameworks, and climate change adaptation strategies. These factors can include available financial resources, technical expertise, governance structures, availability of information, access to adaptation and new technologies, stakeholder engagement, and the presence of local flood and storm surge defences.

- **Weighting and scoring**: Assign weights or scores to each adaptive capacity factor based on their relative importance and contribution to the overall adaptive capacity. This step involves subjective judgement and expert input to determine the significance of each factor.
- **Calculation of adaptive capacity index**: Aggregate the weighted scores or calculate an overall index value to represent the adaptive capacity of the company. The specific formula for calculating the adaptive capacity index may vary depending on the chosen approach and methodology.

$$\text{Adaptive Capacity Index} = (w1 * F1) + (w2 * F2) + \quad + (wn * Fn)$$

where:

Adaptive Capacity Index: The overall adaptive capacity index value.

$w1, w2, ..., wn$: Weights assigned to each adaptive capacity factor.

$F1, F2, ..., Fn$: Scores or ratings assigned to each adaptive capacity factor.

The weights reflect the relative importance of each adaptive capacity factor in determining the overall adaptive capacity. The scores or ratings represent the level of readiness or capability for each factor, which can be determined through expert assessment or quantitative analysis. The adaptive capacity index provides a quantitative measure of a company's ability to adapt to climate change impacts.

By assessing the adaptive capacity index, decision-makers can identify areas of strength and weakness in their company's adaptive capacity and prioritise resources for enhancing resilience. This information supports effective climate change adaptation planning, resource allocation, and risk management.

5.3.5 Loss event frequency (LEF) (E1-1, E1-4, E1-9, E2-6, E3-5, E4-6)

Relevant SDGs: SDG 9: Industry, Innovation, and Infrastructure, SDG 11: Sustainable Cities and Communities, SDG 13: Climate Action

Loss event frequency (LEF) refers to the number of times a specific event or loss occurs within a defined period of time. It is used as a measure to assess the frequency of occurrence of a particular event and to calculate the probability of that event happening. LEF provides valuable information for risk assessment, insurance underwriting, and decision-making related to potential losses. The methodology for determining the loss event frequency typically involves the following steps:

- **Event identification**: By referring to historical data, incident reports, insurance claims databases, industry-specific databases, regulatory agencies, and relevant research studies, identify the specific event or loss for which the frequency needs to be assessed. This could include natural disasters, accidents, equipment failures, cybersecurity incidents, or any other event that may result in a loss.
- **Data collection**: Gather relevant historical data on the occurrence of the identified event. This can include records of past incidents, insurance claims data, industry databases, incident reports, or other reliable sources of information.
- **Calculation of loss event frequency**: Calculate the loss event frequency by dividing the total number of occurrences of the event within the defined time period by the duration of that period. This calculation provides the average frequency of the event happening over time.

  LEF = Number of Occurrences/Time Period

where:

LEF: Loss event frequency.

Number of Occurrences: The total number of times the specific event or loss has occurred within the defined time period.

Time Period: The duration of the time period for which the loss event frequency is being calculated.

The resulting value represents the average frequency of the event occurring per unit of time (e.g., per year, per month, per week).

By analysing loss event frequency, organisations can gain insights into the frequency and probability of specific events, which helps in risk management, contingency planning, and designing appropriate risk mitigation measures. It enables decision-makers to allocate resources effectively, implement preventive measures, and take actions to minimise the potential impact of future events.

5.3.6 Loss event magnitude (LEM) (E1-1, E1-4, E1-9, E2-6, E3-5, E4-6)

Relevant SDGs: SDG 9: Industry, Innovation, and Infrastructure, SDG 11: Sustainable Cities and Communities, SDG 13: Climate Action

Loss event magnitude (LEM) refers to the extent of loss or damage caused by a specific event or incident. It is used to measure the severity and impact of an event on a company or organisation. LEM provides valuable information for assessing the financial, operational, and reputational consequences of an event, allowing organisations to understand the magnitude of the loss incurred. The methodology for determining the loss event magnitude typically involves the following steps:

- **Event assessment**: By referring to incident reports, financial records, insurance claims data, post-event assessments, and regulatory agencies, assess the specific event or incident for which the magnitude needs to be determined. This could include natural disasters, accidents, security breaches, financial losses, or any other event that results in tangible or intangible losses.
- **Quantification of loss**: Quantify the loss or damage caused by the event in relevant units such as monetary value (e.g., currency), physical units, or other applicable metrics. This may involve assessing direct costs, indirect costs, business interruption, environmental impact, human casualties, or any other relevant factors.
- **Calculation of loss event magnitude**: Calculate the loss event magnitude by quantifying the total loss incurred as a result of the event, which provides an understanding of the scale and severity of the event's impact. Accordingly, loss event magnitude (LEM) can vary based on the nature of the event and the units of measurement employed. For example, if the loss is measured in monetary terms, the formula can be as simple as:

  $$\text{LEM} = \text{Total Monetary Loss}$$

where:

LEM: Loss event magnitude.

Total Monetary Loss: The total financial loss or damage caused by the event.

If other units of measurement are used, the formula can be adjusted accordingly to reflect the appropriate quantification method.

By analysing loss event magnitude, organisations can understand the severity of an event's impact, assess its implications on various aspects of their operations, and develop strategies to mitigate future losses. This information aids in risk management, insurance planning, and decision-making related to resilience-building measures.

5.3.7 Annualised loss expectancy (ALE) (E1-1, E1-4, E1-9, E2-6, E3-5, E4-6)

Relevant SDGs: SDG 9: Industry, Innovation, and Infrastructure, SDG 11: Sustainable Cities and Communities, SDG 13: Climate Action

Annualised loss expectancy (ALE) is a metric used to estimate the expected financial loss for a specific hazard or event per year. It takes into account both the occurrence frequency of the hazard and the expected damage or loss per event. ALE provides valuable insights into the potential financial impact of a hazard and helps organisations understand the level of risk they face. The methodology for calculating annualised loss expectancy typically involves the following steps:

- **Hazard assessment**: By referring to historical data, risk assessments, insurance claims data, expert opinions, and industry-specific studies, identify the specific hazard or event for which the ALE needs to be determined. This could include natural disasters, cybersecurity breaches, equipment failures, or any other event that may result in financial losses.
- **Occurrence frequency**: Determine the occurrence frequency of the hazard, typically expressed as the number of times the event is expected to occur within a year. This can be based on historical data, probabilistic models, or expert judgement.
- **Expected damage per event**: Assess the expected damage or loss that would result from a single occurrence of the hazard. This includes quantifying the financial impact, such as direct costs, indirect costs, business interruption, reputational damage, and other relevant factors.
- **Calculation of ALE**: Calculate the annualised loss expectancy by multiplying the occurrence frequency by the expected damage per event. This provides an estimate of the average annual financial loss associated with the hazard.

$$\text{ALE} = \text{Occurrence Frequency} \times \text{Expected Damage per Event}$$

where:

ALE: Annualised loss expectancy.

Occurrence Frequency: The number of times the hazard is expected to occur within a year.

Expected Damage per Event: The estimated financial loss or damage that would result from a single occurrence of the hazard.

The resulting value represents the expected average annual financial loss associated with the hazard. By calculating the ALE, organisations can assess the potential financial impact of a hazard and make informed decisions regarding risk mitigation, insurance coverage, and allocation of resources for resilience measures. It helps organisations prioritise efforts and allocate resources effectively to minimise the financial consequences of hazards.

5.3.8 Exposure at risk (EAR) (E1-1, E1-9, E2-6, E3-5, E4-6)

Relevant SDGs: SDG 9: Industry, Innovation, and Infrastructure, SDG 11: Sustainable Cities and Communities, SDG 13: Climate Action

Exposure at risk (EAR) is a metric used to measure the extent to which an organisation's assets and liabilities are exposed to a specific event or hazard. It provides an understanding of the total potential loss that could be incurred as a result of the event. EAR is particularly useful for assessing the financial impact and vulnerability of an organisation to various risks. The methodology for calculating exposure at risk typically involves the following steps:

- **Identify the event**: Determine the specific event or hazard for which the exposure needs to be assessed. This could include natural disasters, market fluctuations, geopolitical risks, or any other event that may have financial implications for the organisation.
- **Assess assets and liabilities**: By referring to asset inventories, financial statements, risk assessments, insurance data, market analyses, and expert opinions, identify and quantify the assets and liabilities that may be affected by the event. This includes tangible assets (e.g., property, equipment), financial assets (e.g., investments, cash reserves), intangible assets (e.g., intellectual property), and liabilities (e.g., debts, obligations).

- **Determine exposure**: Evaluate the potential impact of the event on each identified asset and liability. This may involve considering factors such as physical vulnerability, market volatility, operational disruptions, legal and regulatory implications, and reputational risks.
- **Calculate exposure at risk**: Aggregate the estimated potential losses associated with each exposed asset and liability to derive the exposure at risk. This provides a comprehensive measure of the organisation's total potential loss from the event.

  EAR = Σ (Potential Loss of Exposed Asset/Liability)

where:

EAR: Exposure at risk.

Σ: Summation symbol.

Potential Loss of Exposed Asset/Liability: The estimated potential loss that could be incurred for each identified exposed asset or liability.

By quantifying the exposure at risk, organisations can assess their vulnerability to different events and make informed decisions regarding risk mitigation, insurance coverage, and contingency planning. EAR helps organisations identify areas of high exposure and prioritise efforts to minimise potential losses, enhance resilience, and protect their financial stability.

5.3.9 Business downtime (E1-1, E1-9, E2-6, E3-5, E4-6, S1-4)

Relevant SDGs: SDG 9: Industry, Innovation, and Infrastructure, SDG 11: Sustainable Cities and Communities, SDG 13: Climate Action

Business downtime refers to the period during which normal business operations are disrupted or temporarily halted due to various factors, such as equipment failures, natural disasters, cyber-attacks, or other unforeseen events. It encompasses the impacts on business operations, including the duration of downtime, repair costs, productivity losses, and potential increases in production costs. The methodology for assessing business downtime typically involves the following considerations:

- **Identify downtime events**: By referring to historical records of downtime events, maintenance logs, equipment performance data, incident reports, employee productivity records, and financial statements, identify the events or factors that can cause disruptions to business operations. These can include equipment breakdowns, power outages, supply chain disruptions, technological failures, natural disasters, or any other event that can significantly impact the continuity of operations.

- **Measure downtime duration**: Quantify the length of time during which business operations are affected or completely halted. This can be measured in hours, days, or any appropriate unit of time. Accurate data on downtime duration is essential for estimating the financial and operational impacts.
- **Evaluate repair costs**: Assess the costs associated with repairing or restoring the affected systems, equipment, or infrastructure. This includes expenses for repair parts, labour, contractor services, and any additional resources required to bring the operations back to normal.
- **Calculate productivity loss**: Estimate the loss of productivity resulting from the business interruption. This can involve analysing the number of employees affected, the average wage rates, and the expected reduction in output or service delivery during the downtime period.
- **Consider increased production costs**: Account for any additional costs incurred to expedite production or service delivery after the downtime event. These costs may include overtime wages, rush orders, expedited shipping, or other expenses related to restoring normal operations as quickly as possible.

Source of information:

Information sources for assessing business downtime can include historical records of downtime events, maintenance logs, equipment performance data, incident reports, employee productivity records, financial statements, and other relevant sources of information.

6

E1: Climate Change

In this chapter: Assessing the organisation's impact on the climate and identifying key sources. Scenario and resilience analysis. Target setting and transition planning. Monitoring and tracking process. Example indicators and metrics.

The transition to a sustainable economy and the reduction of greenhouse gas (GHG) emissions have become critical imperatives for organisations worldwide. In particular, the transition towards a net zero future poses unique challenges for organisations, as it potentially requires extensive operational adaptation and investment.

In line with global efforts to combat climate change and limit global warming to 1.5 °C, the Corporate Sustainability Reporting Directive (CSRD) has established a comprehensive standard for organisations to report their transition towards a net zero future. The Science-Based Targets initiative (SBTi) framework provides a robust methodology for organisations to set GHG emission reduction targets that are aligned with the goals of the Paris Agreement and thus the CSRD/ESRS. The SBTi framework ensures that these targets are compatible with limiting global warming to 1.5 °C and achieving climate neutrality by 2050. It also emphasises the importance of assessing and managing material climate-related impacts, risks, and opportunities across an organisation's operations and value chain.

Under the CSRD/ESRS, organisations are obligated to disclose their performance, strategies, and actions related to climate change mitigation and adaptation. These disclosures include information on GHG emission reduction targets, policies, actions, and resources dedicated to climate change initiatives. Additionally, organisations must provide transparency regarding their energy consumption, energy mix, and exposure to coal-, oil-, and gas-related activities.

ESRS E1: CLIMATE CHANGE		
GOVERNANCE & STRATEGY	**IMPACTS, RISKS, & OPPORTUNITIES**	**METRICS & TARGETS**
• E1-1: Comprehensive *transition plan* for climate change *mitigation* INCLUDES CROSS-CUTTING DISCLOSURES: • Related to ESRS 2 GOV1: The role of the *administrative, management, and supervisory bodies* in decarbonisation, climate change adaptation, and mitigation • Related to ESRS 2 SBM 3: Climate resilience of *strategy and business model* • Related to ESRS 2 GOV-3: Integration of climate change *strategies* and performance in *incentive schemes*	• E1-2: *Policies* related to climate change mitigation and adaptation • E1-3: *Action plans* and *resources* in relation to climate change policies and targets INCLUDES CROSS-CUTTING DISCLOSURES: • ESRS 2 IRO-1: Description of *processes to identify and assess* material climate-related impacts, risks, and opportunities	• E1-4: *GHG emission reduction targets* • E1-5: *Energy consumption* and mix (inc intensity) • E1-6: Gross Scopes 1, 2, 3 and Total *GHG emissions* (inc intensity in high climate impact sectors) • E1-7: *GHG removals* and *GHG mitigation* projects financed through carbon credits • E1-8:Internal *carbon pricing* • E1-9: *Potential financial effects* from material physical risks, material transition risks, and climate-related opportunities

Before diving into detailed metrics and targets, begin with describing your materiality assessment and organisational governance initiatives that directly interact with climate change topics, most importantly, incentivisation, resilience analysis, and screening processes and methodologies that link your materiality analysis to transition planning (see **6.1 E1-1 Comprehensive transition plan for climate change mitigation**).

TIPS AND TRICKS

Being Low Carbon is Profitable: Case Study, Nudie Jeans

Company name: Nudie Jeans

Sector: Apparel industry

Problem: Vision to become world's most sustainable brand. Nudie Jeans wanted to gain full transparency on all their supply chain tiers, insights into fibres used and create a circular business model

Solution: Sustainability and social impact were embedded into the strategy of Nudie Jeans from the very beginning. Company worked closely with its suppliers to achieve full transparency on all of their supply chain tiers including CO_2 emissions and water usage. Nowadays, forever repairs are directly embedded into the pricing of Nudie Jeans and company encourages returning back old jeans that people do not wear anymore, in exchange for discount vouchers on a new pair. This way, they promote consumer behaviour leading to circular business and ultimately to circular economy. In 2023, they have repaired over 73,000 pairs of jeans and their sales of second-hand jeans have been growing. All of Nudie Jeans are made of 100% certified organic cotton. For over five years, they have been tracking their scope 1, scope 2, and scope 3 GHG emissions to cover the total invested in carbon offsetting. Their Climate Action Roadmap is fully aligned to SBTs 1.5 °C pathway. Nudie Jeans continues growing and attracting new conscious customers, through sales of their new jeans' models but also through growing a second-hand segment, where the affordability attracts younger consumers. Last year, they have opened stores in New York and in Indonesia, and they are partnering with new repair centres and have plans to further expand, enabling their existing and new customers to create their own tomorrow's vintage.

Table 40 Indicators and metrics related to E1 – Climate change incentives, resilience and screening processes.

ESRS	DR	Paragraph	Name	Data type	Tips and tricks	Related CDR	VSMEs	ISSB / IFRS S1 & S2	GRI
E1	E1.GOV-3	13	Disclosure of how climate-related considerations are factored into remuneration of members of administrative, management, and supervisory bodies	Narrative	5.2.5 GOV-3: Integration of sustainability-related performance in incentive schemes			IFRS S1.21(b), IFRS S2.29(g)(i), IFRS S2.29(g)(ii)	GRI 2-19
E1	E1.GOV-3	13	Percentage of remuneration recognised that is linked to climate related considerations	Per cent				IFRS S1.21(b), IFRS S2.29(g)(i), IFRS S2.29(g)(ii)	GRI 2-19
E1	E1.GOV-3	13	Explanation of climate-related considerations that are factored into remuneration of members of administrative, management, and supervisory bodies	Narrative				IFRS S1.21(b), IFRS S2.29(g)(i), IFRS S2.29(g)(ii)	GRI 2-19
E1	E1.SBM-3	18	Type of climate-related risk	Semi-narrative	Chapter 3: Single and Double Materiality Assessment, including resilience and scenario analysis including classification of climate-related hazards, see Table 36			IFRS S2.10(b)	GRI 201-2
E1	E1.SBM-3	19a	Description of scope of resilience analysis	Narrative				IFRS S2.22(b)(i)(7), IFRS S2.22(b)(i)*	
E1	E1.SBM-3	19b	Disclosure of how resilience analysis has been conducted	Narrative				IFRS S2.22(b)(i)*, IFRS S2.22(b)(iii), IFRS S2.25(b)	
E1	E1.SBM-3	19b	Date of resilience analysis	Date				IFRS S2.22(b)(i)*, IFRS S2.22(b)(iii), IFRS S2.25(b)	
E1	E1.SBM-3	AR 7b	Time horizons applied for resilience analysis	Semi-narrative				IFRS S2.36.(e)(iii)	

E1	E1.SBM-3	19c	Description of results of resilience analysis	Narrative				IFRS S2.22(b)(i)*, IFRS S2.25(b), IFRS S2.22(a)(i), IFRS S2.22(a)(ii), IFRS S2.22(a)(iii)(1–3)	
E1	E1.SBM-3	AR 8b	Description of ability to adjust or adapt strategy and business model to climate change	Narrative				IFRS S2.14(a)(i), IFRS S2.22(a)(iii)(1–3)	
E1	E1.IRO-1	20a, AR 9	Description of process in relation to impacts on climate change	Narrative					
E1	E1.IRO-1	20b	Description of process in relation to climate-related physical risks in own operations and along value chain	Narrative					
E1	E1.IRO-1	AR 11a	Climate-related hazards have been identified over short-, medium-, and long-term time horizons	Semi-narrative					
E1	E1.IRO-1	AR 11a	Undertaking has screened whether assets and business activities may be exposed to climate-related hazards	Semi-narrative	See 6.1 E1-1: Description of transition plan for climate change mitigation including key drivers in scenario analysis and subsequent link(s) to policies. Can also use TCFD, ISO 14091:2021 or other recognised industry standard.				
E1	E1.IRO-1	AR 11b	Short-, medium-, and long-term time horizons have been defined	Semi-narrative				IFRS S2.10(d)4	
E1	E1.IRO-1	AR 11c	Extent to which assets and business activities may be exposed and are sensitive to identified climate-related hazards has been assessed	Semi-narrative					

(Continued)

Table 40 (*Continued*).

ESRS	DR	Paragraph	Name	Data type	Tips and tricks	Related CDR	VSMEs	ISSB / IFRS S1 & S2	GRI
E1	E1.IRO-1	AR 11d	Identification of climate-related hazards and assessment of exposure and sensitivity are informed by high emissions climate scenarios	Semi-narrative	If significant OpEx and CapEx are required, then also relate to R&D expenditure and risk and opportunity metrics, see 5.3			IFRS S2.22(b)(i)(2)*, IFRS S2.22(b)(i)(3)*	
E1	E1.IRO-1	21	Explanation of how climate-related scenario analysis has been used to inform identification and assessment of physical risks over short, medium, and long term	Narrative				IFRS S2.22(b)(i)(1-6)*, IFRS S2.25(a)(ii)*, IFRS S2.25(b)*	
E1	E1.IRO-1	20c	Description of process in relation to climate-related transition risks and opportunities in own operations and along value chains	Narrative				IFRS S2.25(b)*, IFRS S2.22(b)(i)(4)*	
E1	E1.IRO-1	AR 12a	Transition events have been identified over short-, medium-, and long-term time horizons	Semi-narrative					
E1	E1.IRO-1	AR 12a	Undertaking has screened whether assets and business activities may be exposed to transition events	Semi-narrative					

E1	E1.IRO-1	AR 12b	Extent to which assets and business activities may be exposed and are sensitive to identified transition events has been assessed	Semi-narrative					
E1	E1.IRO-1	AR 12c	Identification of transition events and assessment of exposure has been informed by climate-related scenario analysis	Semi-narrative				IFRS S2.22(b)(i)(1-6)*, IFRS S2.25(a)(ii)*, IFRS S2.25(b)*	
E1	E1.IRO-1	AR 12d	Assets and business activities that are incompatible with or need significant efforts to be compatible with transition to climate-neutral economy have been identified	Semi-narrative					
E1	E1.IRO-1	AR 15	Explanation of how climate scenarios used are compatible with critical climate-related assumptions made in financial statements	Narrative				IFRS S1.23	

- Indicates that additional relevant information can be found in Section 3. ESRS to IFRS S2 (climate): information that an entity starting with ESRS needs to know when also applying ISSB Standards to enable compliance with both sets of standards.

Source: CDR (EU) 2023/2772 supplementing Directive 2013/34/EU, *IFRS S1 General Requirements for Disclosure of Sustainability-related Financial Information, IFRS S2 Climate-related Disclosures, GRI 1, 2, 3: 11, 12, 13, 14, 306, 101, 201, 202, 203, 204, 205, 206, 207,301, 302, 303, 304, 305, 306, 308,401, 402, 403, 404, 405, 406, 407, 408, 409, 410, 411, 413, 414, 415, 416, 417, 418*, and the author.

6.1 Comprehensive Transition Plan for Climate Change Mitigation (E1-1, 2, and 3)

Climate change transition plans are a mandatory component of all ESG reporting standards and must be aligned with international frameworks based on the GHG Protocol. The following methodology for developing and reporting a climate change transition plan complies with IFRS, SFDR, GRI, CDSB, SASB, ISO, and CSRD/ESRS requirements and should be used in combination with an international framework verifier, for example, SBTi or TPT.

The key components of the methodology are outlined in the process diagram below, and each component is explained in detail in the following sections and on the support website https://esgreportpro.com along with examples of some key metrics.

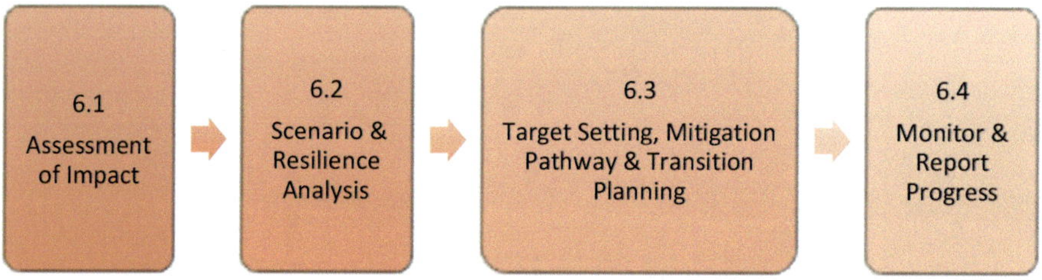

Graphic 19 Climate change assessment methodology components. Source: Author.

Climate change is a critical reporting issue in the CSRD/ESRS and directly relates to your organisation's approach to governance, strategy, and risk management, specifically:

- Related to ESRS 2 GOV1: The role of the *administrative, management, and supervisory bodies* in decarbonisation, climate change adaptation, and mitigation.
- Related to ESRS 2 SBM 3: Climate resilience of *strategy and business model*.
- Related to ESRS 2 GOV-3: Integration of climate change *strategies* and performance in *incentive schemes*.

Assessing the organisation's impact on the climate involves mapping its greenhouse gas (GHG) emissions and identifying the sources that contribute to its carbon footprint. This assessment provides a foundation for measuring progress and identifying areas for improvement. A detailed breakdown of the four steps involved in the assessment is as follows.

6.1.1 Creating a greenhouse gas inventory

Creating a greenhouse gas inventory involves first identifying and categorising the sources and scale of GHG emissions within the organisation's operations and external value chain. This process should begin as part of your stakeholder engagement process (see Graphic 20) (**see Stage 3: Stakeholder Engagement and Implementation**) and include links to your asset register, and maintenance and servicing records. GHG emissions must be categorised as Scope 1, Scope 2, and Scope 3 and cross-referenced by relevant emission source per location, using these definitions (see Graphic 21 for approximate comparisons by industry of Scope 1 + 2 versus Scope 3 emission relative amounts):

- **Scope 1 emissions**: Direct emissions from sources owned or controlled by the organisation, such as combustion of fossil fuels (for example, fuel for vehicles owned by the organisation), on-site energy generation (for example, diesel generators), and industrial processes (for example, refrigeration coolants used in air conditioners and refrigerators, fire suppression systems, and certain industrial chemicals and processes).
- **Scope 2 emissions**: Indirect emissions from purchased energy for organisational use including electricity or natural gas that is purchased from a local power utility to power a building or facility, and/or energy consumption and usage for offices and sites that are purchased from a utility, and/or energy generation from any leased, non-renewable sources, and/or other purchased emissions from rented or leased equipment not already accounted for in sites and facilities.
- **Scopes 1 and 2 locked-in GHG emissions of key assets:** This will be assessed as the sum of the estimated Scopes 1 and 2 GHG emissions over the operating lifetime of the active and firmly planned key assets. Key assets are those owned or controlled by the organisation, and they consist of existing or planned assets (such as stationary or mobile installations, facilities, and equipment) that are sources of either significant direct or energy-indirect GHG emissions. Firmly planned key assets are those that the undertaking will most likely deploy within the next five years.
- **Scope 3 emissions**: Indirect emissions that occur throughout the value chain, including upstream and downstream activities, business travel (including accommodation and events), employee commuting, waste creation/management, purchased goods and services, capital goods, transportation and distribution from suppliers and to customers, use of services like email and hosting/data centres, leased assets, franchises, investments,

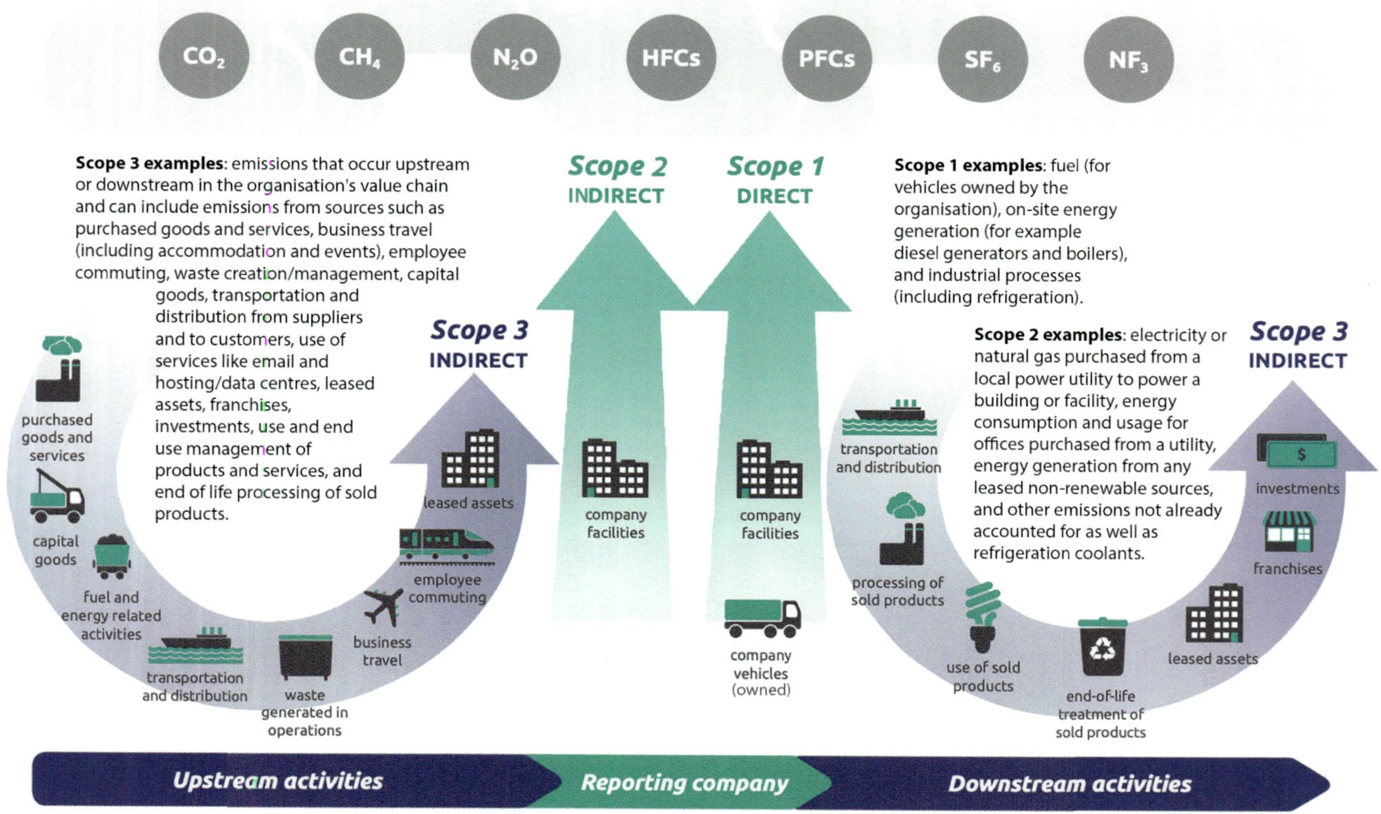

Graphic 20 GHG Protocol Scopes 1, 2, and 3. Source: Adapted from GHG Protocol, https://ghgprotocol.org/sites/default/files/standards/Scope3_Calculation_Guidance_0.pdf

use and end use management of products and services, and end of life processing of sold products. Organisations that identify the Scope 3 category 'use of sold products' (see table 41) as significant must consider the cumulative locked-in GHG emissions associated with the direct use-phase GHG emissions of sold products in tCO2eq by assessing the sales volume of products in the reporting year multiplied by the sum of estimated direct use-phase GHG emissions over their expected lifetime.

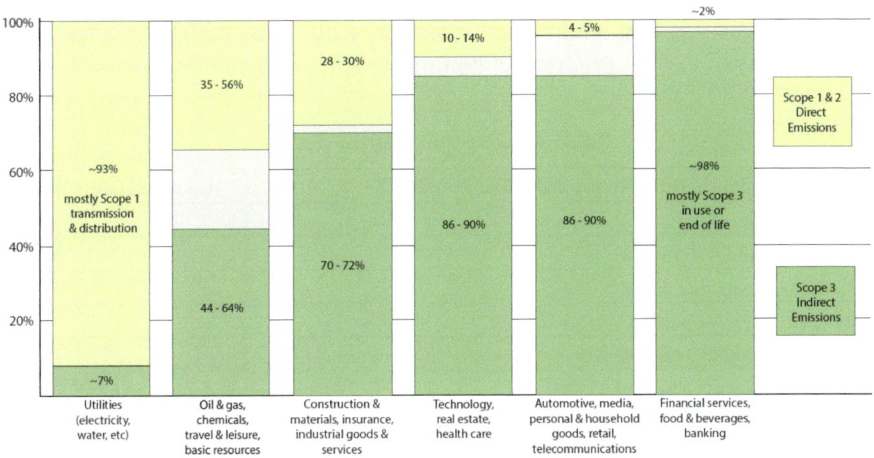

Graphic 21 Example ratios of Scope 1 and 2 vs. Scope 3 emissions for various sectors. Source: GHG Protocol and the author.

Table 41 GHG Protocol Scope 3 categories.

Scope 3 categories	8 Upstream leased assets
1 Purchased goods and services	9 Downstream transportation and distribution
2 Capital goods	10 Processing of sold products
3 Fuel- and energy-related activities	11 Use of sold products
4 Upstream transportation and distribution	12 End-of-life treatment of sold products
5 Waste generated in operations	13 Downstream leased assets
6 Business travel	14 Franchises
7 Employee commuting	15 Investments

Source: GHG Protocol, for description of categories, see Annex 3: Scope 3 categories based on GHG corporate protocol.

Exclude carbon credits and avoided emissions:

In the process of measuring and reducing emissions, it is important to exclude certain factors from being counted as reductions towards meeting science-based targets. This includes carbon credits and avoided emissions:

Carbon credits:

Carbon credits are tradable permits that represent the reduction or removal of greenhouse gas emissions elsewhere. They are generated through projects or activities that offset emissions by supporting emission reduction initiatives. When setting and tracking targets, it is crucial not to count carbon credits as direct reductions. This is because relying solely on purchased offsets does not contribute to an actual reduction in emissions from the organisation's own operations or value chain. Instead, the focus should be on implementing internal measures that lead to real emission reductions within the organisation. However, carbon credits can be reported as a separate action.

Table 42 Reporting carbon credits.

Carbon credits cancelled in the reporting year 20XX	**Comparative**	***N***
Total (tCO2e)		
Share from removal projects (%)		
Share from reduction projects (%)		
Recognised quality standard 1 (%)		
Recognised quality standard 2 (%)		
Recognised quality standard 3 (%)		
...		
Share from projects within the EU (%)		
Share of carbon credits that quality as corresponding adjustments (%)		

Carbon credits planned to be cancelled in the future	**Amount until (period)**
Total (tCO2e)	

Source: CDR (EU) 2023/2772 supplementing Directive 2013/34/EU.

Avoided emissions:

Avoided emissions refer to emissions that would have occurred under a business-as-usual scenario but are avoided due to specific actions or interventions. It is important to note that avoided emissions occur outside of the product's lifecycle or the direct control of the organisation. They are often the result of external factors or actions taken by other parties.

- When assessing Scope 1, 2, and 3 emissions, avoided emissions *should not* be counted as reductions. This is because the organisation did not directly cause or control these emission reductions.
- Instead, the organisation should focus on reducing emissions within its own operations and value chain, taking direct responsibility for the emissions it generates.

Once the GHG inventory is finalised, it is time to apply a double-materiality assessment (**see Chapter 3: Single and Double Materiality**) to prioritise and scale emission sources based on their significance and potential for emissions reduction, noting that emissions could be financially and physically material, or neither. Materiality assessments should ask questions regarding the significance of the emission source to the organisational strategy, goals, and ongoing business; for example, should the source be replaced, then what impacts, risks, and opportunities might arise? And, how significant would replacement be to future operations?

6.1.2 Quantification and reporting

Once all material emission sources have been identified along with their double materiality assessment and associated impacts, risks, and opportunities, it is then time to quantify and report them. This involves two steps:

- Quantify the emissions from each identified source and then calculate the scale of each emission source using appropriate emission factors, methodologies, and indicators.
- Produce a comprehensive report that includes a breakdown of emissions by source, scope, and location – for the report, include visual aids such as charts, graphs, or diagrams to present the emissions data in a clear and understandable manner.

6.1.3 Establishing the baseline year

Having now identified and quantified emissions, it is necessary to select a baseline year, which is a reference point from which all future emissions, reduction targets, and mitigation efforts will be compared, such that:

- you can provide a complete record of all material emissions;
- is representative of the organisation's typical operations; and
- has occurred since 2015.

Common baseline years that are pre- and post-COVID for medium to large organisations are 2018, 2019, 2022, or 2023.

Common baseline years that are pre- and post-COVID for small to medium organisations are 2018, 2019, 2020, or 2021.

Now review your GHG inventory and report for emissions in your selected baseline year. Be careful to note any differences in the parameters or boundaries of each emission type. For example, investments or divestments may make a significant difference to values.

You should now have two complete GHG inventories, one for your baseline year and one for your reporting year, as well as two corresponding reports.

6.1.4 Leakage and offsetting assessment

The final step in creating a greenhouse gas inventory involves identifying possible leakages and mitigation measures. For each item in both your baseline and reporting years, describe:

- your assessment of any potential leakage of emissions due to changes in activities or operations;
- opportunities identified for emissions reduction and energy efficiency improvements to mitigate leakage risks; and/if and/or how you have considered offsetting strategies to compensate for unavoidable emissions through credible carbon offset projects or investments.

6.2 Scenario and Resilience Analysis

A climate change scenario and resilience analysis is an assessment of your organisation's vulnerability to climate-related challenges and identifies opportunities for adaptation and mitigation. The Locate, Evaluate, Assess, Prepare (LEAP) framework is used to guide this analysis, and ensure that you consider the following:

- **Plausibility**: Plausible scenarios are based on credible narratives and are more likely to occur. While they may not represent the most probable future outcomes, they are the most feasible paths for achieving the target of limiting global warming to 1.5 °C or well below 2 °C. Plausible scenarios demonstrate internal consistency and logical coherence, avoiding assumptions or parameters that contradict current evidence without a rational explanation. Benchmarking is advisable and when necessary use statistical techniques, such as calculating medians and ranges from a sample of scenarios that meet specific criteria, to assess the plausibility of scenarios.
- **Responsiveness**: Responsible scenarios are built upon the principle of minimising the risk of failing to meet the objectives of the Paris Agreement. These scenarios are objective and unbiased, focusing on risk management rather than organisational preferences. Risk mitigation strategies can include diversification to reduce dependency on

specific future developments and adopting a portfolio-based approach. By taking proactive measures and avoiding delays in action, risks associated with climate change can be better managed.

6.2.1 Locate your interface, evaluate dependencies, and impacts

For each operational or financial activity associated with a GHG inventory item in a specific location (including through the supply chain), describe how activities may be dependent or influenced by climate-related factors. Factors should include the hazards that we identified in **5.2.10 IRO-1: Identify and assess material impacts, risks, and opportunities** as well as regulatory frameworks, reliance on climate-sensitive resources and inputs, such as water, energy, raw materials, or agricultural products.

Having identified each interface, evaluate and describe your dependencies and potential impacts (for example, the likelihood, quality and cost implications, and assessment reliability) to your organisation and value chain. It is unlikely that your organisation and/or your value chain are not affected by any climate-related factors, so stakeholder engagement to confirm your assessment is advisable.

6.2.2 Assessment of risks and opportunities

Conduct a comprehensive risk assessment to identify climate-related risks and opportunities that may affect each of the organisation's activities, including operations, value chain, infrastructure, financial performance, or reputation. Examples of climate-related risks include:

- *Physical risks* (e.g., infrastructure damage, supply chain disruptions) are risks associated with climate-induced events such as extreme weather events, sea-level rise, and changes in temperature or precipitation patterns.
- *Transition risks* (e.g., policy changes, shifts in consumer preferences) are risks arising from the transition to a low-carbon economy, including policy changes, regulatory shifts, market disruptions, and changes in consumer preferences.
- *Liability risks* (e.g., legal and regulatory actions) are risks related to legal or regulatory actions due to environmental damages, non-compliance with climate-related regulations, or failure to meet stakeholder expectations.

Table 43 Examples of climate-related risks and potential impacts.

Identified risk	Potential impact	Likelihood	Cost implications	Reliability
Governance				
• GHG emissions cost increases • Increased GHG emission reporting obligations • Regulation and control of existing products • Litigation exposure	• Increased operating costs for compliance, premiums, fines • Policy changes incur current asset write-offs/impairment/early retirement	High in EU Moderate North America and Asia Pacific	Low to moderate <5% revenue	To be expected >80% probability
Technology and innovation				
• New lower emission products • Unsuccessful new products • Transitioning and adaptation challenges	• Stock write-offs and/or reduced demand for older products • Increased R&D costs and capital investment	Moderate to high in all plant locations	Moderate to high >10% revenue	Uncertain 40%–60% probability
Market and pricing				
• Changing customer preferences • Market uncertainty • Raw material volatility	• Fluctuating costs and variable demand reduce performance • Increased price competition • Production constraints	Low to moderate in all major markets	Low <2% revenue	Less than likely <40%
Physical impacts				
Acute: • Increasingly severe weather events • Chronic: • Weather pattern variability • Rising temperatures • Rising sea levels	• Value chain disruption • Own workforce disruption or displacement • Damage to plant, equipment, stock • Higher cost of capital and/or premiums	Low to moderate in all plant locations	Low to moderate <5% revenue	More than likely 60%–80%

Source: Author.

Table 44 Example of climate-related opportunities and potential impacts.

Identified opportunity	Potential impact	Likelihood	Cost implications	Reliability
Operations				
• More energy efficient distribution • Reduced consumption and waste • More energy efficient locations	• Reduced operating costs • Reduced production costs • Increased production capacity and storage	High in EU Moderate North America and Asia Pacific	Low <2% revenue	To be expected >80% probability
Products and services				
• Lower product lifetime footprints • Higher integration with suppliers and customers • Greener sales mix	• Reduced exposure to GHG emission costs • Co-creation and collaborative development • Enhanced reputation	Moderate to high in all markets	Moderate to high >10% revenue	Less than likely <40%
Markets				
• New product and markets • Early entrant advantage • Increased differentiation	• New service offerings • Increased revenue and channels • Improved competitive advantage	High in EU Moderate to high in North America and Asia Pacific	High >15% revenue	Uncertain 40%–60% probability
Resilience				
• Energy consumption reduction • Circular resource use	• Access to new financing tools and investors • Stronger value chain and increased loyalty			More than likely 60%–80%

Source: Author.

- **Reputational risks**: Risks associated with negative public perception, stakeholder backlash, or reputational damage due to insufficient climate action or irresponsible practices.
- **Supply chain risks**: Risks stemming from climate impacts on suppliers, including disruptions in the availability, quality, or cost of inputs.

Simultaneously, identify opportunities that arise from climate-related changes, such as developing new sustainable products and services, improving resource efficiency, or accessing new markets driven by climate mitigation and adaptation measures.

6.2.3 Prepare to respond to climate-related risks and opportunities, and report

Using your GHG inventory and the above Locate, Evaluate, Assess and Prepare (LEAP) approach, describe:

- a climate resilience strategy that outlines measures to adapt to climate change and capitalise on opportunities;
- how you intend to implement risk management strategies and action plans to mitigate identified risks;
- plan to enhance adaptive capacity by integrating climate considerations into decision-making processes, supply chain management, infrastructure design, and business continuity planning; and
- you have or will establish mechanisms for ongoing monitoring, evaluation, and reporting of climate-related performance and progress towards resilience goals.

6.3 Target Setting, Mitigation Pathway, and Transition Planning

6.3.1 Target setting

Target setting involves establishing an emissions target for each organisational activity and then totalling the target emissions per year, that is, for each GHG inventory item and considering your baseline year, establish targets for reducing GHG emissions by 2030 and then by 2050. An example of an organisational target is *to reduce GHG emissions by 50% by 2030, compared to a baseline year of 2019, across all operations and supply chain, and by 95% by 2050*. Throughout this process, it is important to:

- **Set specific, ambitious yet realistic and measurable targets**: For each emission source, set annual targets with the aim to be net zero by 2030 and as close to zero emissions by 2050, with any residual emissions offset each year. Consider sector-specific criteria and other relevant factors, such as your organisation's capacity for mitigation of emissions, available resources and technology, and industry benchmarks. Best practice targets should focus on remaining within planetary boundaries and encourage peer buy-in, especially through your value chain. Ultimately, your targets should be challenging enough to drive meaningful and resilient change while remaining attainable.
- **Stakeholder engagement**: When setting targets involve relevant stakeholders, including employees, investors, customers, and communities. Seek their input, align with their expectations, and communicate your organisation's commitment to sustainability.

TIPS AND TRICKS

Planetary boundaries are key to environmental best practice:

In September 2023, a team of scientists quantified, for the first time, all nine processes that regulate the stability and resilience of the Earth system, planetary boundaries, which are a set of nine environmental limits within which humanity can operate safely. They include boundaries for climate change, biodiversity loss, ocean acidification, land-system changes, freshwater changes, and more. They are crucial for organisations and ESG reporting as they provide a framework for understanding the environmental impact of human (including business) activities, thus guiding companies towards more sustainable practices and helping investors assess a company's sustainability performance. Adhering to planetary boundaries is key for long-term business resilience and ensuring a sustainable future.

For more information, see https://www.stockholmresilience.org/research/planetary-boundaries.html

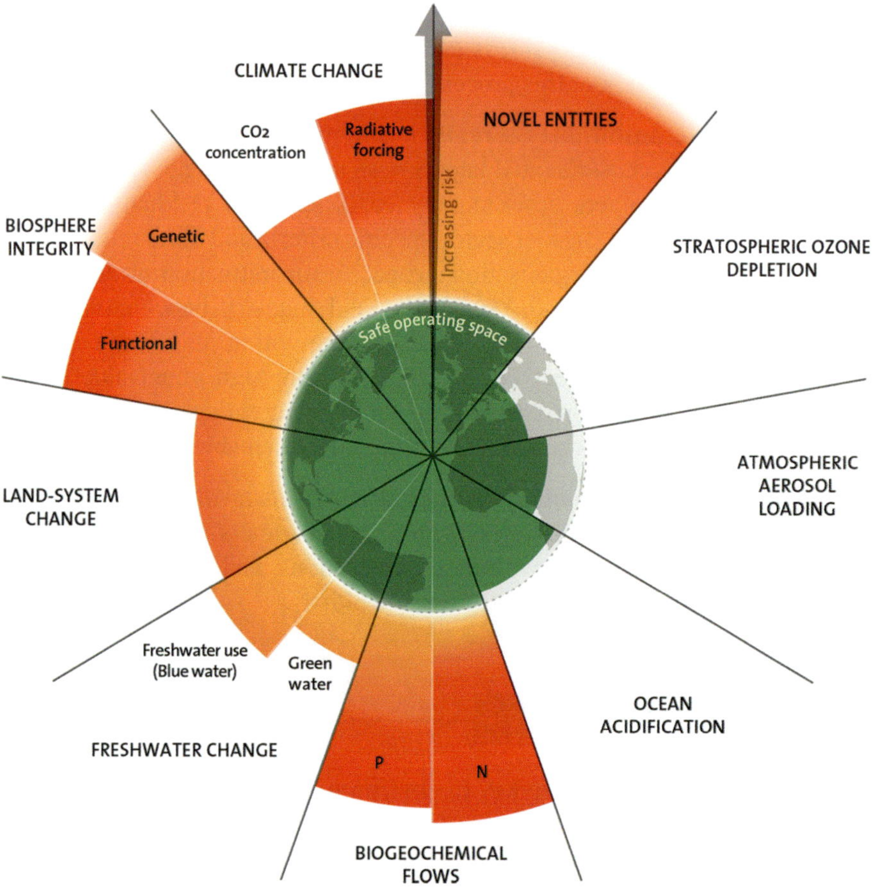

Graphic 22 The 2023 update to the planetary boundaries. Source: Azote for Stockholm Resilience Centre, based on analysis in Richardson et al. (2023).

6.3.2 Estimating emissions mitigation pathway

After setting targets, your organisation needs to explain how it intends to meet them through a mitigation pathway using one of the following methodologies:

- **Absolute emission reduction**: Set targets for reducing absolute emissions independently of business growth.
- **Intensity-based emission reduction**: Measure emissions relative to a specific metric while allowing for potential growth in business activities.

- **Sector-based decarbonisation**: Set targets based on decarbonisation pathways specific to the organisation's sector, considering sectoral benchmarks and best practice based on planetary boundaries.

Among the several methodologies prescribed by the SBTi framework, I strongly recommend using the *absolute contraction approach (ACA)*, which has been implemented widely because of its simplicity and applicability in heterogeneous sectors. The absolute contraction approach (ACA) is a methodology based on the principle that global warming can be limited to 1.5 °C if all organisations worldwide reduce their absolute GHG emissions in accordance with emission scenarios aligned with this temperature threshold. In other words, if we all aim for the same target and reduce emissions at the corresponding rate, we should be able to limit global warming to 1.5 °C. The ACA can be applied to companies in all sectors, particularly in cases where a sector-specific approach is not mandated by the GHG Protocol. There are two key aspects that you need to include in your mitigation pathway:

- **Calculation of emission reduction rate**: The GHG Protocol and SBTi considered 20 scenarios aligned with limiting global warming to 1.5 °C and 28 scenarios aligned with even more ambitious temperature thresholds to determine the required absolute emission reduction rate of limiting global warming to the 1.5 °C target. Their conclusion is that GHG emissions must be reduced by an average linear reduction of at least 4.2% per year relative to a specified base year, with net reduction targets of 42% by 2030 and 90% by 2050.

- **Target milestones:** The emission reduction milestones depend on the chosen base year. For instance, if the base year is 2018, the ACA requires a minimum emission reduction of 29.4% by 2025 and 50.4% by 2030 to align with the linear reduction trajectory of 4.2% per year.

6.3.3 Transition plan

A transition plan describes how your organisation will adjust its strategy and business model to meet its emissions targets based on your emissions mitigation pathway. So, the goal of the transition plan is to limit global warming to 1.5 °C by achieving net zero emissions by 2030 and climate neutrality by 2050, and should include the following aspects:

- **Emission reduction target pathway actions**: For each target, identify decarbonisation levers and key actions, such as changes in product portfolio, operational adaptions, lower emission alternatives, and adoption of new technologies.

- **Investments and funding**: The company explains its investments and funding initiatives supporting the transition plan's implementation, focusing on climate change mitigation actions. This includes disclosure of significant operational expenditures (OpEx) and/or capital expenditures (CapEx) required for implementation of the action plan.
- **Cumulative locked-in GHG emissions**: The company assesses cumulative locked-in GHG emissions associated with key assets and sold products, ensuring responsible management and considering their impact over their expected lifetime.
- **Management of GHG-intensive assets**: How the company plans to transform, decommission, or phase-out GHG-intensive assets related to coal, oil, and gas, thus minimising climate risk and aligning with sustainability goals.
- **EU Paris-aligned benchmarks**: The company discloses whether and why it may be excluded from the EU Paris-aligned benchmarks.
- **Integration with business strategy and financial planning**: How the transition plan is embedded within the company's overall business strategy and financial planning, approved by relevant administrative and management bodies.
- **Reporting progress of implementation**: The company reports on its progress in implementing the transition plan, demonstrating milestones achieved and steps taken.
- **Integration of sustainability in incentive schemes**: The company considers integrating sustainability-related performance indicators into its incentive schemes to promote and reward sustainability efforts.

Typically, a transition plan outlines the steps and associated impacts of mitigation pathway activities graphically and tabulates progress targets over time, see example in Graphic 23 and Table 45 below.

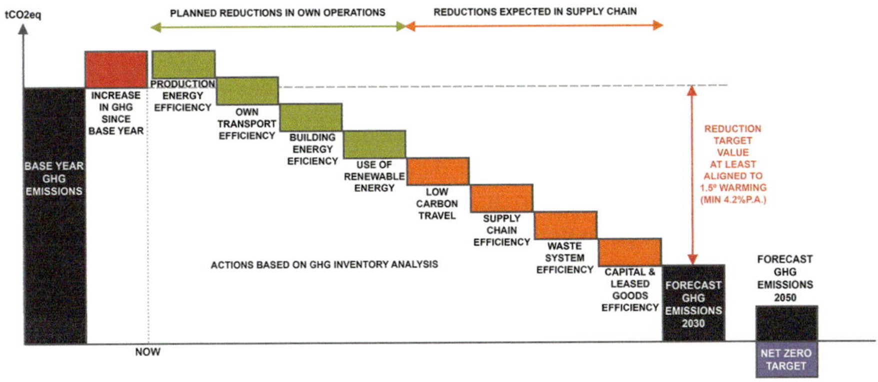

Graphic 23 Transition plan example using absolute contraction approach (ACA). Source: Author, adapted from GHG Protocol.

Table 45 Transition plan example using absolute contraction approach (ACA).

	Baseline year	**2030 target**	**2035 target**	**....**	**2050 target**
Total GHG emissions	100	58	42		...
Production energy efficiency		–10	–3		
Own transport efficiency		–7	–2		
Building energy efficiency		–2	–2		
Use of renewable energy		–12	–3		
Low carbon travel		–3	–		
Supply chain efficiency		–4	–2		
Waste system efficiency		–1			
Capital and leased goods efficiency		–2	–4		
Other		–1			

Source: Author.

6.4 Monitoring and Reporting Progress

A transition plan is meaningless without appropriate embedded mechanisms for monitoring, tracking, and reporting progress in implementing climate-related actions and achieving mitigation targets. By effectively monitoring and tracking progress, your organisation can assess the effectiveness of its strategies, identify areas for continuous improvement, and ensure transparency in reporting. Subsequent reporting allows for broad access to stakeholders and industry peers and regulators, as well as promoting responsible practices, continuous improvement, and transparency.

The monitoring and reporting process must explain how your organisation's transition plan is aligned with the overall business strategy and financial planning, and how the plan and ongoing implementation is approved by administrative, management, and supervisory bodies.

6.4.1 Data collection and management

In addition to the processes described in **Chapter 4: Data Collection and Analysis**, your Climate Change reporting system will need to gather the following relevant information on climate-related activities, emissions, and progress towards targets:

- Select and report on key performance indicators (KPIs) and metrics to measure the organisation's climate performance, such as GHG emissions intensity, energy consumption, renewable energy adoption, and waste reduction.
- Describe how your organisation ensures data accuracy, consistency, and transparency by implementing standardised measurement and reporting protocols, and relevant industry guidelines.

6.4.2 Reporting and public disclosure

- Describe your regular reporting and disclosure framework to communicate the organisation's climate-related performance internally and externally.
- Describe how your organisation engages and seeks input from internal and external stakeholders in the monitoring and tracking process to ensure buy-in, collaboration, and shared responsibility, as well as accountability and transparency.
- Possibly through your Green Team, describe how the climate change transitioning process involves collaboration between relevant departments, such as sustainability, finance, operations, and supply chain, to

integrate climate monitoring and tracking into existing processes and decision-making frameworks.

6.4.3 Performance evaluation and continuous improvement

Included in your transition plan should be the results and conclusions of regular evaluations and reviews of the organisation's climate performance that assess the effectiveness of mitigation strategies and identify areas for improvement. This should include:

- Analyse data trends, evaluate progress against targets, and identify any deviations or gaps.
- The outcome of periodic audits or independent verifications to ensure the accuracy and reliability of reported climate-related information.

Finally, describe how your organisation intends to use insights gained from monitoring and tracking to refine and enhance climate-related strategies and actions. This might involve:

- Identifying opportunities for innovation and efficiency improvements to drive further emission reductions and enhance resilience to climate-related risks.
- Staying informed about evolving best practices, emerging technologies, and regulatory developments to adapt and improve your organisation's climate performance over time.

In summary, your report should clearly demonstrate your organisation's commitment to climate action, enhanced transparency and accountability, and ambition to drive continuous improvement in your climate performance. Outcomes and insights should then enable informed decision-making, and support the organisation's long-term sustainability goals.

Table 46 Indicators and metrics related to E1-1, 2, and 3 – Transition plan, policies, and actions relating to Climate Change.

ESRS	DR	Paragraph	Name	Data type	Tips and tricks	Related CDR	VSMEs	ISSB/IFRS S1 & S2	GRI
ESRS 2		62	Disclosures to be reported in case the undertaking has not adopted policies or actions or targets	Narrative	See 5.1.5				
E1	E1-1	14	Disclosure of transition plan for climate change mitigation	Narrative	Summary of Chapter 6, ideally using a SBTi or similar framework	CL	V	IFRS S2.14(a)(iv)*	
E1	E1-1	16a	Explanation of how targets are compatible with limiting of global warming to one and half degrees Celsius in line with the Paris Agreement	Narrative	6.3.2 Mitigation pathway		V	IFRS S2.14(a)(iv)*	
E1	E1-1	16b	Disclosure of decarbonisation levers and key action	Narrative	6.3.3 Transition Plan Plus related to financial impacts from IRO-1			IFRS S2.14(a)(iv)*	
E1	E1-1	16c	Disclosure of significant operational expenditures (OpEx) and/or capital expenditures (CapEx) required for implementation of action plan	Narrative				IFRS S2.14(a)(iv)*, IFRS S2.29(e)	GRI 201-2
E1	E1-1	16c	Financial resources allocated to action plan (OpEx)	Monetary				IFRS S2.14(a)(iv)*, IFRS S2.29(e)	GRI 201-2
E1	E1-1	16c	Financial resources allocated to action plan (CapEx)	Monetary				IFRS S2.14(a)(iv)*, IFRS S2.29(e)	GRI 201-2

E1	E1-1	16d	Explanation of potential locked-in GHG emissions from key assets and products and of how locked-in GHG emissions may jeopardise achievement of GHG emission reduction targets and drive transition risk	Narrative				IFRS S2.14(a)(iv)*	
E1	E1-1	16e	Explanation of any objective or plans (CapEx, CapEx plans, OpEx) for aligning economic activities (revenues, CapEx, OpEx) with criteria established in Commission Delegated Regulation 2021/2139	Narrative				IFRS S2.14(a)(iv)*, IFRS S2.29(e)	
E1	E1-1	16f	Significant CapEx for coal-related economic activities	Monetary				IFRS S2.14(a)(iv)*, IFRS S2.29(e)	
E1	E1-1	16f	Significant CapEx for oil-related economic activities	Monetary				IFRS S2.14(a)(iv)*, IFRS S2.29(e)	
E1	E1-1	16f	Significant CapEx for gas-related economic activities	Monetary				IFRS S2.14(a)(iv)*, IFRS S2.29(e)	
E1	E1-1	16g	Undertaking is excluded from EU Paris-aligned benchmarks	Semi-narrative		PILLAR 3 + Benchmark		IFRS S2.14(a)(iv)*	
E1	E1-1	16h	Explanation of how transition plan is embedded in and aligned with overall business strategy and financial planning	narrative	6.4 Monitoring and reporting progress and related to MDR-P, and SBM-1			IFRS S2.14(a)(iv)*	
E1	E1-1	16i	Transition plan is approved by administrative, management, and supervisory bodies	Semi-narrative				IFRS S2.14(a)(iv)*	
E1	E1-1	16j	Explanation of progress in implementing transition plan	Narrative				IFRS S2.14(c)	

(Continued)

Table 46 (*Continued*).

ESRS	DR	Paragraph	Name	Data type	Tips and tricks	Related CDR	VSMEs	ISSB/IFRS S1 & S2	GRI
E1	E1-1	17	Date of adoption of transition plan for undertakings not having adopted transition plan yet	GYear					
E1	E1-2	24	Policies in place to manage its material impacts, risks, and opportunities related to climate change mitigation and adaptation	MDR-P				IFRS S2.25(a), IFRS S2.25(b)*	GRI 305-3-3
E1	E1-2	25	Sustainability matters addressed by policy for climate change	Semi-narrative					GRI 302-3-3
E1	E1-3	28	Actions and resources related to climate change mitigation and adaptation	MDR-A	5.1.2 MDR-A – Minimum Disclosure and 6.7 Energy mix			IFRS S2.14(a)(ii), IFRS S2.14(a)(iii), IFRS S2.14(b)	GRI 201-2, GRI 302-3-3, GRI 305-3-3, GRI 305-1.2
E1	E1-3	29a	Decarbonisation lever type	Semi-narrative					
	E1-3	AR19d	Adaptation solution type(s)	Semi-narrative	6.3.3 Transition plan, Indicators and Metrics Related to GHG Removals and GHG Mitigation Projects Financed Through Carbon Credits, and GHG Intensity Reduction through Removals and Reductions		V		
E1	E1-3	29b	Achieved GHG emission reductions	GHG emissions					GRI 305-5
E1	E1-3	29b	Expected GHG emission reductions	GHG emissions					GRI 305-5
E1	E1-3	AR21	Explanation of extent to which ability to implement action depends on availability and allocation of resources	Narrative					

E1	E1-3	29ci	Explanation of relationship of significant CapEx and OpEx required to implement actions taken or planned to relevant line items or notes in financial statements	Narrative	And then related to financial impacts from IRO-1 in CapEx and OpEx				
E1	E1-3	29cii, 16c	Explanation of relationship of significant CapEx and OpEx required to implement actions taken or planned to key performance indicators required under Commission Delegated Regulation (EU) 2021/2178***	Narrative				IFRS S2.29(e)**	
E1	E1-3	29ciii, 16c	Explanation of relationship of significant CapEx and OpEx required to implement actions taken or planned to CapEx plan required by Commission Delegated Regulation (EU) 2021/2178***	Narrative				IFRS S2.29(e)**	
E1	E1-3	AR22	Explanation of any potential differences between significant OpEx and CapEx and key performance indicators disclosed under the EU Taxonomy, Commission Delegated Regulation (EU) 2021/2178***	Narrative					

*Indicates that additional relevant information can be found in Section 3. ESRS to IFRS S2 (climate): information that an entity starting with ESRS needs to know when also applying ISSB Standards to enable compliance with both sets of standards; **Indicates that additional relevant information can be found in Section 4. IFRS S2 (climate) to ESRS: information that an entity starting with ISSB Standards needs to know when also applying ESRS to enable compliance with both sets of standards.

*** See where relevant Annex 2 to 11 for reporting templates for KPIs: https://eur-lex.europa.eu/legal-content/EN/TXT/?uri=CELEX%3A32021R2178

Source: CDR (EU) 2023/2772 supplementing Directive 2013/34/EU, *IFRS S1 General Requirements for Disclosure of Sustainability-related Financial Information, IFRS S2 Climate-related Disclosures, GRI 1, 2, 3: 11, 12, 13, 14, 306, 101, 201, 202, 203, 204, 205, 206, 207,301, 302, 303, 304, 305, 306, 308,401, 402, 403, 404, 405, 406, 407, 408, 409, 410, 411, 413, 414, 415, 416, 417, 418*, and the author.

Table 47 Indicators and metrics related to E1-4 GHG emission targets.

ESRS	DR	Paragraph	Name	Data type	Tips and tricks	Related CDR	VSMEs	ISSB / IFRS S1 & S2	GRI
E1	E1-4	32	Tracking effectiveness of policies and actions through targets	MDR-T	See 5.1.4			IFRS S2.33	GRI 302-3-3, GRI 305-3-3, GRI 305-1.2
E1	E1-4	33	Disclosure of how GHG emissions reduction targets and/or any other targets have been set to manage material climate-related impacts, risks, and opportunities	Narrative	GHG Intensity Reduction through Removals and Reductions			IFRS S2.33(b)	GRI 302-3-3, GRI 305-3-3, GRI 305-1.2
E1	E1-4	34a + 34 b	Tables: Multiple dimensions (base-line year and targets; GHG types, Scope 3 categories, decarbonisation levers, entity-specific denominators for intensity value)	Table	All of 34a & 34b can be set out on a single table – see Table 52: Reporting Scope 1, 2, and 3 Emissions			IFRS S2.33, IFRS S2.33(g)**, IFRS S2.36 (a, b)	
E1	E1-4	34a + 34 b	Absolute value of total greenhouse gas emissions reduction	Table/GHG emissions		SFDR+PILLAR 3+ BENCHMARK		IFRS S2.33, IFRS S2.33(g)**, IFRS S2.36 (a, b)	
E1	E1-4	34a + 34 b	Percentage of total greenhouse gas emissions reduction (as of emissions of base year)	Table/per cent		SFDR+PILLAR 3+ BENCHMARK		IFRS S2.33, IFRS S2.33(g)**, IFRS S2.36 (a, b)	
E1	E1-4	34a + 34 b	Intensity value of total greenhouse gas emissions reduction	Table/decimal		SFDR+PILLAR 3+ BENCHMARK		IFRS S2.33, IFRS S2.33(g)**, IFRS S2.36 (a, b)	
E1	E1-4	34a + 34 b	Absolute value of Scope 1 greenhouse gas emissions reduction	Table/GHG emissions		SFDR+PILLAR 3+ BENCHMARK	V	IFRS S2.33, IFRS S2.33(g)**, IFRS S2.36 (a, b)	

E1	E1-4	34a + 34 b	Percentage of Scope 1 greenhouse gas emissions reduction (as of emissions of base year)	Table/per cent		SFDR+PILLAR 3+ BENCHMARK	V	IFRS S2.33, IFRS S2.33(g)**, IFRS S2.36 (a, b)	
E1	E1-4	34a + 34 b	Intensity value of Scope 1 greenhouse gas emissions reduction	Table/decimal		SFDR+PILLAR 3+ BENCHMARK		IFRS S2.33, IFRS S2.33(g)**, IFRS S2.36 (a, b)	
E1	E1-4	34a + 34 b	Absolute value of location-based Scope 2 greenhouse gas emissions reduction	Table/GHG emissions		SFDR+PILLAR 3+ BENCHMARK	V	IFRS S2.33, IFRS S2.33(g)**, IFRS S2.36 (a, b)	
E1	E1-4	34a + 34 b	Percentage of location-based Scope 2 greenhouse gas emissions reduction (as of emissions of base year)	Table/per cent		SFDR+PILLAR 3+ BENCHMARK	V	IFRS S2.33, IFRS S2.33(g)**, IFRS S2.36 (a, b)	
E1	E1-4	34a + 34 b	Intensity value of location-based Scope 2 greenhouse gas emissions reduction	Table/decimal		SFDR+PILLAR 3+ BENCHMARK		IFRS S2.33, IFRS S2.33(g)**, IFRS S2.36 (a, b)	
E1	E1-4	34a + 34 b	Absolute value of market-based Scope 2 greenhouse gas emissions reduction	Table/GHG emissions		SFDR+PILLAR 3+ BENCHMARK	V	IFRS S2.33, IFRS S2.33(g)**, IFRS S2.36 (a, b)	
E1	E1-4	34a + 34 b	Percentage of market-based Scope 2 greenhouse gas emissions reduction (as of emissions of base year)	Table/per cent		SFDR+PILLAR 3+ BENCHMARK	V	IFRS S2.33, IFRS S2.33(g)**, IFRS S2.36 (a, b)	
E1	E1-4	34a + 34 b	Intensity value of market-based Scope 2 greenhouse gas emissions reduction	Table/decimal		SFDR+PILLAR 3+ BENCHMARK		IFRS S2.33, IFRS S2.33(g)**, IFRS S2.36 (a, b)	
E1	E1-4	34a + 34 b	Absolute value of Scope 3 greenhouse gas emissions reduction	Table/GHG emissions		SFDR+PILLAR 3+ BENCHMARK	V	IFRS S2.33, IFRS S2.33(g)**, IFRS S2.36 (a, b)	

(Continued)

Table 47 (*Continued*).

ESRS	DR	Paragraph	Name	Data type	Tips and tricks	Related CDR	VSMEs	ISSB / IFRS S1 & S2	GRI
E1	E1-4	34a + 34 b	Percentage of Scope 3 greenhouse gas emissions reduction (as of emissions of base year)	Table/per cent		SFDR+PILLAR 3+ BENCHMARK	V	IFRS S2.33, IFRS S2.33(g)**, IFRS S2.36 (a, b)	
E1	E1-4	34a + 34 b	Intensity value of Scope 3 greenhouse gas emissions reduction	Table/decimal		SFDR+PILLAR 3+ BENCHMARK		IFRS S2.33, IFRS S2.33(g)**, IFRS S2.36 (a, b)	
E1	E1-4	34b	Explanation of how consistency of GHG emission reduction targets with GHG inventory boundaries has been ensured	Narrative	Statement from assessor or verification body, e.g., SBTi, and combined with relevant indicators and metrics. See 6.5–6.17		V	IFRS S2.36(b), IFRS S2.36(a)	GRI 305-3-3, GRI 305-1.2
E1	E1-4	34c	Disclosure of past progress made in meeting target before current base year	Narrative			V	IFRS S2.33(e)	GRI 305-1, GRI 305-2, GRI 305-3, GRI 305-5
E1	E1-4	AR 25 a	Description of how it has been ensured that baseline value is representative in terms of activities covered and influences from external factors	Narrative				IFRS S2.33(e)	

E1	E1-4	AR 25 b	Description of how a new baseline value affects new target, its achievement, and presentation of progress over time	Narrative				IFRS S2.33(e)	GRI 305-5
E1	E1-4	34e,16a	GHG emission reduction target is science based and compatible with limiting global warming to one and a half degrees Celsius	Semi-narrative				IFRS S2.33(h), IFRS S2.34(a)12, IFRS S2.36(d), IFRS S2.14(a)(iv)*	
E1	E1-4	34f,16b	Description of expected decarbonisation levers and their overall quantitative contributions to achieve GHG emission reduction target	Narrative				IFRS S2.14(a)(iv)*	
E1	E1-4	AR 30c	Diverse range of climate scenarios have been considered to detect relevant environmental, societal, technology, market and policy-related developments, and determine decarbonisation levers	Narrative					GRI 302-1

*Indicates that additional relevant information can be found in Section 3. ESRS to IFRS S2 (climate): information that an entity starting with ESRS needs to know when also applying ISSB Standards to enable compliance with both sets of standards; **Indicates that additional relevant information can be found in Section 4. IFRS S2 (climate) to ESRS: information that an entity starting with ISSB Standards needs to know when also applying ESRS to enable compliance with both sets of standards.

Source: CDR (EU) 2023/2772 supplementing Directive 2013/34/EU, *IFRS S1 General Requirements for Disclosure of Sustainability-related Financial Information, IFRS S2 Climate-related Disclosures, GRI 1, 2, 3: 11, 12, 13, 14, 306, 101, 201, 202, 203, 204, 205, 206, 207,301, 302, 303, 304, 305, 306, 308,401, 402, 403, 404, 405, 406, 407, 408, 409, 410, 411, 413, 414, 415, 416, 417, 418*, and the author.

Table 48 Indicators and metrics related to E1-5 – Energy consumption and mix.

ESRS	DR	Paragraph	Name	Data Type	Tips and Tricks	Related CDR	VSMEs	ISSB / IFRS S1 & S2	GRI
E1	E1-5	37	Total energy consumption related to own operations	Energy	See 6.7 Energy Mix and Consumption and relate to applicable metrics from 6.5 to 6.17	SFDR	V		GRI 302-1
E1	E1-5	37a	Total energy consumption from fossil sources	Energy		SFDR	V		GRI 302-1
E1	E1-5	37b	Total energy consumption from nuclear sources	Energy		SFDR			GRI 302-1
E1	E1-5	AR 34	Percentage of energy consumption from nuclear sources in total energy consumption	Per cent					
E1	E1-5	37c	Total energy consumption from renewable sources	Energy		SFDR			GRI 302-1
E1	E1-5	37ci	Fuel consumption from renewable sources	Energy		SFDR			GRI 302-1
E1	E1-5	37cii	Consumption of purchased or acquired electricity, heat, steam, and cooling from renewable sources	Energy		SFDR			GRI 302-1
E1	E1-5	37ciii	Consumption of self-generated non-fuel renewable energy	Energy		SFDR			GRI 302-1
E1	E1-5	AR 34, AR 71	Percentage of renewable sources in total energy consumption	Per cent					
E1	E1-5	38a	Fuel consumption from coal and coal products	Energy		SFDR			GRI 302-1
E1	E1-5	38b	Fuel consumption from crude oil and petroleum products	Energy		SFDR			GRI 302-1
E1	E1-5	38c	Fuel consumption from natural gas	Energy		SFDR			GRI 302-1

E1	E1-5	38d	Fuel consumption from other fossil sources	Energy		SFDR			GRI 302-1
E1	E1-5	38e	Consumption of purchased or acquired electricity, heat, steam, or cooling from fossil sources	Energy		SFDR			GRI 302-1
E1	E1-5	AR 34	Percentage of fossil sources in total energy consumption	Per cent					
E1	E1-5	39	Non-renewable energy production	Energy					
E1	E1-5	39	Renewable energy production	Energy					
E1	E1-5	40	Energy intensity from activities in high climate impact sectors (total energy consumption per net revenue)	Per cent	Coal, crude oil, natural gas, and other fossil fuels and purchased or acquired electricity, heat, steam, or cooling from fossil fuels, see 6.7 and 6.8	SFDR			GRI 302-3
E1	E1-5	41	Total energy consumption from activities in high climate impact sectors	Energy		SFDR			GRI 302-3
E1	E1-5	42	High climate impact sectors used to determine energy intensity	Semi-narrative		SFDR			GRI 302-3
E1	E1-5	43	Disclosure of reconciliation to relevant line item or notes in financial statements of net revenue from activities in high climate impact sectors	Narrative		SFDR			
E1	E1-5	AR 38b	Net revenue from activities in high climate impact sectors	Monetary			V		
E1	E1-5	AR 38b	Net revenue from activities other than in high climate impact sectors	Monetary			V		

Source: CDR (EU) 2023/2772 supplementing Directive 2013/34/EU, *IFRS S1 General Requirements for Disclosure of Sustainability-related Financial Information, IFRS S2 Climate-related Disclosures, GRI 1, 2, 3: 11, 12, 13, 14, 306, 101, 201, 202, 203, 204, 205, 206, 207,301, 302, 303, 304, 305, 306, 308,401, 402, 403, 404, 405, 406, 407, 408, 409, 410, 411, 413, 414, 415, 416, 417, 418*, and the author.

Table 49 Indicators and metrics related to E1-6 – GHG emissions Scopes 1, 2, and 3.

Organisations not exceeding on their balance sheet dates the average number of 750 employees during the financial year (on a consolidated basis where applicable) may omit the data points under E1-6 on Scope 3 emissions and total GHG emissions for the first year of preparation of their sustainability statement.

ESRS	DR	Paragraph	Name	Data type	Tips and tricks	Related CDR	VSMEs	ISSB / IFRS S1 & S2	GRI
E1	E1-6	44	Gross Scopes 1, 2, and 3 and total GHG emissions – GHG emissions per scope	Table	All paragraphs concerning emissions by scope can be consolidated onto one or two tables. See 6.5 and 6.6 Note that there is a phase-in option for: AR46d, AR50, AR52, 51 44+52, 44+52a, 44+52b, 52a) 52b) See Annex 2 for more details			IFRS S2.29(a)(i)(1–3)**, IFRS S2.B38–B57	GRI 305-1, GRI 305-2, GRI 305-3
E1	E1-6	50	Gross Scopes 1, 2, and 3 and total GHG emissions – financial and operational control	Table				IFRS S2.29(a)(iv)*	GRI 305-1, GRI 305-2, GRI 305-3
E1	E1-6	AR 41	GHG emissions – by country, operating segments, economic activity, subsidiary, GHG category, or source type	Table					
E1	E1-6	AR 46 d	Gross Scopes 1, 2, and 3 and total GHG emissions – Scope 3 GHG emissions (GHG Protocol) [table]	Table					GRI 305-1, GRI 305-2, GRI 305-3
E1	E1-6	AR 50	Gross Scopes 1, 2, and 3 and total GHG emissions – Scope 3 GHG emissions (ISO 14064-1) [table]	Table					GRI 305-1, GRI 305-2, GRI 305-3

E1	E1-6	AR 52	Gross Scopes 1, 2, and 3 and total GHG emissions – total GHG emissions – value chain [table]	Table					GRI 305-1, GRI 305-2, GRI 305-3
E1	E1-6	48 a	Gross Scope 1 greenhouse gas emissions	Table/GHG emissions		SFDR+PILLAR 3+ BENCHMARK	V	IFRS S2.29(a)(v), IFRS S2.B30, IFRS S2.B31	GRI 305-1
E1	E1-6	48 b	Percentage of Scope 1 GHG emissions from regulated emission trading schemes	Table/per cent		SFDR+PILLAR 3+ BENCHMARK	V		GRI 305-1
E1	E1-6	49 a	Gross location-based Scope 2 greenhouse gas emissions	Table/GHG emissions		SFDR+PILLAR 3+ BENCHMARK	V	IFRS S2.B30	GRI 305-2
E1	E1-6	49 b	Gross market-based Scope 2 greenhouse gas emissions	Table/GHG emissions		SFDR+PILLAR 3+ BENCHMARK	V		GRI 305-2
E1	E1-6	51	Gross Scope 3 greenhouse gas emissions	Table/GHG emissions		SFDR+PILLAR 3+ BENCHMARK	V	IFRS S2.29(a)(vi)(1)	GRI 305-3
E1	E1-6	44+52	Total GHG emissions	Table/GHG emissions		SFDR+PILLAR 3+ BENCHMARK	V		
E1	E1-6	44+52a	Total GHG emissions location-based	Table/GHG emissions		SFDR+PILLAR 3+ BENCHMARK	V		
E1	E1-6	44+52b	Total GHG emissions market-based	Table/GHG emissions		SFDR+PILLAR 3+ BENCHMARK			
E1	E1-6	52 a)	Scope 2 location-based	Table/GHG emissions		SFDR+PILLAR 3+ BENCHMARK	V		GRI 305-2
E1	E1-6	52 b)	Scope 2 market-based	Table/GHG emissions		SFDR+PILLAR 3+ BENCHMARK			GRI 305-2

(*Continued*)

Table 49 (*Continued*).

ESRS	DR	Paragraph	Name	Data type	Tips and tricks	Related CDR	VSMEs	ISSB / IFRS S1 & S2	GRI
E1	E1-6	47	Disclosure of significant changes in definition of what constitutes reporting in the organisation and its value chain and explanation of their effect on year-to-year comparability of reported GHG emissions	Narrative	Not relevant in first report				
E1	E1-6	AR 39 b	Disclosure of methodologies, significant assumptions, and emissions factors used to calculate or measure GHG emissions	Narrative	Include link to tool(s) used			IFRS S2.B38–B57, IFRS S2.29(a)(iii)(1–3)	GRI 305-1
E1	E1-6	AR 42 c	Disclosure of the effects of significant events and changes in circumstances (relevant to its GHG emissions) that occur between the reporting dates of the organisations in its value chain and the date of the organisation's general purpose financial statements	Narrative	Not relevant in first report				

E1	E1-6	AR 45d	Percentage of contractual instruments, Scope 2 GHG emissions	Per cent	See 6.5			IFRS S2.B31	
E1	E1-6	AR 45d	Disclosure of types of contractual instruments, Scope 2 GHG emissions	Per cent				IFRS S2.B31	
E1	E1-6	AR 45d	Percentage of market-based Scope 2 GHG emissions linked to purchased electricity bundled with instruments	Per cent			V	IFRS S2.B31	
E1	E1-6	AR 45d	Percentage of contractual instruments used for sale and purchase of energy bundled with attributes about energy generation in relation to Scope 2 GHG emissions	Per cent				IFRS S2.B31	
E1	E1-6	AR 45d	Percentage of contractual instruments used for sale and purchase of unbundled energy attribute claims in relation to Scope 2 GHG emissions	Per cent				IFRS S2.B31	
E1	E1-6	AR 45d	Disclosure of types of contractual instruments used for sale and purchase of energy bundled with attributes about energy generation or for unbundled energy attribute claims	Narrative				IFRS S2.B31	

(Continued)

Table 49 (*Continued*).

ESRS	DR	Paragraph	Name	Data type	Tips and tricks	Related CDR	VSMEs	ISSB / IFRS S1 & S2	GRI
E1	E1-6	AR 45e	Biogenic emissions of CO_2 from combustion or bio-degradation of biomass not included in Scope 2 GHG emissions	GHG emissions					
E1	E1-6	AR 46 g	Percentage of GHG Scope 3 calculated using primary data	Per cent	See 6.6			IFRS S2.B38–B57, IFRS S2.B56(a)	GRI 305-3
E1	E1-6	AR 46 i	Disclosure of why Scope 3 GHG emissions category has been excluded	Narrative				IFRS S2.B56(a)	GRI 305-3
E1	E1-6	AR 46 i	List of Scope 3 GHG emissions categories included in inventory	Semi-narrative				IFRS S2.B56(a)	GRI 305-3
E1	E1-6	AR 46 i	List of Scope 3 GHG emissions categories included in inventory	Semi-narrative					GRI 305-3
E1	E1-6	AR 46 h	Disclosure of reporting boundaries considered and calculation methods for estimating Scope 3 GHG emissions	Narrative					GRI 305-3

E1	E1-6	53	GHG emissions intensity, location-based (total GHG emissions per net revenue)	Intensity	See 6.8	SFDR+PILLAR 3+ BENCHMARK			GRI 305-4
E1	E1-6	53	GHG emissions intensity, market-based (total GHG emissions per net revenue)	Intensity		SFDR+PILLAR 3+ BENCHMARK			GRI 305-4
E1	E1-6	55	Disclosure of reconciliation to financial statements of net revenue used for calculation of GHG emissions intensity	Narrative	From financial statements	SFDR+PILLAR 3+ BENCHMARK			
E1	E1-6	55	Disclosure of reconciliation to relevant line item or notes in financial statements of net revenue amounts	Narrative		SFDR+PILLAR 3+ BENCHMARK			
E1	E1-6	AR 55	Net revenue	Monetary					
E1	E1-6	AR55	Net revenue used to calculate GHG intensity	Monetary					
E1	E1-6	AR55	Net revenue other than used to calculate GHG intensity	Monetary					

*Indicates that additional relevant information can be found in Section 3. ESRS to IFRS S2 (climate): information that an entity starting with ESRS needs to know when also applying ISSB Standards to enable compliance with both sets of standards; **Indicates that additional relevant information can be found in Section 4. IFRS S2 (climate) to ESRS: information that an entity starting with ISSB Standards needs to know when also applying ESRS to enable compliance with both sets of standards.

Source: CDR (EU) 2023/2772 supplementing Directive 2013/34/EU, *IFRS S1 General Requirements for Disclosure of Sustainability-related Financial Information, IFRS S2 Climate-related Disclosures, GRI 1, 2, 3: 11, 12, 13, 14, 306, 101, 201, 202, 203, 204, 205, 206, 207,301, 302, 303, 304, 305, 306, 308,401, 402, 403, 404, 405, 406, 407, 408, 409, 410, 411, 413, 414, 415, 416, 417, 418*, and the author.

6.5 Scope 1 and 2 GHG Emissions (ESRS2-IRO1, E1-4, E1-6, E1-7)

Relevant SDGs: SDG 13: Climate Action.

This GHG emissions indicator measures an organisation's direct (Scope 1) and indirect (Scope 2) greenhouse gas emissions, which are defined in **6.1.1, Creating a GHG Inventory**. Scope 1 emissions include the direct emissions from sources that are controlled or owned by the organisation, such as emissions from fuel combustion in boilers, furnaces, and vehicles. Scope 2 emissions encompass the indirect emissions associated with the purchase of energy sources, for example, electricity, steam, heat, or cooling. The ultimate goal is to achieve net zero GHG emissions. This means eliminating both the direct emissions from organisation-controlled sources and the indirect emissions from purchased energy. The sustainability threshold or norm for an organisation's GHG emissions is set at zero (i.e., net zero). This indicator assesses the sustainability of an organisation based on whether the actual GHG emissions (Scopes 1 and 2) of a specific year are less than or equal to zero (net zero) or greater than zero (unsustainable).

While calculating Scope 1 and 2 emissions, review your GHG inventory and transition plan insights that relate to your materiality assessment (see Chapter 3: Single and Double Materiality Assessment), notably**:**

- Consider the financial impact of emissions on the organisation, as well as the broader environmental and social impacts.
- Assess emissions based on their materiality in terms of financial risks, reputational risks, and potential for value creation.
- Include emissions sources that are considered material from both financial and environmental perspectives, even if they fall outside the boundaries defined by the GHG Protocol method.

Specific disclosures relating to CSRD/ESRS:

The disclosure on gross *Scope 1 GHG emissions* required shall include:

a) the gross Scope 1 GHG emissions in metric tonnes of CO2eq; and

b) the percentage of Scope 1 GHG emissions from regulated emission trading schemes.

The disclosure on gross *Scope 2 GHG emissions* required shall include:

a) the gross location-based Scope 2 GHG emissions in metric tonnes of CO2eq; and

b) the gross market-based Scope 2 GHG emissions in metric tonnes of CO2eq.

The undertaking shall disaggregate the information, separately disclosing emissions from:

a) the consolidated accounting group (the parent and subsidiaries); and

b) investees such as associates, joint ventures, or unconsolidated subsidiaries that are not fully consolidated in the financial statements of the consolidated accounting group, as well as contractual arrangements that are joint arrangements not structured through an entity (i.e., jointly controlled operations and assets), for which it has operational control.

When calculating GHG emissions, your organisation must make the following considerations:

- Ensure that your emissions inventory covers at least 95% of company-wide Scope 1 and 2 GHG emissions.
- Include all seven greenhouse gases or classes of GHGs covered by the UNFCCC/Kyoto and Montreal Protocols.

Table 50 Common GHG refrigerants.[1]

Greenhouse gases	**Refrigerants**
Carbon dioxide (CO_2)	HFC-32
Methane (CH_4)	HFC-134a
Nitrous oxide (N2O)	R-407A
Hydrofluorocarbons (HFCs)	R-404A
Perfluorocarbons (PFCs)	R-410A
Sulphur hexafluoride (SF6)	R-507
Ozone (O_3)	R-422D
	R427-A

Source: GHG Protocol.

To calculate Scope 1 and 2 emissions:

Identify and quantify direct GHG emissions from sources that are owned or controlled by the organisation. For Scope 1, this includes emissions from

[1] For a comprehensive e list see: GHG Protocol, Refrigeration and Air-Conditioning Equipment, https://ghgprotocol.org › default › files › hfc-pfc_1

European Environment Agency, Ozone-depleting substances 2021 (aka Montreal Protocol), https://www.eea.europa.eu/themes/climate/ozone-depleting-substances-and-climate-change/ozone-depleting-substances-2021-pdf-version

activities such as combustion of fossil fuels in company-owned vehicles or onsite boilers, as well as process emissions from industrial operations. Use appropriate emission factors for calculating Scope 1 emissions using the abbreviated general formula:

$$\text{Emissions} = \text{Energy Consumption} \times \text{Emission Factor}$$

where:

i. 'Emissions' represents the carbon emissions in kilograms of CO2e.

ii. 'Energy Consumption' refers to the amount of energy used in a specific unit of measurement, such as kilowatt-hours (kWh) for electricity or cubic meters (m^3) for natural gas.

iii. 'Emission Factor' is the conversion factor that represents the amount of carbon emissions associated with each unit of energy consumed. It is typically expressed in kilograms of CO2e per unit of energy.

To calculate Scope 2 emissions:

Begin with determining operational boundaries for Scope 2 emissions by choosing one of three control criteria: operational control, financial control, or equity share, based on the level of control or influence over emissions sources:

- **Operational control**: Include emissions from sources that your organisation owns or operates.
- **Financial control**: Include emissions from sources where your organisation has financial control, such as subsidiaries or joint ventures.
- **Equity share**: Include emissions based on your organisation's share of equity in organisations where it has partial ownership.

Within your chosen boundary, determine and measure indirect GHG emissions associated with the consumption of purchased electricity, heat, or steam. Use appropriate emission factors for calculating Scope 2 emissions using the abbreviated general formula:

$$\text{Emissions} = \text{Energy Consumption} \times \text{Emission Factor}$$

where:

iv. 'Emissions' represents the carbon emissions in kilograms of CO2e.

v. 'Energy Consumption' refers to the amount of energy used in a specific unit of measurement, such as kilowatt-hours (kWh) for electricity or cubic meters (m^3) for natural gas.

vi. 'Emission Factor' is the conversion factor that represents the amount of carbon emissions associated with each unit of energy consumed. It is typically expressed in kilograms of CO2e per unit of energy.

Calculate separate emissions reported outside the GHG inventory:
Energy sources do not always fall within the boundaries of a GHG inventory, for example, biomass and bioenergy. However, organisations must report direct CO_2 emissions from biomass combustion, processing, and distribution, as well as land-use emissions and removals associated with bioenergy feedstock.

6.6 Scope 3 GHG Emissions (ESRS 2 IRO 1, E1-4, E1-6, E1-7)

Relevant SDGs: SDG 13: Climate Action.

GHG emissions (Scope 3) refer to the indirect greenhouse gas emissions generated by an organisation's activities outside of its direct operations. These emissions are associated with the entire value chain, including suppliers, customers, and other third parties. Measuring and addressing Scope 3 emissions is crucial for organisations to understand and mitigate their overall environmental impact.

Unless a business has significant real estate holdings and energy consumption, Scope 3 typically represents the most significant contributor to a company's carbon footprint because it includes things like:

- Business travel (e.g., air travel)
- Employee commuting
- Waste generated in operations and waste disposal
- Purchased goods and services
- Transportation and distribution tied to suppliers and customers
- Capital goods, investments, and franchises
- Leased assets

- Emissions from the use of a product or service sold
- Product's end of life (when it is no longer useful)

The organisation shall include GHG emissions in metric tonnes of CO2eq from each significant *Scope 3 category* (i.e., each Scope 3 category that is a priority for the undertaking).

Companies selling or distributing fossil fuels must report use-phase emissions associated with those fossil fuels in Scope 3 category 11 (use of sold products). The following categories may be referred to while reporting the use-phase emissions associated with the products/services sold:

- **For products that consume energy (such as fuels or electricity) during use**: This process includes analysing the use-phase of the product, determining the emissions per product, and aggregating the total emissions.
- **For fuels and feedstocks**: This step involves gathering data on fuel usage and multiplying it by emission factors representative of the specific type of fuel being used.
- **For products that contain or produce greenhouse gases (GHGs) during use**: This entails collecting data on the amount of GHGs present in the product and multiplying it by the percentage of GHGs released during use, along with the corresponding GHG emission factors.

Additional sources of information:

- Suppliers, customers, and third parties: Obtain Scope 3 GHG emissions data from various stakeholders in the organisation's value chain. This data can be collected through surveys, reporting requirements, or engagement with suppliers, customers, and other relevant third parties.
- Authentic sustainability assessment: A user manual for the sustainable development performance indicators. 2022. UNRISD.

To calculate Scope 3 emissions:

- While building your GHG inventory, identify all three items by relevant categories of upstream and downstream emission sources. Then determine the appropriate treatment of indirect use-phase emissions, which occur during the use of products indirectly, and should only be estimated if deemed significant and if there are levers to address them.

- Consult sector-specific guidance (GRI, SBTi, etc.) and peers for best practices in inventory and target boundary-setting, emissions accounting, and target calculation.
- Review policies, actions, and targets both internally and with stakeholders to identify possible gaps and overlaps in Scope 3 items.

For each Scope 3 GHG inventory item, determine and measure indirect GHG emissions using the appropriate emission factors and the abbreviated general formula:

$$\text{Emissions} = \text{Energy Consumption} \times \text{Emission Factor}$$

where:

vii. 'Emissions' represents the carbon emissions in kilograms of CO2e.

viii. 'Energy Consumption' refers to the amount of energy used in a specific unit of measurement, such as kilowatt-hours (kWh) for electricity or cubic meters (m^3) for natural gas.

ix. 'Emission Factor' is the conversion factor that represents the amount of carbon emissions associated with each unit of energy consumed. It is typically expressed in kilograms of CO2e per unit of energy.

TIPS AND TRICKS

Example: Policy for reducing GHG emissions from business travel

Reduce your total and relative per day carbon emissions by:

- For air travel, no international flights are offered for distances less than 1000 km.
- For flight trips between 1000 and 3000 km, the duration of stay in the destination is at least 10 days.
- For travel for more than 3000 km, the duration of stay in the destination is at least 14 days.
- Flights are only economy and not first or business class.
- Sustainability certified hotels are used at least 40% of the time.
- No high carbon impact activities take place (helicopter flights, etc.).
- Carbon emissions are calculated for the full trip.
- Food provided (if part of the package) does not include pork or beef (chicken and fish is allowed).

6.7 Energy Mix and Consumption (ESRS2 IRO-1, E1-4, E1-5)

Relevant SDGs: SDG 13: Climate Action.

This indicator assesses the proportion of renewable energy consumed within the organisation compared to the total energy consumption. It aims to measure the organisation's reliance on renewable fuel sources and its commitment to sustainable energy practices.

Methodology:

1. Calculation of numerator (renewable energy consumed, REC): Your organisation determines the amount of renewable energy consumed, which includes renewable fuel sources (such as biofuels), solar energy, biomass, hydropower, geothermal energy, ocean energy, heat from renewable sources, and electricity from renewable sources.
2. Calculation of denominator (total energy consumption, TEC): Your organisation calculates the total energy consumption within the organisation, which includes non-renewable fuel consumed, renewable fuel consumed, electricity, heating, cooling, and steam purchased for consumption, self-generated electricity, heating, cooling, and steam that are not consumed, and subtracts the electricity, heating, cooling, and steam sold.
3. Conversion factors: To ensure comparability, both the numerator and the denominator are expressed in the same units of measurement, for example kWhrs, joules (J).
4. Calculation of proportion of renewable energy: The proportion of renewable energy (PRE) is calculated by dividing the renewable energy consumed (REC) by the total energy consumption (TEC).
5. Sustainability threshold: A sustainability norm or threshold for renewable energy is set at 100%. PRE scores of 1.0 or greater are considered sustainable, indicating a high reliance on renewable energy, while scores below 1.0 are considered unsustainable.

It is expressed as:

Table 51 Energy consumption and mix.

Energy consumption and mix	**Comparative**	**Year N**
Fuel consumption from coal and coal products (MWh)		
Fuel consumption from crude oil and petroleum products (MWh)		
Fuel consumption from natural gas (MWh)		
Fuel consumption from other non-renewable sources (MWh)		
Consumption from nuclear products (MWh)		
Consumption of acquired or purchased electricity, heat, steam, and cooling from non-renewable sources (MWh)		
Total of non-renewable energy sources (MWh) – sum total of 1–6.		
Share of non-renewable sources in total energy consumption (%)		
Fuel consumption for renewable sources (including biomass, biogas, non-fossil fuel waste, renewable hydrogen, etc.) (MWh)		
Consumption of purchased or acquired electricity, heat, steam, and cooling from renewable sources (MWh)		
The consumption of self-generated non-fuel based renewable energy (MWh)		
Total renewable energy consumption (MWh) – sum-total of 8–10		
Share of renewable sources in total energy consumption %		
Total energy consumption (MWh) – sum total of 7 and 11		

Source: CDR (EU) 2023/2772 supplementing Directive 2013/34/EU.

6.8 Emissions Intensity per Net Revenue/Unit of Production/Service (ESRS 2 IRO-1, E1-6)

Relevant SDGs: SDG 13: Climate Action.

The choice of relevant activity will depend on the specific industry or sector. For example:

- In manufacturing, it can be the quantity of goods produced or units manufactured.
- In services, it can be the number of services provided, customer interactions, or any other appropriate metric.
- In hospitality, it can be guest nights or room occupancy.
- In construction, it can be the square footage of constructed buildings or infrastructure projects completed.

Table 52 Reporting Scope 1, 2, and 3 emissions.

	Retrospective				Milestones and target years			
	Base year	**Comparative**	***N***	**% *N*/*N* – 1**	**2025**	**2030**	**2050**	**Annual % target/base year**
Scope 1 GHG emissions								
Gross Scope 1 GHG emissions (tCO2e)								
% of Scope 1 GHG emissions from regulated ETS								
Scope 2 GHG emissions								
Gross location-based Scope 2 emissions (tCO2e)								
Gross market-based Scope 2 GHG emissions (tCO2e)								
Scope 3 GHG emissions (only include those assessed as material)								
Total gross indirect Scope 3 GHJG emissions (tCO2e)								
Purchased goods and services								
Cloud computing and data services								
Capital goods								
Fuel and energy-related activities								
Upstream leased assets								
Waste generated in operations								
Processing of sold products								
Use of sold products								
End of life treatment of sold products								
Downstream leased assets								
Franchises								
Upstream transportation and distribution								
Downstream transportation and distribution								
Business travel								
Employee commuting								
Financial investments								
TOTAL GHG emissions								
Total GHG emissions (location-based) (tCO2e)								
Total GHG emissions (market-based) (tCO2e								

Source: CDR (EU) 2023/2772 supplementing Directive 2013/34/EU.

The emissions intensity per unit of production can be calculated by dividing the total emissions by the total activity. The following formula can be referred to:

Emissions Intensity = Total GHG Emissions/Net Revenue or Total Activity or Total Service

where:

Total GHG Emissions: The sum of greenhouse gas (GHG) emissions (Scope 1, 2, and 3) produced by the organisation.

Total Activity: The quantity of relevant activity performed or delivered by the organisation.

It is expressed as:

Table 53 Energy intensity per net revenue.

Energy intensity per net revenue	**Comparative**	*N*	**% *N/N* – 1**
Total energy consumption from activities in high climate impact sectors per net revenue (MWh/monetary unit)			

Source: CDR (EU) 2023/2772 supplementing Directive 2013/34/EU.

6.9 GHG Reduction Targets (E1-1, E1-4)

Relevant SDGs: SDG 13: Climate Action.

Setting to reduce GHG emissions over a specific timeframe first involves establishing a baseline year for emissions (see **6.1.3, Establishing the Baseline Year**), then determining a target reduction percentage, and implementing measures to achieve and report the targets (see **6.3, Target Setting, Mitigation Pathway and Transition Planning**). The formula for calculating the reduction target percentage is as follows:

Reduction Target % = ((Baseline Emissions – Target Emissions)/ Baseline Emissions) × 100

where:

Baseline Emissions: The total greenhouse gas emissions in the baseline year.

Target Emissions: The desired level of greenhouse gas emissions to be achieved.

It is expressed as:

Table 54 Cross-sector (ACA) reductions pathway.

	2030	**2050**
Cross-sector (ACA) reductions pathway based in the year 2020 as the reference year	−42%	−90%

Source: CDR (EU) 2023/2772 supplementing Directive 2013/34/EU.

6.10 Energy Efficiency Improvements (E1-3, E1-5)

Relevant SDGs: SDG 7: Affordable and Clean Energy

Measurement of improvements in energy efficiency resulting in reduced energy consumption and lower emissions. The methodology for evaluating energy efficiency improvements involves comparing energy consumption data before and after implementing energy-saving measures. The formula for calculating energy efficiency improvements is as follows:

Energy Efficiency Improvement (%) = ((Initial Energy Consumption – Final Energy Consumption)/Initial Energy Consumption) × 100

where:

Initial Energy Consumption: The energy consumption before implementing energy-saving measures.

Final Energy Consumption: The energy consumption after implementing energy-saving measures.

6.11 GHG Emissions Intensity Reduction (E1-1, E1-3, E1-4, E1-7)

Relevant SDGs: SDG 13: Climate Action.

According to the GHG Protocol, the methodology for calculating greenhouse gas emissions intensity reduction involves dividing the reduction in greenhouse gas emissions by the reduction in output or production. It is essential that your organisation verify all reduction and removals with an established third party through a fully documented process. The formula for calculating emissions intensity reduction is as follows:

Emissions Intensity Reduction (%) = ((Initial Emissions Intensity – Final Emissions Intensity)/Initial Emissions Intensity) × 100

Table 55 Indicators and metrics related to E1-7 – GHG removals and GHG mitigation projects financed through carbon credits.

ESRS	DR	Paragraph	Name	Data type	Tips and tricks	Related CDR	VSMEs	ISSB / IFRS S1 & S2	GRI
E1	E1-7	56a	Disclosure of GHG removals and storage resulting from projects developed in own operations or contributed to in upstream and downstream value chains	Narrative	All removals, carbon credits, etc. should be reported separately, see 6.1.4 of Transition Plan and use a table to capture all relevant data in this section	CL			GRI 305-5
E1	E1-7	56b	Disclosure of GHG emission reductions or removals from climate change mitigation projects outside value chain financed or to be financed through any purchase of carbon credits	Narrative		CL		IFRS S2.36(e)(iii)	GRI 305-3-3, GRI 305-1.2, GRI 305-5
E1	E1-7	58	Removals and carbon credits are used	Semi-narrative					
E1	E1-7	58a	GHG removals and storage activity by undertaking scope and by removal and storage activity	Table					
E1	E1-7	58a	Total GHG removals and storage	Table/GHG emissions					
E1	E1-7	AR 58f	GHG emissions associated with removal activity	Table/GHG emissions					
E1	E1-7	58b	Disclosure of calculation assumptions, methodologies, and frameworks applied (GHG removals and storage)	Narrative	Link to tool(s)				

(Continued)

Table 55 (*Continued*).

ESRS	DR	Paragraph	Name	Data type	Tips and tricks	Related CDR	VSMEs	ISSB / IFRS S1 & S2	GRI
E1	E1-7	AR 58e	Removal activity has been converted into carbon credits and sold on to other parties on voluntary market	Semi-narrative	Certificates from third party				
E1	E1-7	59a	Total amount of carbon credits outside value chain that are verified against recognised quality standards and cancelled	GHG emissions	See 6.11 and then tabulate			IFRS S2.36(e)(ii)**	
E1	E1-7	59b	Total amount of carbon credits outside value chain planned to be cancelled in future	GHG emissions				IFRS S2.36(e)(ii)**, IFRS S2.36(e)(i)	
E1	E1-7	AR 60	Reversals	GHG emissions					
E1	E1-7	AR 61	Disclosure of extent of use and quality criteria used for carbon credits	Narrative					
E1	E1-7	AR 62a	Percentage of reduction projects	Per cent					
E1	E1-7	AR 62a	Percentage of removal projects	Per cent					
E1	E1-7	AR 62b	Type of carbon credits from removal projects	Semi-narrative				IFRS S2.36(e)(iii)	
E1	E1-7	AR 62c	Percentage for recognised quality standard	Per cent				IFRS S2.36(e)(ii)**	
E1	E1-7	AR 62d	Percentage issued from projects in European Union	Per cent					
E1	E1-7	AR 62e	Percentage that qualifies as corresponding adjustment	Per cent					
E1	E1-7	AR 64	Date when carbon credits outside value chain are planned to be cancelled	Date	From offsetting partner				

E1	E1-7	60	Explanation of scope, methodologies, and frameworks applied and how residual GHG emissions are intended to be neutralised	Narrative	See 6.1.1, Creating a Greenhouse Gas Inventory				
E1	E1-7	61	Public claims of GHG neutrality that involve the use of carbon credits have been made	Semi-narrative	Ensure claims are checked by legal experts			IFRS S2.36(e)(i)	
E1	E1-7	61a	Public claims of GHG neutrality that involve the use of carbon credits are accompanied by GHG emission reduction targets	Semi-narrative					
E1	E1-7	61b	Claims of GHG neutrality and reliance on carbon credits neither impede nor reduce achievement of GHG emission reduction targets or net zero target	Semi-narrative					
E1	E1-7	61 a, b	Explanation of how public claims of GHG neutrality that involve use of carbon credits are accompanied by GHG emission reduction targets and how claims of GHG neutrality and reliance on carbon credits neither impede nor reduce achievement of GHG emission reduction targets or net zero target	Narrative					
E1	E1-7	61c	Explanation of credibility and integrity of carbon credits used	Narrative				IFRS S2.36(e)(ii) , IFRS S2.36(e)(iv)*	GRI 305-3-3, GRI 305-1.2
E1	E1-7	AR 62	Percentage for recognised quality standards	Per cent					

*Indicates that additional relevant information can be found in Section 3. ESRS to IFRS S2 (climate): information that an entity starting with ESRS needs to know when also applying ISSB Standards to enable compliance with both sets of standards; **Indicates that additional relevant information can be found in Section 4. IFRS S2 (climate) to ESRS: information that an entity starting with ISSB Standards needs to know when also applying ESRS to enable compliance with both sets of standards.

Source: CDR (EU) 2023/2772 supplementing Directive 2013/34/EU, *IFRS S1 General Requirements for Disclosure of Sustainability-related Financial Information, IFRS S2 Climate-related Disclosures, GRI 1, 2, 3: 11, 12, 13, 14, 306, 101, 201, 202, 203, 204, 205, 206, 207,301, 302, 303, 304, 305, 306, 308,401, 402, 403, 404, 405, 406, 407, 408, 409, 410, 411, 413, 414, 415, 416, 417, 418*, and the author.

Table 56 GHG intensity reduction through removals and reductions.

GHG intensity reduction and removal	Comparative	*N*	% *N/N* – 1
GHG removal activity 1 (e.g., forest restoration, etc.)			
GHG removal activity 2 (e.g., direct air capture, etc.)			
…			
Total GHG removals from own operations (tCO2e)			
GHG reduction from activity 1 (e.g., shift to renewable energy, etc.)			
GHG reduction from activity 2 (e.g., divestment in XYZ asset, etc.)			
Total GHG reductions from own operations (tCO2e)			
GHG removal activity 1 (e.g., forest restoration, etc.)			
GHG removal activity 2 (e.g., direct air capture, etc.)			
Total GHG removals from value chain (tCO2e)			
GHG reduction activity 1 (e.g., forest restoration, etc.)			
GHG reduction activity 2 (e.g., direct air capture, etc.)			
Total GHG reduction from value chain (tCO2e)			

Source: CDR (EU) 2023/2772 supplementing Directive 2013/34/EU.

6.12 Internal Carbon Pricing (Shadow Pricing) (E1-8)

The shadow pricing methodology aims to internalise the cost of carbon when making investment decisions within an organisation. It assigns a theoretical value to carbon emissions, considering their lifecycle environmental and financial costs, and is normally used for:

a) the assessment of the useful life and residual value of its assets (intangibles, property, plant, and equipment);

b) the impairment of assets; and

c) the fair value measurement of acquired assets during business acquisitions.

A detailed explanation of the shadow pricing methodology is provided as follows.

1. Calculate lifecycle environmental and financial costs:

To calculate the lifecycle environmental and financial costs associated with carbon emissions:

Table 57 Indicators and metrics related to E1-8 – Internal carbon pricing.

ESRS	DR	Paragraph	Name	Data type	Tips and tricks	Related CDR	VSMEs	ISSB / IFRS S1 & S2	GRI
E1	E1-8	63 a	Carbon pricing scheme by type	Table	Explain whether and how the carbon prices used in internal carbon pricing schemes are consistent with those used in financial statements and then tabulate results. See 6.12			IFRS S2.29(f)	
E1	E1-8	63 a	Type of internal carbon pricing scheme	Table/narrative				IFRS S2.29(f)	
E1	E1-8	63b	Description of specific scope of application of carbon pricing scheme	Table/narrative					
E1	E1-8	63c	Carbon price applied for each metric tonne of greenhouse gas emission	Table/monetary				IFRS S2.29(f)	
E1	E1-8	63c	Description of critical assumptions made to determine carbon price applied	Table/narrative				IFRS S2.29(f)	
E1	E1-8	63d	Percentage of gross Scope 1 greenhouse gas emissions covered by internal carbon pricing scheme	Table/per cent					
E1	E1-8	63d	Percentage of gross Scope 2 greenhouse gas emissions covered by internal carbon pricing scheme	Table/per cent					
E1	E1-8	63d	Percentage of gross Scope 3 greenhouse gas emissions covered by internal carbon pricing scheme	Table/per cent					
E1	E1-8	AR 65	Disclosure of how carbon price used in internal carbon pricing scheme is consistent with carbon price used in financial statements	Table/narrative				IFRS S2.29(d)*	

Source: CDR (EU) 2023/2772 supplementing Directive 2013/34/EU, *IFRS S1 General Requirements for Disclosure of Sustainability-related Financial Information, IFRS S2 Climate-related Disclosures, GRI 1, 2, 3: 11, 12, 13, 14, 306, 101, 201, 202, 203, 204, 205, 206, 207,301, 302, 303, 304, 305, 306, 308,401, 402, 403, 404, 405, 406, 407, 408, 409, 410, 411, 413, 414, 415, 416, 417, 418*, and the author.

- Divide the lifecycle of the investment into stages: extraction, manufacturing, transportation, use/operation, and disposal.
- Determine the key inputs and outputs for each stage and their associated energy consumption.
- Quantify GHG emissions using Scope 1, 2, and 3 methodologies as previously mentioned. This should include waste generation and consideration of potential environmental damages.
- Calculate financial costs, which could be based on estimation, so make sure you know the accuracy of your quantities and that they match your desired reporting period:
- Financial cost = Quantity of emissions/energy/waste × Cost per unit
- Analyse the data and interpret the results to understand the environmental and financial costs.
- Identify potential mitigation measures to reduce impacts and associated costs.

2. Discount rate selection:

Through discussion with your investors and financiers, select an appropriate discount rate that reflects the time value of money and your organisation's risk preferences. The discount rate represents the rate at which future costs and benefits are discounted to their net present value. It accounts for the opportunity cost of capital and the uncertainty associated with future cash flows:

- **Time value of money**: The discount rate should reflect the idea that costs and benefits occurring in the future are less valuable than those occurring in the present. A common approach is to use your organisation's cost of capital or the rate of return required by investors for similar projects.
- **Risk preferences**: Take into account your organisation's risk tolerance and the level of uncertainty associated with future cash flows. Higher discount rates can be used if your organisation is more risk-averse or if the project has higher uncertainty. Lower discount rates may be appropriate for low-risk projects or when there is high confidence in the future cash flows.
- **Regulatory guidance**: Consider any specific guidance or regulations provided by relevant authorities or regulatory bodies. They may

recommend or require the use of specific discount rates or methodologies for carbon shadow pricing.

3. Define the shadow price:

The shadow price is the value assigned to carbon emissions per tonne, reflecting the estimated environmental and financial costs. It is usually denoted by λ (lambda).

In the process of defining the shadow price, it is essential to consider the assessment of existing environmental impacts, targets, and transition plan. This involves understanding the current environmental footprint of the organisation, setting emissions reduction targets, and developing a comprehensive plan for transitioning to a low-carbon future.

3.1. Calculation of shadow price:

To determine the shadow price, the following equation can be referred to:

Price on Carbon ($/tCO2e) = Yearly funding required for initiatives ($)/ Annual GHG emissions in boundary ($/tCO2e)

This equation reflects the financial value attributed to each metric tonne of carbon emissions, derived from the funding needed to support emission reduction initiatives divided by the annual greenhouse gas emissions within the organisation's boundary.

4. Incorporate shadow price in investment funding analysis:

Include the shadow price in the financial analysis of investment projects. The shadow price is multiplied by the carbon emissions (E) associated with the project to calculate the environmental cost component. This can be represented as:

$$\text{Environmental Cost} = \lambda \times E$$

5. Net present value (NPV) calculation:

Calculate the net present value of the investment by discounting the future cash flows, including the environmental cost component determined in Step 4. The NPV represents the present value of the project's expected cash inflows minus the present value of its expected cash outflows. The equation for NPV can be expressed as:

$$\text{NPV} = \Sigma\, [(Ct - ECt)/(1 + r)^{\wedge}t] - \text{Initial Investment}$$

where:

Ct is the cash flow in period *t*,
ECt is the environmental cost in period *t*,
r is the discount rate,
t is the time period,
Initial Investment represents the initial cost of the investment.

6. Decision rule:

Evaluate the NPV calculated in Step 5. If the NPV is positive, the investment is considered financially viable, considering the environmental cost component. If the NPV is negative, the investment is not economically feasible, taking into account the environmental impact.

By incorporating the shadow pricing methodology into investment analysis, organisations can make informed decisions that consider the environmental costs associated with carbon emissions. It incentivises decision-makers to select investments that are not only financially attractive but also environmentally sustainable, promoting energy-efficient infrastructure and practices.

Table 58 Internal carbon pricing.

Type of internal carbon pricing	Volume at stake (tCO2e)	Prices applied (€/tCO2e)	Perimeter description
CapEx shadow price			
Research and development investment shadow price			
Internal carbon fee or fund			
Carbon prices for impairment testing			
Other…			

Source: CDR (EU) 2023/2772 supplementing Directive 2013/34/EU.

6.13 Green Investment (E1-2, E1-3, E1-9)

Relevant SDGs: SDG 13: Climate Action.

Green investment refers to the expenditures made by an organisation in environmentally sustainable activities and technologies. It can be measured using frameworks and checklists developed by international organisations such as the High-Level Expert Group on Sustainable Finance, the European Patent Office, the Organisation for European Co-operation and Development (OECD), and the European Union's Classification of Environmental Protection Activities (CEPA) list.

Table 59 Indicators and metrics related to E1-9 – Potential financial effects related to Climate Change.

E1-9 is subject to phase-in for the first three years – see Annex 2: List of Phased-in Disclosure Requirements.

ESRS	DR	Paragraph	Name	Data type	Tips and tricks	Related CDR	VSMEs	ISSB / IFRS S1 & S2	GRI
E1	E1-9	66a	Assets at material physical risk before considering climate change adaptation actions	Monetary	Tabulate results based on materiality assessment and risk and opportunity metrics, see 5.2.10 IRO-1: Identify and assess material impacts, risks, and opportunities Disaggregated table to use latest NUTS codes within EU	PILLAR 3	V	IFRS S2.29(c)**	GRI 201-2
E1	E1-9	66a	Assets at acute material physical risk before considering climate change adaptation actions	Monetary		PILLAR 3	V	IFRS S2.29(c)**	GRI 201-2
E1	E1-9	66a	Assets at chronic material physical risk before considering climate change adaptation actions	Monetary		PILLAR 3	V	IFRS S2.29(c)**	GRI 201-2
E1	E1-9	66a	Percentage of assets at material physical risk before considering climate change adaptation actions	Per cent		PILLAR 3	V	IFRS S2.29(c)**	GRI 201-2
E1	E1-9	66a	Disclosure of location of significant assets at material physical risk	Narrative		PILLAR 3	V	IFRS S2.29(c)**	GRI 201-2
E1	E1-9	AR 70 c i)	Disclosure of location of its significant assets at material physical risk (disaggregated by NUTS codes[1])	Table/ narrative		PILLAR 3	V		GRI 201-2
E1	E1-9	66b	Percentage of assets at material physical risk addressed by climate change adaptation actions	Per cent			V		GRI 201-2

[1] For more information see https://ec.europa.eu/eurostat/web/nuts

(Continued)

Table 59 (*Continued*).

ESRS	DR	Paragraph	Name	Data type	Tips and tricks	Related CDR	VSMEs	ISSB / IFRS S1 & S2	GRI
E1	E1-9	66d	Net revenue from business activities at material physical risk	Monetary			V	IFRS S2.29(c)**	GRI 201-2
E1	E1-9	66d	Percentage of net revenue from business activities at material physical risk	Per cent				IFRS S2.29(c)**	GRI 201-2
E1	E1-9	AR 69a	Disclosure of how anticipated financial effects for assets and business activities at material physical risk have been assessed	Narrative					
E1	E1-9	AR 69b	Disclosure of how assessment of assets and business activities considered to be at material physical risk relies on or is part of the process to determine material physical risk and to determine climate scenarios	Narrative					
E1	E1-9	AR 71b	Disclosure of risk factors for net revenue from business activities at material physical risk	Narrative			V		
E1	E1-9	AR 71b	Disclosure of magnitude of anticipated financial effects in terms of margin erosion for business activities at material physical risk	Narrative			V		
E1	E1-9	67a	Assets at material transition risk before considering climate mitigation actions	Monetary				IFRS S2.29(b)**	

E1	E1-9	67a	Percentage of assets at material transition risk before considering climate mitigation actions	Per cent				IFRS S2.29(b)**	
E1	E1-9	67b	Percentage of assets at material transition risk addressed by climate change mitigation actions	Per cent					
E1	E1-9	67c	Total carrying amount of real estate assets by energy efficient classes	Table/ monetary	Financial report	PILLAR 3	V		
E1	E1-9	AR 72a, AR 73a	Disclosure of how potential effects on future financial performance and position for assets and business activities at material transition risk have been assessed	Narrative	see 5.2.10 IRO-1: Identify and assess material impacts, risks, and opportunities			IFRS S2.17**	
E1	E1-9	AR 72b	Disclosure of how assessment of assets and business activities considered to be at material transition risk relies on or is part of the process to determine material transition risks and to determine scenarios	Narrative					
E1	E1-9	AR 73a	Estimated amount of potentially stranded assets	Monetary				IFRS S2.17**	
E1	E1-9	AR 73a	Percentage of estimated share of potentially stranded assets of total assets at material transition risk	Per cent	Chapter 3, Single and Double Materiality Assessment	SFRD		IFRS S2.17**	
E1	E1-9	AR 73b	Total carrying amount of real estate assets for which energy consumption is based on internal estimates	Monetary		SFRD			

(Continued)

Table 59 (*Continued*).

ESRS	DR	Paragraph	Name	Data type	Tips and tricks	Related CDR	VSMEs	ISSB / IFRS S1 & S2	GRI
E1	E1-9	67d	Liabilities from material transition risks that may have to be recognised in financial statements	Monetary	Financial report	SFRD			
E1	E1-9	AR 74c	Number of Scope 1 GHG emission allowances within regulated emission trading schemes	Integer			V		
E1	E1-9	AR 74c	Number of emission allowances stored (from previous allowances) at the beginning of reporting period	Integer			V		
E1	E1-9	AR 74d	Potential future liabilities, based on existing contractual agreements, associated with carbon credits planned to be cancelled in near future	Monetary			V		
E1	E1-9	AR 74e	Monetised gross Scope 1 and 2 GHG emissions	Monetary			V		
E1	E1-9	AR 74e	Monetised total GHG emissions	Monetary			V		
E1	E1-9	67e	Net revenue from business activities at material transition risk	Monetary	Cross-reference to the related line item or disclosure in the financial statements if these amounts are identifiable			IFRS S2.29(b)**	
E1	E1-9	67e	Net revenue from customers operating in coal-related activities	Monetary				IFRS S2.29(b)**	
E1	E1-9	67e	Net revenue from customers operating in oil-related activities	Monetary				IFRS S2.29(b)**	

E1	E1-9	67e	Net revenue from customers operating in gas-related activities	Monetary				IFRS S2.29(b)**	
E1	E1-9	67e	Percentage of net revenue from customers operating in coal-related activities	Per cent				IFRS S2.29(b)**	
E1	E1-9	67e	Percentage of net revenue from customers operating in oil-related activities	Per cent				IFRS S2.29(b)**	
E1	E1-9	67e	Percentage of net revenue from customers operating in gas-related activities	Per cent				IFRS S2.29(b)**	
E1	E1-9	AR 76, 67e	Percentage of net revenue from business activities at material transition risk	Per cent			V		
E1	E1-9	AR 76b	Disclosure of risk factors for net revenue from business activities at material transition risk	Narrative			V		
E1	E1-9	AR 76b	Disclosure of magnitude of anticipated financial effects in terms of margin erosion for business activities at material transition risk	Narrative			V		
E1	E1-9	68a	Disclosure of reconciliations with financial statements of significant amounts of assets and net revenue at material physical risk	Narrative					
E1	E1-9	68b	Disclosure of reconciliations with financial statements of significant amounts of assets, liabilities, and net revenue at material transition risk	Narrative					

(Continued)

Table 59 (*Continued*).

ESRS	DR	Paragraph	Name	Data type	Tips and tricks	Related CDR	VSMEs	ISSB / IFRS S1 & S2	GRI
E1	E1-9	69a	Expected cost savings from climate change mitigation actions	Monetary		BENCHMARK			
E1	E1-9	69a	Expected cost savings from climate change adaptation actions	Monetary		BENCHMARK			
E1	E1-9	69b	Potential market size of low-carbon products and services or adaptation solutions to which undertaking has or may have access	Monetary	Estimations of opportunities	BENCHMARK			
E1	E1-9	69b	Expected changes to net revenue from low-carbon products and services or adaptation solutions to which undertaking has or may have access	Monetary		BENCHMARK			

**Indicates that additional relevant information can be found in Section 4. IFRS S2 (climate) to ESRS: information that an entity starting with ISSB Standards needs to know when also applying ESRS to enable compliance with both sets of standards.

Source: CDR (EU) 2023/2772 supplementing Directive 2013/34/EU, *IFRS S1 General Requirements for Disclosure of Sustainability-related Financial Information, IFRS S2 Climate-related Disclosures, GRI 1, 2, 3: 11, 12, 13, 14, 306, 101, 201, 202, 203, 204, 205, 206, 207,301, 302, 303, 304, 305, 306, 308,401, 402, 403, 404, 405, 406, 407, 408, 409, 410, 411, 413, 414, 415, 416, 417, 418*, and the author.

1. Frameworks and checklists:

- Use frameworks or checklists provided by international organisations to identify and classify green investments. These frameworks cover various areas such as environmental management, renewable energy, combustion technologies, climate change mitigation, transportation, and buildings.
- Consider the expenditures related to environmental protection activities, capital formation, purchase of environmental protection products, and investments in adapted goods.

2. Calculation of indicators:

Total amount of green investments: The total amount of green investments incurred during a specific reporting period can be calculated using the following equation:

Total Amount of Green Investments = $\sum$ (Cost indicated on corresponding invoices),

where $\sum$ represents the summation of costs indicated on all corresponding invoices.

3. Green investment ratio:

The green investment ratio, which represents the proportion of green investment relative to the overall financial position of your organisation, can be calculated using the following equation:

Green Investment Ratio (in percentage) = (Green Investment in period t/ Total Assets or Revenue in period t) $\times$ 100,

where Green Investment in period t represents the amount of green investment incurred during period t, and Total Assets or Revenue in period t represents the total assets or revenue of your organisation during period t.

Sourcing relevant information for green investment ratio can be difficult with limited transparency. Look for information in the income statement (P&L statement) as part of production costs or selling expenses if the expenses are not capitalised. Capitalised green investments can be found in internal management reports such as capital budgets and are included in the fixed assets section of the balance sheet.

Note: Given the lack of a shared definition of green investment across industries and contexts, it is important to provide a consistent explanation of why an investment has been categorised as green. The use of frameworks and checklists helps in identifying, classifying, and calculating the total amount of green investments, providing valuable information for reporting purposes.

6.14 Carbon Tax (E1-1, E1-4, E1-8, E1-9)

Relevant SDGs: SDG 13: Climate Action.

The carbon tax is a pricing mechanism that aims to reduce greenhouse gas emissions by placing a monetary cost on each unit of carbon dioxide equivalent (CO2e) emitted. It is calculated by multiplying the tax rate per unit of CO2e emitted (in USD/tCO2e) by the quantity of emissions (in metric tonnes of CO2e). This tax provides an economic incentive for industries and individuals to reduce their carbon footprint and transition to cleaner alternatives.

The methodology for calculating the carbon tax amount is straightforward. It involves two variables: the tax rate per unit of CO2e emitted and the quantity of emissions. To calculate the tax amount, the tax rate is multiplied by the emissions. The tax rate is typically set by the government or the regulatory body responsible for implementing the carbon tax.

$$\text{Carbon Tax Amount} = \text{Tax Rate (USD/tCO2e)} \times \text{Emissions (tCO2e)}$$

Example: If the tax rate is \$10 per tCO2e and the emissions are 1000 tCO2e, the carbon tax amount would be \$10,000.

The source of reference for specific tax rates, regulations, and implementation details varies depending on the country or region. It is recommended to consult official government websites, environmental agencies, or relevant legislation for accurate and up-to-date information on carbon tax.

6.15 Cap-and-Trade System (E1-1, E1-4, E1-8, E1-9)

Relevant SDGs: SDG 13: Climate Action.

In a cap-and-trade system, the calculation involves determining the quantity of emissions allowances or permits allocated to each participant and the corresponding market price.

The formula for calculating the total cost of emissions is:

$$\text{Total Cost of Emissions} = \text{Number of Allowances} \times \text{Market Price per Allowance}$$

The number of allowances represents the emissions limit set by the cap, and the market price per allowance is determined by the supply and demand dynamics of the trading market. Additionally, for participants in a cap-and-trade system, the compliance cost can be calculated by multiplying the quantity of emissions (in metric tonnes of CO2e) by the market price per allowance:

Compliance Cost = Emissions (tCO2e) × Market Price per Allowance

This formula determines the financial cost incurred by participants for emitting a certain quantity of CO2e.

The specific formulas and calculations involved in carbon pricing depend on the design and implementation details of the chosen policy instrument or financial mechanism. These formulas help determine the financial impact and incentives for reducing emissions, ensuring that the cost of carbon pollution is appropriately incorporated into economic decisions.

6.16 Stranded Asset Risk (E1-1, E1-9)

Relevant SDGs: SDG 7: Affordable and Clean Energy, SDG 9: Industry, Innovation, and Infrastructure, SDG 13: Climate Action.

Stranded asset risk refers to the potential financial risk that a company's assets, such as infrastructure, facilities, or investments, will lose value or become obsolete due to changes in regulations, technology advancements, or shifts in consumer preferences. This risk arises when assets that were once considered economically viable and productive become unprofitable or redundant in a rapidly changing energy landscape.

Estimating and assessing stranded asset risk helps companies evaluate the potential financial impacts of changes in the energy market and regulations on their operations and assets. It assists in identifying assets that may face reduced value or become stranded, enabling proactive decision-making and mitigation strategies to navigate such risks effectively.

Assessing stranded asset risk involves evaluating various factors that may impact the value of company assets. While there is no specific mathematical formula, the following steps are commonly involved:

- **Identify at-risk assets**: Identify assets that may be exposed to potential devaluation or obsolescence due to changes in regulations, emerging technologies, or shifts in consumer preferences. This may involve analysing factors such as the asset's lifespan, regulatory environment, market trends, and technological advancements.
- **Conduct scenario analysis**: Develop scenarios based on potential future conditions and evaluate how each scenario could impact the

value of at-risk assets. This analysis considers factors such as changes in carbon pricing, renewable energy adoption, policy shifts, market dynamics, and emerging risks and opportunities.

- **Quantify potential financial impacts**: Estimate the potential financial impacts of stranded asset risk by assigning monetary values to the affected assets. This can include evaluating the costs of asset write-downs, impairments, or potential lost revenues over a specified time frame.

- **Continuously monitor and update assessments**: Stranded asset risk is an evolving issue, and it is crucial to continuously monitor and update assessments as new information becomes available, market conditions change, or regulations evolve.

6.17 Return Period (E1-1, E1-4, E1-9)

Relevant SDGs: SDG 11: Sustainable Cities and Communities, SDG 13: Climate Action, SDG 15: Life on Land.

The return period is a metric used to estimate the average time between occurrences of a specific hazard, such as a 100-year flood. It is commonly expressed as the number of years within which an event with a certain probability of occurrence is expected to happen. For example, a 100-year flood has a 1% probability of occurring in any given year. The return period provides insight into the likelihood of a hazard happening in a particular location and helps in assessing the associated risks and planning appropriate mitigation measures.

The return period is determined based on historical data, statistical analysis, and probability calculations. The following steps are typically involved:

- **Data collection**: Gather relevant data related to the occurrence of the specific hazard over a specified period. This can include historical records, observations, or climate data.

- **Frequency analysis**: Perform statistical analysis on the collected data to assess the frequency of the hazard event. Common methods include probability distributions, such as the Gumbel distribution or Weibull distribution, to estimate the return period.

- **Probability calculation**: Calculate the probability of the hazard event occurring within a specific time frame. This is usually expressed as a percentage or a fraction. The formula for calculating the probability is:

- Probability = 1/(Return Period in Years)
- **Return period estimation**: Determine the average time between occurrences of the hazard by taking the reciprocal of the probability calculated in the previous step. For instance, a hazard with a 1% probability of occurring in any given year corresponds to a return period of 100 years.

By estimating the return period, decision-makers and planners can assess the likelihood of hazards in a given location and incorporate this information into risk assessments, land-use planning, infrastructure design, and emergency management strategies.

7

E-2 Pollution

In this chapter: Assessment of pollution impacts and actions to prevent and mitigate negative impacts. Adaption to a sustainable economy. Risks and opportunities related to pollution impacts and their financial effects. Common indicators and metrics.

Understanding the potential impacts of your organisation on air, water, and soil pollution is crucial for effective environmental management. The following pollution assessment methodology, indicators, and metrics enable a comprehensive assessment of these impacts, allowing for the identification of specific sources of pollution, emissions, effluents, and waste generated during operations. By evaluating the magnitude, frequency, and potential harm associated with these impacts, it provides valuable insights into your organisation's pollution footprint.

In addition to assessing the negative impacts, this methodology recognises the significance of material positive impacts resulting from pollution reduction measures and environmentally friendly practices. It encourages the identification and quantification of such positive impacts to promote sustainable practices and recognise the efforts made towards pollution mitigation. By considering both positive and negative impacts, the methodology provides a holistic view of the organisation's pollution profile, facilitating informed decision-making and targeted actions to address pollution challenges.

Overall, this pollution assessment methodology serves as a valuable tool for undertakings to assess their pollution-related impacts comprehensively. It helps identify risks and opportunities associated with pollution, enables the development of effective pollution prevention and control strategies, and supports the organisation's transition towards a sustainable economy. By implementing this methodology, undertakings can actively contribute to the reduction of pollution, protect environmental resources, adapt to sustainable practices, and manage financial and environmental risks, ultimately contributing to the creation of a cleaner, healthier, and more sustainable environment.

In the following chapter:

- None of the data points are applicable to your ESG report if the topic is deemed not material, with the exception of IRO1 that are to be disclosed irrespective of the outcome of your materiality assessment.
- Not all DPs are to be reported if the undertaking has not adopted the respective policies, implemented the respective actions, or set the respective targets in relation to a sustainability matter that has been assessed to be material.
- Some data points may be phased-in for the first three years, Annex 2: List of Phased-in Disclosure Requirements.

ESRS E2: POLLUTION		
GOVERNANCE & STRATEGY	**IMPACTS, RISKS, & OPPORTUNITIES**	**METRICS & TARGETS**
INCLUDES CROSS-CUTTING DISCLOSURE: • IRO-1: Description of *processes to identify and assess* material pollution-related impacts, risks, and opportunities	• E2-1: *Policies* related to pollution • E2-2: Pollution *actions* and *resources*	• E2-3: *Targets* related to pollution and *tracking effectiveness* of policies and actions through *targets* • E2-4: Pollution of *air, water, and soil* • E2-5: *Substances of concern* and substances of very high concern • E2-6: *Potential financial effects* from pollution-related impacts, risks, and opportunities

Before diving into detailed metrics and targets, begin with describing your materiality assessment and organisational governance initiatives that directly interact with pollution topics, most importantly, the processes and methodologies to identify and assess impacts, risks, and opportunities related to pollution (see **Chapter 3: Single and Double Materiality Assessment and 7.1 Pollution Assessment Methodology**).

Table 60 Pollution assessment indicators relating to impacts, risks, and opportunities.

ESRS	DR	Paragraph	Name	Data type	Tips and tricks	Related CDR	VSMEs	GRI
E2	E2.IRO-1	11 a	Information about methodologies, assumptions, and tools used to screen site locations and business activities in order to identify actual and potential pollution-related impacts, risks, and opportunities in own operations and upstream and downstream value chain	Narrative	Chapter 3: Single and Double Materiality Assessment and 7.1 Pollution Assessment Methodology using TNFD, EU BREF standard and/or Do No Significant Harm framework			
E2	E2.IRO-1	11 b	Disclosure of whether and how consultations have been conducted relating to pollution	Narrative				
E2	E2.IRO-1	AR9	Disclosure of results of materiality assessment relating to pollution	Narrative				GRI 305-3-3, GRI 305-1.2

Source: CDR (EU) 2023/2772 supplementing Directive 2013/34/EU, *GRI 1, 2, 3: 11, 12, 13, 14, 306, 101, 201, 202, 203, 204, 205, 206, 207,301, 302, 303, 304, 305, 306, 308,401, 402, 403, 404, 405, 406, 407, 408, 409, 410, 411, 413, 414, 415, 416, 417, 418*, and the author.

7.1 Pollution Assessment Methodology

This pollution assessment methodology should be incorporated into your materiality assessment process as it is designed to help you evaluate the impacts of your organisation on air, water, and soil pollution and assess the material positive and negative effects resulting from your organisation's activities. This methodology supports the transition to a sustainable economy by taking into account the actual and potential impacts of pollution, aiming to prevent, control, and eliminate pollution to create a toxic-free environment in alignment with the EU Action Plan 'Towards a Zero Pollution for Air, Water, and Soil'. Additionally, it considers your organisation's capacity to adapt its strategies and business models in line with the needs for pollution prevention, control, and elimination.

7.1.1 Assessment of pollution impacts

Examination of the specific ways in which your organisation affects air, water, and soil pollution, including the emissions, effluents, waste, and other pollutants released during its operations.

- Identification and quantification of material positive impacts resulting from pollution reduction measures and environmentally friendly practices implemented by your organisation and value chain.
- Analysis of material negative impacts, encompassing the emissions, effluents, or waste that contribute to pollution, with an evaluation of their magnitude, frequency, and potential harm.
- Consideration of potential future impacts that may arise from your organisation and value chain's activities and processes, taking into account associated risks and their potential consequences on air quality, water resources, and soil health.

7.1.2 Actions to prevent and mitigate negative impacts

Identification of measures undertaken to prevent or mitigate actual or potential material negative impacts on pollution, including the implementation of pollution control technologies and the adoption of best management practices.

- Evaluation of the effectiveness of pollution prevention measures, waste management systems, and environmental controls implemented by your organisation and value chain.

- Assessment of outcomes and results achieved through these actions in terms of pollution reduction and mitigation, including the quantification of emissions, effluents, or waste reductions.
- Monitoring of compliance with relevant environmental regulations and standards to ensure adherence to pollution control requirements.

7.1.3 Adaptation to a sustainable economy

Evaluation of your organisation's plans and capacity to align its strategy and business models with the transition to a sustainable economy.

- Identification of opportunities to integrate pollution prevention and control measures into operations, including the exploration of innovative technologies and the promotion of resource efficiency.
- Assessment of the organisation's commitment to supporting the EU Action Plan 'Towards a Zero Pollution for Air, Water, and Soil' and its contribution towards achieving the outlined objectives.
- Analysis of strategies employed to address the needs for pollution prevention, control, and elimination in order to create a toxic-free environment.

7.1.4 Risks and opportunities related to pollution impacts

Identification and assessment of the nature, type, and extent of material risks and opportunities associated with your organisation's pollution-related impacts and dependencies.

- Evaluation of potential opportunities to develop innovative pollution control technologies, sustainable practices, or eco-friendly products either internally or with value chain collaboration.
- Assessment of your organisation's risk management strategies, including proactive measures, monitoring systems, and compliance with regulations.

7.1.5 Financial effects of pollution impacts

Evaluation of the short-, medium-, and long-term financial effects resulting from material risks and opportunities associated with pollution impacts.

- Analysis of potential costs associated with pollution control, mitigation, and elimination measures, as well as the potential savings resulting from the implementation of sustainable practices.
- Consideration of financial implications related to regulatory compliance and the adoption of pollution prevention technologies.

TIPS AND TRICKS

The methodology and sources of information for these indicators may vary depending on the specific organisation's practices, industry, and regulatory requirements. It is important to consider relevant local and international environmental regulations and standards when assessing potential financial effects from pollution-related impacts, risks, and opportunities.

IMPORTANT: The indicators and metrics in this chapter may have direct and indirect impacts beyond the scope of pollution, for example on biodiversity, and should therefore be considered in parallel and combination with all relevant impacts.

7.2 Hazardous Waste Treatment (ESRS 2 IRO 1, E2-5, E5-2)

Relevant SDGs: SDG 3: Good Health and Well-being, SDG 11: Sustainable Cities and Communities.

This indicator assesses the management and treatment of hazardous waste generated by your organisation during a reporting period. It includes measuring the total amount of hazardous waste generated and disclosing the proportion of hazardous waste that is treated to reduce or eliminate its danger to people and the environment.

1. Measurement of total hazardous waste: Calculate the total amount of hazardous waste generated during the reporting period, including all types of waste listed in the **Basel Convention**[1] definitions and should be measured in kilos and metric tonnes.
2. Proportion of hazardous waste treated: Determine the proportion of hazardous waste that undergoes treatment to reduce or eliminate its

[1] For more information, see: https://www.basel.int/Implementation/HazardCharacteristics/Overview/tabid/3931/Default.aspx

Table 61 Indictors and metrics relating to policies, actions, resources, and targets concerning pollution.

ESRS	DR	Paragraph	Name	Data type	Tips and tricks	Related CDR	VSMEs	GRI
ESRS 2		62	Disclosures to be reported if the undertaking has not adopted policies and/or actions and/or targets	Narrative				
E2	E2-1	14	Policies to manage its material impacts, risks, and opportunities related to pollution	MDR-P	Combine TNFD (or similar), materiality assessment, pollution assessment methodology, and 5.1.1 MDR-P			GRI 303-3-3, GRI 305-3-3, GRI 305-1.2
E2	E2-1	15 a	Disclosure of whether and how policy addresses mitigating negative impacts related to pollution of air, water, and soil	Narrative				
E2	E2-1	15 b	Disclosure of whether and how policy addresses substituting and minimising use of substances of concern and phasing out substances of very high concern	Narrative				
E2	E2-1	15 c	Disclosure of whether and how policy addresses avoiding incidents and emergency situations, and if and when they occur, controlling and limiting their impact on people and environment	Narrative				
E2	E2-1	AR 12	Disclosure of contextual information on relations between policies implemented and how policies contribute to EU Action Plan Towards Zero Pollution for Air, Water, and Soil	Narrative			V	GRI 305-3-3, GRI 305-1.2

E2	E2-2	18	Actions and resources in relation to pollution	MDR-A	Combine TNFD (or similar), materiality assessment, pollution assessment methodology, and 5.1.2 MDR-A			GRI 303-3-3, GRI 305-3-3, GRI 305-1.2
E2	E2-2	19	Layer in mitigation hierarchy to which action can be allocated to pollution	Semi-narrative				GRI 303-3-3, GRI 305-3-3
E2	E2-2	AR 13	Action related to pollution extends to upstream/downstream value chain engagements	Semi-narrative				
E2	E2-2	19	Layer in mitigation hierarchy to which resources can be allocated to pollution	Semi-narrative			V	
E2	E2-2	AR 15	Information about action plans that have been implemented at site-level pollution	Narrative			V	
E2	E2-3	22	Tracking effectiveness of policies and actions through targets	MDR-T	Combine TNFD (or similar), materiality assessment, pollution assessment methodology, and 5.1.4 MDR-T,			GRI 303-3-3, GRI 305-3-3, GRI 305-1.2
E2	E2-3	23 a	Disclosure of whether and how target relates to prevention and control of air pollutants and respective specific loads	Narrative				
E2	E2-3	23 b	Disclosure of whether and how target relates to prevention and control of emissions to water and respective specific loads	Narrative				
E2	E2-3	23 c	Disclosure of whether and how target relates to prevention and control of pollution to soil and respective specific loads	Narrative				
E2	E2-3	23 d	Disclosure of whether and how target relates to prevention and control of substances of concern and substances of very high concern	Narrative				

(Continued)

Table 61 (*Continued*).

ESRS	DR	Paragraph	Name	Data type	Tips and tricks	Related CDR	VSMEs	GRI
E2	E2-3	24	Ecological thresholds and entity-specific allocations were taken into consideration when setting pollution-related target	Semi-narrative			V	GRI 303-2
E2	E2-3	24 a	Disclosure of ecological thresholds identified and methodology used to identify ecological thresholds relating to pollution	Narrative			V	GRI 303-2
E2	E2-3	24 b	Disclosure of how ecological entity-specific thresholds were determined for pollution	Narrative			V	GRI 303-2
E2	E2-3	24 c	Disclosure of how responsibility for respecting identified ecological thresholds is allocated for pollution	Narrative			V	GRI 303-2
E2	E2-3	25	Pollution-related target is mandatory (required by legislation) or voluntary	Semi-narrative				
E2	E2-3	AR 17	Pollution-related target addresses shortcomings related to Substantial Contribution criteria for Pollution Prevention and Control	Semi-narrative			V	
E2	E2-3	AR 18	Information about targets that have been implemented at site-level (pollution)	Narrative			V	

Source: CDR (EU) 2023/2772 supplementing Directive 2013/34/EU, *GRI 1, 2, 3: 11, 12, 13, 14, 306, 101, 201, 202, 203, 204, 205, 206, 207,301, 302, 303, 304, 305, 306, 308,401, 402, 403, 404, 405, 406, 407, 408, 409, 410, 411, 413, 414, 415, 416, 417, 418*, and the author.

Table 62 Indictors and metrics relating to pollution of air, water, and soil and potential financial effects.

Note that E2-6 is subject to phase-in provisions, see Annex 2: List of Phased-in Disclosure Requirements.

ESRS	DR	Paragraph	Name	Data type	Tips and tricks	Related CDR	VSMEs	GRI
E2	E2-4	28 a	Pollution of air, water, and soil in multiple dimensions: at site level or by type of source, by sector or by geographical area	Table	Where assessed as material and using pollution assessment methodology, select appropriate metrics from 7.2 to 7.11 and then tabulate results with explanations of methodologies		V	GRI 305-7
E2	E2-4	28 a	Emissions to air by pollutant	Table/mass		SFDR	V	GRI 305-7
E2	E2-4	28 a	Emissions to water by pollutant [+ by sectors/ geographical area/type of source/site location]	Table/mass		SFDR	V	GRI 305-7
E2	E2-4	28 a	Emissions to soil by pollutant [+ by sectors/ geographical area/type of source/site location]	Table/mass		SFDR	V	GRI 305-7
E2	E2-4	28 b	Microplastics generated	Mass				
E2	E2-4	28 b	Microplastics used	Mass				
E2	E2-4	30 a	Description of changes over time (pollution of air, water, and soil)	Narrative				
E2	E2-4	30 b	Description of measurement methodologies (pollution of air, water, and soil)	Narrative				GRI 305-7
E2	E2-4	30 c	Description of process(es) to collect data for pollution-related accounting and reporting	Narrative				GRI 305-7
E2	E2-4	AR 23 c	Percentage of total emissions of pollutants to water occurring in areas at water risk	Per cent			V	
E2	E2-4	AR 23 c	Percentage of total emissions of pollutants to water occurring in areas of high-water stress	Per cent			V	
E2	E2-4	AR 23 c	Percentage of total emissions of pollutants to soil occurring in areas at water risk	Per cent			V	

(*Continued*)

Table 62 (*Continued*).

ESRS	DR	Paragraph	Name	Data type	Tips and tricks	Related CDR	VSMEs	GRI
E2	E2-4	AR 23 c	Percentage of total emissions of pollutants to soil occurring in areas of high-water stress	Per cent			V	
E2	E2-4	31	Disclosure of reasons for choosing inferior methodology to quantify emissions	Narrative				GRI 305-7
E2	E2-4	AR 25a	Disclosure of list of installations operated that fall under IED and EU BAT conclusions	Narrative			V	
E2	E2-4	AR 25b	Disclosure of list of any non-compliance incidents or enforcement actions necessary to ensure compliance in case of breaches of permit conditions	Narrative			V	GRI 2-27
E2	E2-4	AR 25c	Disclosure of actual performance and comparison of environmental performance against emission levels associated with best available techniques (BAT-AEL) as described in EU-BAT conclusions	Narrative			V	
E2	E2-4	AR 25d	Disclosure of actual performance against environmental performance levels associated with best available techniques (BAT-AEPLs) applicable to sector and installation	Narrative			V	
E2	E2-4	AR 25e	Disclosure of list of any compliance schedules or derogations granted by competent authorities according to Article 15(4) IED that are associated with implementation of BAT-AELs	Narrative			V	
E2	E2-5	34	Total amount of substances of concern that are generated or used during production or that are procured by main hazard classes of substances of concern	Table	Where assessed as material and using pollution assessment methodology, tabulate all substances of high concern and of very high concern			

E2	E2-5	34	Total amount of substances of concern that are generated or used during production or that are procured	Table/mass				
E2	E2-5	34	Total amount of substances of concern that leave facilities as emissions, as products, or as part of products or services	Table/mass				
E2	E2-5	34	Quantity of substances of concern that leave facilities as emissions by main hazard classes of substances of concern	Table/mass				
E2	E2-5	34	Quantity of substances of concern that leave facilities as products by main hazard classes of substances of concern	Table/mass				
E2	E2-5	34	Quantity of substances of concern that leave facilities as part of products by main hazard classes of substances of concern	Table/mass				
E2	E2-5	35	Quantity of substances of concern that leave facilities as services	Table/mass				
E2	E2-5	35	Total amount of substances of very high concern that are generated or used during production or that are procured by main hazard classes of substances of concern	Table/mass				
E2	E2-5	35	Total amount of substances of very high concern that leave facilities as emissions, as products, or as part of products or services by main hazard classes of substances of concern	Table/mass				
E2	E2-5	35	Quantity of substances of very high concern that leave facilities as emissions by main hazard classes of substances of concern	Table/mass				

(Continued)

Table 62 (*Continued*).

ESRS	DR	Paragraph	Name	Data type	Tips and tricks	Related CDR	VSMEs	GRI
E2	E2-5	35	Quantity of substances of very high concern that leave facilities as products by main hazard classes of substances of concern	Table/mass				
E2	E2-5	35	Quantity of substances of very high concern that leave facilities as part of products by main hazard classes of substances of concern	Table/mass				
E2	E2-5	35	Quantity of substances of very high concern that leave facilities as services by main hazard classes of substances of concern	Table/mass				
E2	E2-6	39 a	Disclosure of quantitative information about anticipated financial effects of material risks and opportunities arising from pollution-related impacts	Narrative	Cross reference with financial reports, 5.3 Risk and Opportunity Indicator Metrics and see 7.12 and 7.13			
E2	E2-6	40 a	Percentage of net revenue made with products and services that are or that contain substances of concern	Per cent				
E2	E2-6	40 a	Percentage of net revenue made with products and services that are or that contain substances of very high concern	Per cent				
E2	E2-6	40 b	Operating expenditures (OpEx) in conjunction with major incidents and deposits (pollution)	Monetary				
E2	E2-6	40 b	Capital expenditures (CapEx) in conjunction with major incidents and deposits (pollution)	Monetary				

E2	E2-6	40 c	Provisions for environmental protection and remediation costs (pollution)	Monetary				
E2	E2-6	39 a	Disclosure of qualitative information about anticipated financial effects of material risks and opportunities arising from pollution-related impacts	Narrative				
E2	E2-6	39 b	Description of effects considered, related impacts, and time horizons in which they are likely to materialise (pollution)	Narrative				
E2	E2-6	39 c	Disclosure of critical assumptions used to quantify anticipated financial effects, sources and level of uncertainty of assumptions (pollution)	Narrative				
E2	E2-6	41	Description of material incidents and deposits whereby pollution had negative impacts on environment and/or is expected to have negative effects on financial cash flows, financial position, and financial performance	Narrative				
E2	E2-6	AR 33	Disclosure of assessment of related products and services at risk and explanation how time horizon is defined, financial amounts are estimated, and which critical assumptions are made (pollution)	Narrative				

Source: CDR (EU) 2023/2772 supplementing Directive 2013/34/EU, *GRI 1, 2, 3: 11, 12, 13, 14, 306, 101, 201, 202, 203, 204, 205, 206, 207,301, 302, 303, 304, 305, 306, 308,401, 402, 403, 404, 405, 406, 407, 408, 409, 410, 411, 413, 414, 415, 416, 417, 418*, and the author.

danger compared to the total waste reported by your organisation. This proportion is expressed as a percentage and verification processes must be included. If direct measurement is challenging, you can use invoices from waste management companies to reconstruct the necessary information.

3. Breakdown of treatment methods: If applicable, your organisation breaks down the total weight of hazardous waste by different treatment methods, such as reuse, recycling, composting, recovery (including energy recovery), incineration (mass burn), deep well injection, landfill, on-site storage, and other specified methods.

7.3 Ozone Depleting Substances and Chemicals (E2-4, E1-6)

Relevant SDGs: SDG 13: Climate Action.

The measurement methodology for ozone-depleting substances[2] (ODS) and chemicals is based on the information provided in the Montreal Protocol and should be considered in parallel with GHG emissions. The ozone-depleting potential (ODP) value represents the impact of a substance on ozone depletion relative to a reference substance, usually trichlorofluoromethane (CFC 11), which is assigned an ODP value of 1. ODP values are expressed in kg CFC 11 equivalents per kg of the respective substance. The dependency of an entity on ODS is determined by considering the production of ODS, purchases of ODS, and stocks of ODS.

1. Production of ODS: Measure the amount of virgin ODS (not recovered, reclaimed, or recycled) added by your organisation. This information should be recorded in the operating information systems of each specific plant as part of the production outcome amounts for a specific reporting period.

2. Purchases and stocks of ODS: Track your organisation's purchases and stocks of ODS. Information regarding ODS purchases for production processes can be found in the accounts payable and operating information systems of each specific plant. The responsible individual for this information is typically the plant manager or purchasing manager.

[2] See https://www.eea.europa.eu/themes/climate/ozone-depleting-substances-and-climate-change/ozone-depleting-substances-2021-pdf-version

3. ODS embodied in equipment: Determine the ODS present in equipment used outside production processes, such as air conditioning or firefighting equipment. This information can be derived from the descriptions of the specific equipment purchased by your organisation at each facility. The responsible individual for this information is usually the facility manager or general services administrator.

7.4 Air Quality Index (E2-4)

Relevant SDGs: SDG 3: Good Health and Well-being, SDG 11: Sustainable Cities and Communities.

The air quality index (AQI) is calculated based on the concentration of various air pollutants, such as particulate matter (PM2.5 and PM10), nitrogen dioxide (NO_2), sulphur dioxide (SO_2), and ozone (O_3). Each pollutant is assigned a value, and the overall AQI is determined by the highest value among the pollutants. Due to the difficulty and equipment required to measure AQI, it is recommended contracting an expert organisation, while noting that it is essential to reference all local laws and regulations governing air quality in the disclosure.

7.5 Wastewater Treatment Efficiency (E2-1, E2-4, E5-1)

Relevant SDGs: SDG 6: Clean Water and Sanitation.

The efficiency of wastewater treatment is calculated by comparing the concentration of pollutants in the influent (incoming wastewater) with the effluent (treated wastewater). The removal efficiency can be determined using the following formula:

Wastewater Treatment Efficiency (%) = ((Influent Pollutant Concentration − Effluent Pollutant Concentration)/Influent Pollutant Concentration) × 100

7.6 Chemical Spills and Releases (E2-4, E2-5)

Relevant SDGs: SDG 6: Clean Water and Sanitation, SDG 14: Life Below Water.

The methodology for tracking chemical spills and releases involves recording incidents, quantifying the volume of chemicals released, and assessing the impact on the environment and surrounding communities.

Disclosure is required of incident reports, emergency response records, regulatory agencies, and any potential fines and/or punitive measures.

7.7 Emissions of Hazardous Air Pollutants (E2-4, E2-5)

Relevant SDGs: SDG 3: Good Health and Well-being, SDG 11: Sustainable Cities and Communities.

Emissions of hazardous air pollutants are typically measured using continuous monitoring systems, and periodic emissions testing may be conducted for specific pollutants. Emissions are typically reported in terms of mass or concentration. Disclosure is required of incident reports, emergency response records, regulatory agencies, and any potential fines and/or punitive measures.

7.8 Water Pollution Incidents (E2-4, E3-2)

Relevant SDGs: SDG 6: Clean Water and Sanitation, SDG 14: Life Below Water.

Measurement of the number and severity of water pollution incidents, such as spills, leaks, or discharges of pollutants into water bodies. Water pollution incidents are documented based on reported incidents, investigations, and monitoring of water quality parameters. The severity of an incident can be assessed based on factors such as the volume and toxicity of the pollutants released. Disclosure is required of incident reports, emergency response records, regulatory agencies, and any potential fines and/or punitive measures.

7.9 Soil Contamination Levels (E2-4, E4-5)

Relevant SDGs: SDG 3: Good Health and Well-being, SDG 15: Life on Land.

Measurement of the concentration of contaminants in soil, which can negatively impact ecosystems and human health. Soil samples should be collected from various locations, and laboratory analysis is performed to determine the concentration of contaminants, such as heavy metals, pesticides, or industrial chemicals. Due to the difficulty and equipment required to measure soil contamination levels, it is recommended contracting an expert organisation, while noting that it is essential to reference all local laws and regulations in the disclosure. Disclosure is required of incident reports, emergency response records, regulatory agencies, and any potential fines and/or punitive measures.

7.10 Emissions of Volatile Organic Compounds (VOCs) (E2-4, E2-5)

Relevant SDGs: SDG 3: Good Health and Well-being, SDG 11: Sustainable Cities and Communities.

Measurement of the emissions of volatile organic compounds, which contribute to air pollution and can have harmful effects on human health. Emissions of VOCs are typically measured using continuous monitoring systems or periodic emissions testing. The emissions are reported in terms of mass or concentration. Disclosure is required of incident reports, emergency response records, regulatory agencies, and any potential fines and/or punitive measures.

7.11 Noise Pollution Levels (E2-4)

Relevant SDGs: SDG 3: Good Health and Well-being, SDG 11: Sustainable Cities and Communities.

Measurement of ambient noise levels in the environment, particularly in urban areas, which can have adverse effects on human well-being. Due to the difficulty and equipment required to measure soil contamination levels, it is recommended contracting an expert organisation, while noting that it is essential to reference all local laws and regulations in the disclosure. Disclosure is required of incident reports, emergency response records, regulatory agencies, and any potential fines and/or punitive measures.

7.12 Evaluating Cost of Pollution Abatement (E2-6)

Relevant SDGs: SDG 3 (Good Health and Well-being), SDG 6 (Clean Water and Sanitation), SDG 7 (Affordable and Clean Energy), SDG 11 (Sustainable Cities and Communities), SDG 12 (Responsible Consumption and Production), SDG 13 (Climate Action), SDG 14 (Life Below Water), SDG 15 (Life on Land).

Evaluating the cost of pollution abatement measures requires the organisation to allocate financial resources to mitigate or eliminate pollution-related impacts, risks, and opportunities. Therefore, this disclosure requires reference to financial statements, budget reports, pollution control and abatement records, environmental impact assessments, and compliance reports.

Begin with an assessment of pollution abatement measures across your organisation's activities, processes, and operations that contribute to pollution, and identify the measures implemented to mitigate or eliminate pollution-related impacts. Then calculate the direct and indirect costs associated with pollution abatement measures, including investments in pollution control technologies, waste management systems, emission reduction initiatives, and environmental impact assessments.

7.13 Financial Provisioning for Pollution Liabilities (E2-6)

Relevant SDGs: SDG 3 (Good Health and Well-being), SDG 6 (Clean Water and Sanitation), SDG 7 (Affordable and Clean Energy), SDG 11 (Sustainable Cities and Communities), SDG 12 (Responsible Consumption and Production), SDG 13 (Climate Action), SDG 14 (Life Below Water), SDG 15 (Life on Land).

Financial provisioning for pollution liabilities assesses your organisation's financial allocation for potential liabilities arising from pollution-related impacts, risks, and opportunities. Therefore, this disclosure requires reference to financial statements, risk management reports, insurance policies, legal claims and disputes records, and environmental audits.

Begin with identifying potential pollution liabilities that may arise from pollution-related impacts, such as legal claims, fines, remediation costs, and compensation expenses. Then estimate the financial provision or reserves set aside by your organisation to cover potential pollution liabilities. This may include dedicated funds, insurance policies, or budgetary allocations.

8

E-3 Water and Marine

In this chapter: Assessment of water and marine impacts and actions to prevent and mitigate negative impacts. Adaption to a sustainable economy. Risks and opportunities related to water and marine impacts and their financial effects. Common indicators and metrics.

Water and marine resources play a crucial role in sustaining life and ecosystems. As we face growing environmental challenges, it is essential to understand and address the impacts of human activities on these vital resources. Comprehensive ESG reporting systems are necessary to evaluate and disclose these impacts and promote sustainability. This methodology aims to provide a robust framework for assessing an organisation's effects on water and marine resources and disclosing actions taken to mitigate negative impacts and promote sustainability.

The methodology comprises key components that offer a comprehensive approach to reporting on water and marine resources by focusing on assessing the material positive and negative impacts of the undertaking on these resources, including factors such as water quality, quantity, and ecosystem health. By analysing these impacts, stakeholders gain insights into the organisation's contributions to the overall state of water and marine resources. Furthermore, the methodology emphasises your organisation's efforts to prevent and mitigate material negative impacts by considering a range of measures, such as pollution reduction, biodiversity conservation, and resource management. Such actions demonstrate your organisation's commitment to protecting and restoring water and marine resources.

Useful resources for water and marine assessments:

- *ENCORE* tool allows to explore water-related dependencies for the direct operations of 169 sectors of the economy at the global level.

- **System of Environmental Economic Accounting (SEEA) technical note**: *Water Accounting* illustrates the main links between water and economic systems that can be grouped into three types of flows between the two systems: (1) physical flows of water and wastewater, (2) monetary flows associated with water-related products and services, (3) assets (stocks of water and their changes during the accounting period).
- The *AWS Standard v2.0* presents a five-step approach to water stewardship. For each step, criteria and indicators are provided to support companies in their water stewardship journey. Additionally, the standard aims to achieve five main outcomes: water governance, sustainable water balance, water quality status, important water-related areas, and safe WASH (water, sanitation, and hygiene) for all.
- The *CERES Aqua Gauge* tool supports companies in developing and evaluating their water strategy – including risk assessment, goal-setting, accountability systems, and procurement, siting and CapEx decisions. Additionally, the Ceres Roadmap 2030 section on water provides steps to improve water management and includes practical suggestions on performance.
- The *water risk filter* (*WRF*, developed by World Wildlife Forum, WWF) is an assessment screening tool to help companies and investors to prioritise action on what and where it matters the most to address water risks for enhancing business resilience and contributing to a sustainable future.
- The *Aqueduct 4.0 Water Risk Atlas (WRA)* is a robust methodology with updated inputs that provides accurate baseline data, as well as future projections data for 2030 and 2050.

In the following chapter:

- None of the data points are applicable to your ESG report if the topic is deemed not material, with the exception of IRO1 that are to be disclosed irrespective of the outcome of your materiality assessment.
- Not all DPs are to be reported if the undertaking has not adopted the respective policies, implemented the respective actions, or set the respective targets in relation to a sustainability matter that has been assessed to be material.
- Some data points may be phased-in for the first three years, Annex 2: List of Phased-in Disclosure Requirements.

ESRS E3: WATER AND MARINE		
GOVERNANCE & STRATEGY	**IMPACTS, RISKS, & OPPORTUNITIES**	**METRICS & TARGETS**
INCLUDES CROSS-CUTTING DISCLOSURE: • IRO-1: Description of *processes to identify and assess* material water and marine-related impacts, risks, and opportunities	• E3-1: *Policies* related to water and marine resources • E3-2: *Actions* and *resources* related to water and marine resources, policies, and actions	• E3-3: *Targets and tracking effectiveness* related to water and marine resources • E3-4: *Water consumption* • E3-5: *Potential financial effects* from water and marine resources-related impacts, risks, and opportunities

Before diving into detailed metrics and targets, begin with describing your materiality assessment and organisational governance initiatives that directly interact with water and marine topics, most importantly, the processes and methodologies to identify and assess impacts, risks, and opportunities related to water and marine issues (see **Chapter 3: Single and Double Materiality Assessment and 8.1 Water and Marine Assessment Methodology**).

Table 63 Water and marine assessment indicators relating to impacts, risks, and opportunities.

ESRS	DR	Paragraph	Name	Data type	Tips and tricks	Related CDR	VSMEs	GRI
E3	IRO-1	8 a)	Disclosure of whether and how assets and activities have been screened in order to identify actual and potential water and marine resources-related impacts, risks and opportunities in own operations and upstream and downstream value chain and methodologies, assumptions, and tools used in screening	Narrative	Chapter 3: Single and Double Materiality Assessment, and 8.1 Water and Marine Assessment Methodology or using TNFD, and/or WWF–WRF tool and/or WRi Aqueduct Atlas tool			GRI 303-1
E3	IRO-1	8 b	Disclosure of how consultations have been conducted regarding water and marine resources	Narrative				
E3	IRO-1	AR 1	Disclosure of results of materiality assessment of water and marine resources	Narrative				

Source: CDR (EU) 2023/2772 supplementing Directive 2013/34/EU, *GRI 1, 2, 3: 11, 12, 13, 14, 306, 101, 201, 202, 203, 204, 205, 206, 207,301, 302, 303, 304, 305, 306, 308,401, 402, 403, 404, 405, 406, 407, 408, 409, 410, 411, 413, 414, 415, 416, 417, 418*, and the author.

8.1 Water and Marine Assessment Methodology

This comprehensive methodology facilitates transparent and accountable disclosure of your organisation's impacts, actions, and strategies concerning water and marine resources and provides stakeholders with a clear understanding of the level of commitment to preserving and restoring these resources while contributing to broader sustainability objectives. This methodology empowers stakeholders to make informed decisions, promotes responsible resource management, and supports the transition to a sustainable and resilient economy.

The methodology comprises several key components as discussed below:

1. **Impact assessment**: Begin by identifying location, type, quality, and quantity of water bodies, including rivers, lakes, and oceans, and the health and biodiversity of marine ecosystems impacted by your organisation and then evaluate the material positive and negative actual or potential impacts.
2. **Mitigation actions**: Detail the specific measures and actions implemented in your organisation to prevent or mitigate material negative impacts on water and marine resources. In particular, focus on reducing pollution, conserving water, and adopting sustainable practices that align with the European Green Deal's ambitions for clean water, healthy soil, and biodiversity.
3. **Compliance and contribution**: Assess your organisation's compliance with relevant European directives and initiatives, such as the EU water framework directive, the EU marine strategy framework, the EU maritime spatial planning directive, and the Sustainable Development Goals (SDGs) 6 (Clean water and sanitation) and 14 (Life below water). Additionally, evaluate the organisation's contribution to the European Green Deal's objectives, ensuring the sustainability of the blue economy and fisheries sectors, and respecting global environmental limits.
4. **Adaptation strategies**: Analyse the plans and capacity of your organisation to adapt its strategy and business models to preserve and restore water and marine resources globally. Identify initiatives and measures that promote water and marine conservation, reduce impacts, and ensure sustainable resource management practices.
5. **Risk and opportunity assessment:** Evaluate the nature, type, and extent of material risks and opportunities arising from your organisation's

impacts and dependencies on water and marine resources. Assess how your organisation identifies, manages, and mitigates these risks while harnessing opportunities associated with sustainable water and marine resource management.

6. **Financial implications:** Consider the short-, medium-, and long-term financial effects of material risks and opportunities associated with your organisation's impacts and dependencies on water and marine resources. Assess the costs and benefits of adopting sustainable practices, investing in water conservation technologies, and exploring potential market opportunities related to water and marine resource management.

8.2 Water Recycling and Reuse (E3-1, E3-2)

Relevant SDGs: SDG 6: Clean Water and Sanitation.

Water recycling and reuse can be measured using two different indicators. The first indicator is the total volume of water recycled and reused, expressed in cubic meters (m^3). The second indicator is the total volume of water recycled and reused as a percentage of the total water withdrawal and total water received from a third party, also expressed in percentage terms.

1. **Total volume of water recycled and reused**: If your organisation has water or flow meters, calculate the indicator at the facility or site level using appropriate documentation and reporting based on direct measurement of the quantity of water recycled and/or reused. Data on the total volume of water recycled and/or reused should be collected with reference to a relevant time unit (e.g., day, week, month) and accumulated over the total reporting period.

2. **Estimation of water recycled and reused**: If your organisation does not have water or flow meters, estimate the volume of recycled and reused water. Estimations can be based on coefficients (area statistics) relating water use to another characteristic, such as the number of employees or production values, and can be applied to site-specific quantities of that characteristic. You could also calculate the volume based on the water demand satisfied by recycled and/or reused water, instead of further withdrawals or supplies from third parties.

3. **Total volume of water recycled and reused as a percentage**: Calculate the percentage by dividing the total volume of water recycled and

Table 64 Indictors and metrics relating to policies, actions, resources, and targets concerning water and marine.

ESRS	DR	Paragraph	Name	Data type	Tips and tricks	Related CDR	VSMEs	GRI
ESRS 2	E3	62	Disclosures to be reported in case the undertaking has not adopted policies and/or actions and/or targets	Narrative				
E3	E3-1	11	Policies to manage its material impacts, risks, and opportunities related to water and marine resources	MDR-P	Combine TNFD (or similar), ENCORE and/or WWF-WRF tool and/or WRi Aqueduct Atlas tool, materiality assessment, water, and marine assessment methodology and 5.1.1 MDR-P	SFDR		GRI 303-3-3
E3	E3-1	12a	Disclosure of whether and how policy addresses water management	Narrative				
E3	E3-1	12a i	Disclosure of whether and how policy addresses the use and sourcing of water and marine resources in own operations	Narrative				
E3	E3-1	12a ii	Disclosure of whether and how policy addresses water treatment	Narrative				
E3	E3-1	12a iii	Disclosure of whether and how policy addresses prevention and abatement of water pollution	Narrative				
E3	E3-1	12 b	Disclosure of whether and how policy addresses product and service design in view of addressing water-related issues and preservation of marine resources	Narrative				
E3	E3-1	12 c	Disclosure of whether and how policy addresses commitment to reduce material water consumption in areas at water risk	Narrative				
E3	E3-1	13	Disclosure of reasons for not having adopted policies in areas of high-water stress	Narrative		SFDR		
E3	E3-1	13	Disclosure of timeframe in which policies in areas of high-water stress will be adopted	Narrative		SFDR	V	
E3	E3-1	14	Policies or practices related to sustainable oceans and seas have been adopted	Semi-narrative		SFDR		

(*Continued*)

Table 64 (*Continued*).

ESRS	DR	Paragraph	Name	Data type	Tips and tricks	Related CDR	VSMEs	GRI
E3	E3-1	AR 18a	The policy contributes to good ecological and chemical quality of surface water bodies and good chemical quality and quantity of groundwater bodies, in order to protect human health, water supply, natural ecosystems and biodiversity, the good environmental status of marine waters, and the protection of the resource base upon which marine related activities depend	Semi-narrative			V	
E3	E3-1	AR 18b	The policy minimises material impacts and risks and implements mitigation measures that aim to maintain the value and functionality of priority services and to increase resource efficiency on own operations	Semi-narrative			V	
E3	E3-1	AR 18c	The policy avoids impacts on affected communities	Semi-narrative			V	
E3	E3-2	17	Actions and resources in relation to water and marine resources	MDR-A	Combine TNFD (or similar), and/or WWF-WRF tool and/or WRi Aqueduct Atlas tool, materiality assessment, water and marine assessment methodology and 5.1.2 MDR-A			GRI 303-3-3
E3	E3-2	18	Layer in mitigation hierarchy to which action and resources can be allocated to (water and marine resources)	Semi-narrative			V	GRI 303-3-3
E3	E3-2	AR20	Information about specific collective action for water and marine resources	Narrative			V	GRI 303-1
E3	E3-2	19	Disclosure of actions and resources in relation to areas at water risk	Narrative				

Source: CDR (EU) 2023/2772 supplementing Directive 2013/34/EU, *GRI 1, 2, 3: 11, 12, 13, 14, 306, 101, 201, 202, 203, 204, 205, 206, 207,301, 302, 303, 304, 305, 306, 308,401, 402, 403, 404, 405, 406, 407, 408, 409, 410, 411, 413, 414, 415, 416, 417, 418*, and the author.

Table 65 Indictors and metrics relating to water and marine, and potential financial effects.

ESRS	DR	Paragraph	Name	Data type	Tips and tricks	Related CDR	VSMEs	GRI
E3	E3-3	22	Tracking effectiveness of policies and actions through targets	MDR-T	Combine TNFD (or similar), and/or WWF-WRF tool and/or WRi Aqueduct Atlas tool, materiality assessment, pollution assessment methodology and 5.1.4 MDR-T from 8.2 to 8.12 Tabulate by location and flag high water risk or stress locations			GRI 303-3-3
E3	E3-3	23 a	Disclosure of whether and how target relates to management of material impacts, risks, and opportunities related to areas at water risk	Narrative				
E3	E3-3	23 b	Disclosure of whether and how target relates to responsible management of marine resources impacts, risks, and opportunities	Narrative				
E3	E3-3	23 c	Disclosure of whether and how target relates to reduction of water consumption	Narrative				
E3	E3-3	24	(Local) ecological threshold and entity-specific allocation were taken into consideration when setting water and marine resources target	Semi-narrative			V	
E3	E3-3	24 a	Disclosure of ecological threshold identified and methodology used to identify ecological threshold for water and marine resources	Narrative			V	
E3	E3-3	24 b	Disclosure of how ecological entity-specific threshold was determined for water and marine resources	Narrative			V	
E3	E3-3	24 c	Disclosure of how responsibility for respecting identified ecological threshold is allocated for water and marine resources	Narrative			V	

(*Continued*)

Table 65 (*Continued*).

ESRS	DR	Paragraph	Name	Data type	Tips and tricks	Related CDR	VSMEs	GRI
E3	E3-3	25	Adopted and presented water and marine resources-related target is mandatory if based on legislation	Semi-narrative				
E3	E3-3	AR 23 a	Target relates to reduction of water withdrawals	Semi-narrative			V	
E3	E3-3	AR 23 b	Target relates to reduction of water discharges	Semi-narrative			V	
E3	E3-4	28 a	Total water consumption	Volume			V	GRI 303-5
E3	E3-4	28 b	Total water consumption in areas at water risk, including areas of high-water stress	Volume			V	GRI 303-5
E3	E3-4	28 c	Total water recycled and reused	Volume		SFDR		
E3	E3-4	28 d	Total water stored	Volume				GRI 303-5
E3	E3-4	28 d	Changes in water storage	Volume				GRI 303-5
E3	E3-4	28 e	Disclosure of contextual information regarding water consumption	Narrative				GRI 303-5
E3	E3-4	29	Water intensity ratio	Per cent		SFDR		
E3	E3-4	AR30	Water consumption – sectors/segments [table]	Table/volume			V	
E3	E3-4	AR 31	Additional water intensity ratio	Per cent			V	

E3	E3-4	AR 32	Total water withdrawals	Volume			V	
E3	E3-4	AR 32	Total water discharges	Volume			V	
E3	E3-5	33 a	Disclosure of quantitative information about potential financial effects of material risks and opportunities arising from water and marine resources-related impacts	Monetary				
E3	E3-5	33 a	Disclosure of qualitative information of potential financial effects of material risks and opportunities arising from water and marine resources-related impacts	Narrative				
E3	E3-5	33 b	Description of effects considered and related impacts of water and marine resources	Narrative				
E3	E3-5	33 c	Disclosure of critical assumptions used in estimates of financial effects of material risks and opportunities arising from water and marine resources-related impacts	Narrative				
E3	E3-5	AR 33	Description of related products and services at risk of water and marine resources	Narrative			V	
E3	E3-5	AR 33	Explanation of how time horizons are defined, financial amounts are estimated, and critical assumptions made water and marine resources	Narrative			V	

Source: CDR (EU) 2023/2772 supplementing Directive 2013/34/EU, *GRI 1, 2, 3: 11, 12, 13, 14, 306, 101, 201, 202, 203, 204, 205, 206, 207,301, 302, 303, 304, 305, 306, 308,401, 402, 403, 404, 405, 406, 407, 408, 409, 410, 411, 413, 414, 415, 416, 417, 418*, and the author.

reused by the sum of the total water withdrawal and total water received from a third party. The total water withdrawal includes all water drawn into your organisation's boundaries from different sources for any use during the reporting period, such as fresh surface water, groundwater, seawater, produced/process water, and third-party water.

8.3 Eutrophication Levels (E3-1)

Relevant SDGs: SDG 6: Clean Water and Sanitation, SDG 14: Life Below Water.

Measurement of the nutrient enrichment and resulting ecological impacts in water bodies, typically caused by excessive nutrient runoff from human activities, involves the following steps:

1. **Collect water quality data**: Gather data on nutrient concentrations (e.g., nitrogen, phosphorus) and other relevant parameters such as chlorophyll-a, dissolved oxygen, and Secchi depth. These measurements provide insights into nutrient enrichment and algal growth. Historical water quality monitoring data should also be included.

2. **Determine reference values**: Establish reference values or thresholds for nutrient concentrations and other parameters based on water quality standards, ecological assessments and guidelines, historical data, and environmental agency reports.

3. **Assess nutrient levels**: Compare the measured nutrient concentrations with the reference values to determine the extent of nutrient enrichment. Higher concentrations beyond the reference values indicate increased eutrophication.

4. **Evaluate ecological impacts**: Assess the ecological impacts of eutrophication, such as harmful algal blooms, oxygen depletion, and shifts in species composition. This can be done through ecological surveys, biodiversity assessments, or modelling approaches.

5. **Develop mitigation strategies**: Based on the assessment results, develop strategies to mitigate eutrophication, such as nutrient management plans, land use controls, and wastewater treatment improvements.

6. **Estimate financial impacts**: Estimate potential costs of mitigation or reparation where possible, as well as possible damages and insurance claims.

8.4 Water Usage Efficiency (E3-4)

Relevant SDGs: SDG 6: Clean Water and Sanitation.

Water usage efficiency can be calculated by determining the ratio of water used to the output or production achieved. The methodology involves the following steps:

1. **Collect data on water consumption and output**: Gather information from water consumption records, production data, and water management systems on the volume of water consumed during a specific period and the corresponding output or production achieved.
2. **Calculate water usage efficiency**: Divide the output by the water consumption to determine the water usage efficiency. The formula for calculating water usage efficiency is as follows:

   Water Usage Efficiency = Output/Water Consumption
3. **Express the result as a percentage**: Multiply the water usage efficiency by 100 to express it as a percentage.

8.5 Water Quality Index (3-1)

Relevant SDGs: SDG 6: Clean Water and Sanitation, SDG 14: Life Below Water.

The water quality index (WQI) is the measurement of the overall water quality based on various physical, chemical, and biological parameters and is calculated by assessing multiple water quality parameters and assigning weights to each parameter based on its relative importance. The methodology involves the following steps:

1. **Select water quality parameters**: Identify the relevant physical, chemical, and biological parameters that are indicative of water quality, such as pH, dissolved oxygen, turbidity, and levels of pollutants. Environmental agencies should be consulted and relevant local regulations and standards disclosed.
2. **Assign weights to parameters**: Assign weights to each parameter based on its significance in determining overall water quality. The weights reflect the relative importance of each parameter.
3. **Measure and normalise parameter values**: Measure the values of each parameter in the water samples and normalise them to a common

scale. This ensures that all parameters contribute proportionately to the overall index.

4. **Calculate sub-index scores**: Calculate sub-index scores for each parameter by multiplying the normalised value by the assigned weight.
5. **Incorporation of location-specific factors**: Consider location-specific factors that may impact water quality, such as local environmental conditions, geographical characteristics, and regional regulations.
6. **Calculate the overall water quality index**: Aggregate the sub-index scores to obtain the overall water quality index. The index provides a single value that represents the overall quality of the water, which can be categorised into different classes (e.g., excellent, good, fair, poor).
7. **Estimate financial impacts**: Estimate potential costs of mitigation or reparation of water quality where possible, as well as possible damages and insurance claims.

8.6 Marine Protected Area Coverage (E3-1, E3-2)

Relevant SDGs: SDG 14: Life Below Water.

Measurement of the extent of designated marine protected areas for conserving and managing marine ecosystems (related to 9.2 Protected Area Coverage) involves the following steps:

1. **Define the boundaries**: Determine the boundaries of the marine area to be considered for measuring the coverage of designated protected areas using marine conservation reports, government databases, and marine spatial planning records.
2. **Identify and map protected areas**: Identify the designated marine protected areas within the defined boundaries and map their locations.
3. **Calculate the total protected area**: Calculate the total area of all designated protected areas within the defined boundaries.
4. **Calculate the coverage percentage**: Divide the total protected area by the total marine area within the defined boundaries and multiply by 100 to obtain the coverage percentage.

8.7 Water Consumption Reduction Targets (E3-3)

Relevant SDGs: SDG 6: Clean Water and Sanitation.

Water consumption reduction targets are goals set by an organisation to reduce water consumption and promote efficient water management and involves the following steps:

1. **Establish a baseline for water consumption**: Collect data on historical water consumption within the organisation by location by referring to sustainability reports, water consumption records, and water management plans.
2. **Determine the target reduction percentage**: Set a target reduction percentage based on your organisation's water management goals and sustainability commitments.
3. **Calculate the target water consumption**: Multiply the baseline water consumption by (1 – target reduction percentage) to determine the target water consumption.
4. **Implement measures to achieve the targets**: Implement water conservation initiatives, technological improvements, and behaviour change programs to reduce water consumption and achieve the set targets.
5. **Estimate financial impacts**: Estimate potential cost savings and positive impacts reparation where possible.

8.8 Coastal Erosion Rates (E3-2, E3-5)

Relevant SDGs: SDG 11: Sustainable Cities and Communities, SDG 14: Life Below Water.

Measurement of the rate at which coastal areas are eroding or losing land due to your organisation's activities involves:

1. **Collect historical data**: Gather historical data on coastal land area for a specific region or area of interest. This can include maps, aerial photographs, satellite imagery, field surveys, coastal monitoring data, and coastal management reports.
2. **Determine reference points**: Identify reference points or landmarks that can serve as stable reference features over time.
3. **Compare past and present land area**: Compare the land area measurements from different time periods using the reference points. This can be done by overlaying maps or images and measuring the change in land area.

4. **Calculate erosion rate**: Divide the change in land area by the time interval to calculate the erosion rate. The erosion rate is typically expressed in meters per year.

5. **Estimate financial impacts**: Estimate potential costs of mitigation or reparation where possible, as well as possible damages and insurance claims.

8.9 Water Scarcity Risk Assessment (E3-1, E3-5)

Relevant SDGs: SDG 6: Clean Water and Sanitation.

Evaluation of the potential risks and vulnerabilities related to water scarcity risk in a particular region or area involves the following steps:

1. **Define the study area**: Determine the geographic scope of the assessment, which could be a specific region, watershed, or country.

2. **Collect relevant data**: Gather data on water availability, population demands, climate conditions, and water management practices. This may include information on precipitation patterns, groundwater levels, population growth projections, and water allocation policies sourced from water resource assessments, hydrological models, regional water plans, the WWF-WRF website, and WRI Aqueduct Atlas.

3. **Assess water availability**: Analyse the current and projected water availability in the study area. This can be done using hydrological models, water balance calculations, or historical data on water sources.

4. **Evaluate water demand**: Estimate the current and projected water demand based on population growth, economic activities, and water consumption patterns.

5. **Analyse water scarcity indicators**: Calculate indicators such as water stress index, water withdrawal-to-availability ratio, and per capita water availability to assess the level of water scarcity.

6. **Identify vulnerabilities and risks**: Identify vulnerable sectors, communities, or ecosystems that may be particularly affected by water scarcity. Assess the potential socio-economic and environmental risks associated with water scarcity.

7. **Develop mitigation strategies**: Based on the assessment results, develop strategies, and measures to mitigate water scarcity risks. These

may include water conservation programs, water reuse initiatives, and improved water management practices.

8. **Estimate financial impacts**: Estimate potential costs of mitigation or reparation where possible, as well as possible damages and insurance claims.

8.10 Fish Stock Health (E3-1, E3-2)

Relevant SDGs: SDG 14: Life Below Water.

Measurement of the health and sustainability of fish populations in a specific aquatic ecosystem or fishery involves the following steps:

1. **Data collection**: Collect data on fish catch, fishing effort, and biological characteristics of the target fish species. This may include information on catch-per-unit-effort (CPUE), size composition, age structure, reproductive capacity and related information from fishery assessments, fishery management reports, and scientific research.
2. **Calculate key indicators**: Calculate indicators such as spawning stock biomass, fishing mortality rate, and recruitment rate. These indicators provide insights into the status of fish stocks and their ability to sustain population levels.
3. **Reference points**: Establish reference points or thresholds for each indicator, such as maximum sustainable yield (MSY) or biomass reference points or, better still, planetary boundaries. These reference points serve as benchmarks for determining whether fish stocks are overfished or experiencing sustainable harvest levels.
4. **Compare indicators with reference points**: Compare the calculated indicators with the established reference points to assess the health of fish stocks. Determine whether stocks are within safe limits or if corrective measures are needed.
5. **Adaptive management**: Use the assessment results to inform fisheries management decisions, such as setting catch limits, implementing fishing gear restrictions, or establishing marine protected areas to promote stock recovery.

8.11 Water Footprint (E3-3, E3-4)

Relevant SDGs: SDG 6: Clean Water and Sanitation.

Measurement of the total volume of freshwater consumed directly or indirectly to produce goods or services involves the following steps:

1. **Identify water-consuming activities**: Identify and categorise the activities or processes that contribute to water consumption in the production of goods or services. This can include agricultural activities, industrial processes, and domestic water use.
2. **Determine water consumption per activity**: Estimate the volume of water consumed for each activity. This can be done through water consumption data, water balance calculations, water use surveys, water footprint assessments, life-cycle assessments, and product sustainability reports.
3. **Account for indirect water consumption**: Consider the water embedded in the supply chain of goods or services. This includes the water used in the production of raw materials, transportation, and packaging.
4. **Incorporation of location-specific factors**: Consider location-specific factors that may impact water quality, such as local environmental conditions, geographical characteristics, and regional regulations.
5. **Normalise and aggregate water consumption**: Normalise the water consumption data to a common unit, such as litres or cubic meters. Aggregate the water consumption across all activities to calculate the total water footprint.
6. **Interpret and analyse results**: Analyse the water footprint results to identify hotspots, areas of high-water consumption, and potential areas for improvement. This can guide water management strategies and promote water-efficient practices.

8.12 Coral Reef Health Index (E3-1)

Relevant SDGs: SDG 14: Life Below Water.

Measurement of the coral reef health index (CRHI), including the overall health and condition of coral reefs, involves the following steps:

1. **Select key indicators**: Identify a set of key indicators that represent different aspects of coral reef health, such as coral cover, species diversity, abundance of reef fish, and water quality parameters.
2. **Collect data**: Collect data on the selected indicators through field surveys, coral reef monitoring programs, reef health assessments, marine

research, or remote sensing techniques. This may involve underwater surveys, coral health assessments, fish counts, and water quality measurements.

3. **Assign indicator scores**: Assign scores to each indicator based on predetermined criteria and reference values. For example, coral cover may be scored based on the percentage of live coral observed, and fish diversity may be scored based on the number of species identified.

4. **Weight the indicators**: Assign weights to each indicator based on their relative importance in determining overall coral reef health. These weights can be determined through expert consensus or stakeholder consultations.

5. **Calculate the CRHI**: Multiply each indicator score by its assigned weight and sum the weighted scores to obtain the overall coral reef health index. The index provides a quantitative measure of the overall health and condition of the coral reef ecosystem.

9

E-4 Biodiversity

In this chapter: Assessment of biodiversity impacts and actions to prevent and mitigate negative impacts. Adaption to a sustainable economy. Risks and opportunities related to biodiversity impacts and their financial effects. Common indicators and metrics.

Biodiversity and ecosystems are fundamental to the health and resilience of our planet. A study by the Green Finance Institute[1] found that the gradual impacts of biodiversity loss on the economy are as detrimental or more so than climate change in the near-term. Recognising the importance of protecting and conserving biodiversity, comprehensive ESG reporting systems provide a means to assess and disclose the impacts of human activities on these vital natural resources. This methodology aims to establish a robust framework for assessing an organisation's effects on biodiversity and ecosystems and disclosing actions taken to prevent negative impacts and promote conservation.

The objective of this methodology is to ensure that sustainability statements provide users with a clear understanding of how your organisation affects biodiversity and ecosystems by focusing on the assessment of both material positive and negative impacts, enabling stakeholders to grasp the full picture of contributions to biodiversity conservation. By evaluating these impacts, stakeholders can gain insights into the potential threats and opportunities arising from your organisation's activities.

By following this methodology, your organisation can ensure a comprehensive and transparent assessment of its biodiversity impacts and dependencies, which demonstrates its commitment to biodiversity conservation, manage

[1] Assessing the Materiality of Nature-Related Financial Risks for the UK. Green Finance Institute, 2024.

associated risks, and seize opportunities for sustainable growth. Additionally, this methodology facilitates comparability and consistency in biodiversity reporting across organisations, enhancing stakeholder understanding and decision-making related to biodiversity and ecosystem conservation.

In the following chapter:

- None of the data points are applicable to your ESG report if the topic is deemed not material, with the exception of IRO1 that are to be disclosed irrespective of the outcome of your materiality assessment,
- Not all DPs are to be reported if the undertaking has not adopted the respective policies, implemented the respective actions, or set the respective targets in relation to a sustainability matter that has been assessed to be material.
- Organisations with less than 750 employees during the financial year of their reporting may omit all data points in this chapter for the first two years of preparation, unless deemed to be material.
- Some data points may be phased-in for the first three years, Annex 2: List of Phased-in Disclosure Requirements.

ESRS E4: BIODIVERSITY & ECOSYSTEMS		
GOVERNANCE & STRATEGY	**IMPACTS, RISKS, & OPPORTUNITIES**	**METRICS & TARGETS**
• E4-1: *Transition plan* for biodiversity and ecosystems INCLUDES CROSS-CUTTING DISCLOSURES: • SBM 3: Resilience of *strategy* **and** *business model* • IRO-1: Description of *processes to identify and assess* material biodiversity- and ecosystem-related impacts, risks, and opportunities	• E4-2**:** *Policies* related to biodiversity and ecosystems • E4-3: *Action plans* and *resources* in relation to biodiversity and ecosystems policies	• E4-4: *Targets and tracking effectiveness* related to biodiversity and ecosystems • E4-5: *Impact metrics* related to biodiversity and ecosystems change • E4-6: *Potential financial effects* from biodiversity- and ecosystem-related risks and opportunities

Before diving into detailed metrics and targets, begin with describing your materiality assessment and organisational governance initiatives that directly interact with biodiversity and ecosystems, most importantly, the processes and methodologies to identify and assess impacts, risks, and opportunities related to biodiversity and ecosystems issues (see Chapter 3, Single and Double Materiality Assessment and 9.1, Biodiversity and Ecosystems Assessment Methodology).

Table 66 Water and marine assessment indicators relating to impacts, risks, and opportunities.

ESRS	DR	Paragraph	Name	Data type	Tips and tricks	Related CDR	VSMEs	GRI
E4	SBM-3	16 a	List of material sites in own operation	Narrative	If deemed to be material, combine EU Nature Restoration Plan[2] and SDGs with TNFD (or similar), ENCORE,[3] EXIOBASE,[4] materiality assessment, biodiversity, and ecosystem assessment methodology and MDR for policies, actions and targets – for more information, see EU Business for Biodiversity[5] and EU Finance for Biodiversity[6]			
E4	SBM-3	16 a i)	Activities related to sites located in or near biodiversity-sensitive areas negatively affect these areas where conclusions or necessary mitigation measures have not been implemented or are ongoing	Narrative		SFDR		GRI 304-1
E4	SBM-3	16 a ii)	Breakdown of material sites located in or near biodiversity-sensitive area	Narrative				
E4	SBM-3	16 a iii)	Disclosure of biodiversity-sensitive areas impacted	narrative				
E4	SBM-3	16 b	Material negative impacts with regard to land degradation, desertification, or soil sealing have been identified	Semi-narrative		SFDR		
E4	SBM-3	16 c	Own operations affect threatened species	semi-narrative		SFDR		
E4	IRO-1	17 a	Disclosure of whether and how actual and potential impacts on biodiversity and ecosystems at own site locations and in value chain have been identified and assessed	Narrative				
E4	IRO-1	17 b	Disclosure of whether and how dependencies on biodiversity and ecosystems and their services have been identified and assessed at own site locations and in value chain	Narrative				
E4	IRO-1	17 c	Disclosure of whether and how transition and physical risks and opportunities related to biodiversity and ecosystems have been identified and assessed	Narrative				
E4	IRO-1	17 d	Disclosure of whether and how systemic risks to own business model have been considered	Narrative				

E4	IRO-1	17 d	Disclosure of whether and how systemic risks to society have been considered in assessment of biodiversity- and ecosystem-related risks	Narrative				
E4	IRO-1	17 e	Disclosure of whether and how consultations with affected communities on sustainability assessments of shared biological resources and ecosystems have been conducted	Narrative				
E4	IRO-1	17 e (i)	Disclosure of whether and how specific sites, raw materials production, or sourcing with negative or potential negative impacts on affected communities	Narrative				
E4	IRO-1	17 e (ii)	Disclosure of whether and how communities were involved in materiality assessment	Narrative				
E4	IRO-1	17 e (iii)	Disclosure of whether and how negative impacts on priority ecosystem services of relevance to affected communities may be avoided	Narrative				
E4	IRO-1	17 e (iii)	Disclosure of plans to minimise unavoidable negative impacts and implement mitigation measures that aim to maintain value and functionality of priority services	Narrative			V	
E4	IRO-1	18	Business model(s) has been verified using a range of biodiversity and ecosystems scenarios, or other scenarios with modelling of biodiversity- and ecosystem-related consequences, with different possible pathways	Narrative			V	

[2] EU Nature Restoration Plan: https://environment.ec.europa.eu/topics/nature-and-biodiversity/nature-restoration-law_en

[3] ENCORE: Source: https://encore.naturalcapital.finance/en

[4] EXIOBASE: Source: https://www.exiobase.eu/

[5] EU Business for Biodiversity: https://dopa.jrc.ec.europa.eu/kcbd/actions-tracker/#Business%20for%20biodiversity

[6] EU Finance for Biodiversity: https://dopa.jrc.ec.europa.eu/kcbd/dashboard/#Financing%20for%20biodiversity

(Continued)

Table 66 (*Continued*).

ESRS	DR	Paragraph	Name	Data type	Tips and tricks	Related CDR	VSMEs	GRI
E4	IRO-1	18 a	Disclosure of why considered scenarios were taken into consideration	Narrative			V	
E4	IRO-1	18 b	Disclosure of how considered scenarios are updated according to evolving conditions and emerging trends	Narrative			V	
E4	IRO-1	18 c	Scenarios are informed by expectations in authoritative intergovernmental instruments and by scientific consensus	Semi-narrative			V	
E4	IRO-1	19a	Undertaking has sites located in or near biodiversity-sensitive areas	Semi-narrative			V	GRI 304-1
E4	IRO-1	19a	Activities related to sites located in or near biodiversity-sensitive areas negatively affect these areas by leading to deterioration of natural habitats and habitats of species and to disturbance of species for which protected area has been designated	Semi-narrative				
E4	IRO-1	19b	Description of how it has been concluded that it is necessary to implement biodiversity mitigation measures	Semi-narrative				

Source: CDR (EU) 2023/2772 supplementing Directive 2013/34/EU, *GRI 1, 2, 3: 11, 12, 13, 14, 306, 101, 201, 202, 203, 204, 205, 206, 207,301, 302, 303, 304, 305, 306, 308,401, 402, 403, 404, 405, 406, 407, 408, 409, 410, 411, 413, 414, 415, 416, 417, 418*, and the author.

9.1 Biodiversity and Ecosystem Assessment Methodology

The methodology for assessing biodiversity impacts and dependencies aims to provide a comprehensive understanding of an organisation's relationship with biodiversity and ecosystems. Modified from the Taskforce on Nature-based Financial Disclosures (TNFD), this methodology enables the undertaking to evaluate its material positive and negative impacts, disclose actions taken to prevent or mitigate negative impacts, and outline its plans and capacity to adapt its strategy and business models in line with biodiversity conservation objectives. The methodology consists of the following components:

9.1.1 Assess biodiversity impacts

Identify the direct and indirect ways in which your organisation interacts and/or affects biodiversity and ecosystems. This includes impacts on indigenous peoples, production and sourcing of raw materials in sensitive areas (for example, habitats for species on the IUCN Red List of Threatened Species, the Birds and Habitats Directive or on national lists of protected species), and the quantification of such raw materials by weight or other relevant characteristics. Begin with establishing a biodiversity baseline[7] and then quantify the material positive and negative impacts in terms of changes to species diversity, habitat quality, ecosystem functions, and ecological resilience. Consider the potential impacts throughout the value chain, including sourcing, production, distribution, and disposal of products or services.

Table 67 Biodiversity areas dependencies and impacts.

Ecosystem service description…	**Actual or potential dependencies**	
…	Change of functionality	Financial loss
…	Limited, moderate, or significant	Limited, moderate, or significant
…	…	…

[7] For further information, see Gullison, R.E., J. Hardner, S. Anstee, M. Meyer. 2015, *Good Practices for the Collection of Biodiversity Baseline Data.*

Site location	Threatened species, protected area, key biodiversity areas	Actual or potential impacts			
...		Frequency of occurrence	Speed of impact	Severity of impact	Potential for mitigation
...		High, medium, or low	<1 year, 1–3 years, or >3 years	High, medium, or low	High, medium, or low
...		...	...	...	...

Source: CDR (EU) 2023/2772 supplementing Directive 2013/34/EU.

9.1.2 Actions to prevent or mitigate negative impacts

Document and disclose the measures taken by your organisation to prevent or mitigate material negative impacts on biodiversity and ecosystems, within appropriate and substantiated thresholds and metrics.[8] Examples of actions may include implementing biodiversity conservation plans, restoring degraded habitats, minimising pollution, and promoting sustainable land and resource management practices. Highlight donations, contributions, and collaborations with local communities, NGOs, governmental agencies, and stakeholders to address biodiversity conservation, habitat restoration, sustainable land and resource management, and other environmental initiatives, showcasing outcomes and impacts. Provide evidence of the outcomes and effectiveness of these actions, highlighting success stories, challenges faced, and lessons learned.

Table 68 Biodiversity actions and targets.

Type of target according to mitigation hierarchy	**Baseline value and year**	**Target value and geographical scope**				**Connected policy or legislation**
		2025	**2030**	**By 2050**		
Avoidance						
Minimisation						
Rehabilitation and restoration						
Compensation or offsets						

Source: CDR (EU) 2023/2772 supplementing Directive 2013/34/EU.

[8] For further details on Biodiversity Metrics, see Lammerant J. et al., 2022, *Assessment of Biodiversity Measurement Approaches for Businesses and Financial Institutions, Update Report 4* on behalf of the EU Business @ Biodiversity Platform.

9.1.3 Adapting strategy and business models

Outline your organisation's plans and capacity to adapt its strategy and business models to align with biodiversity conservation objectives. Consider the integration of biodiversity considerations into corporate policies, decision-making processes, and risk management frameworks (SBM3 and IRO1). Address your organisation's commitment to supporting global and regional biodiversity strategies and frameworks, such as the Post-2020 Global Biodiversity Framework and the EU Biodiversity Strategy.

9.1.4 Managing material risks and opportunities

In parallel or combination with your **Single or Double Materiality Assessment (Chapter 3)**, identify and assess the nature, type, and extent of your organisation's material risks and opportunities related to its impacts or dependencies on biodiversity and ecosystems. Design and implement robust management practices to address these risks and capitalise on opportunities, considering aspects such as regulatory compliance, stakeholder engagement, and biodiversity-related innovation. Provide transparency on how the undertaking monitors, evaluates, and manages biodiversity-related risks and opportunities.

9.1.5 Financial effects

Evaluate the short-, medium-, and long-term financial effects of material risks and opportunities arising from your organisation's impacts and dependencies on biodiversity and ecosystems.[9] Consider the potential costs associated with negative impacts, such as fines, litigation, and reputational damage, as well as the potential benefits of positive impacts, such as enhanced brand reputation, access to new markets, and stakeholder trust.

[9] For further information, see NGFS, 2023, *Recommendations toward the development of scenarios for assessing nature-related economic and financial risks.*

Table 69 Indictors and metrics relating to biodiversity and ecosystems, and potential financial effects.

Note that E4-6 is subject to phase-in provisions. See Annex 2: List of Phased-in Disclosure Requirements.

ESRS	DR	Paragraph	Name	Data type	Tips and tricks	Related CDR	VSMEs	GRI
ESRS 2	E4	62	Disclosures to be reported in case the undertaking has not adopted policies and/or actions and/or targets	Narrative				
E4	E4-1	13 a	Disclosure of resilience of current business model(s) and strategy to biodiversity- and ecosystem-related physical, transition, and systemic risks and opportunities	Narrative	For each material impact, combine EU Nature Restoration Plan[10] and SDGs with TNFD (or similar), ENCORE,[11] EXIOBASE,[12] materiality assessment, biodiversity, and ecosystem assessment methodology and MDR for policies, actions and targets – for more information see EU Business for Biodiversity[13] and EU Finance for Biodiversity[14]			
E4	E4-1	13 b	Disclosure of scope of resilience analysis along own operations and related upstream and downstream value chain	Narrative				
E4	E4-1	13 c	Disclosure of key assumptions made for biodiversity and ecosystems	Narrative				
E4	E4-1	13 d	Disclosure of time horizons used for analysis for biodiversity and ecosystems	Narrative				
E4	E4-1	13 e	Disclosure of results of resilience analysis for biodiversity and ecosystems	Narrative				
E4	E4-1	13 f	Disclosure of involvement of stakeholders for biodiversity and ecosystems	Narrative				
E4	E4-1	15	Disclosure of transition plan to improve and achieve alignment of its business model and value chain	Narrative			V	

E4	E4-1	AR 1 a	Explanation of how strategy and business model will be adjusted to improve and, ultimately, achieve alignment with relevant local, national, and global public policy goals	Narrative			V	
E4	E4-1	AR 1 b	Include information about its own operations and explain how it is responding to material impacts in its related value chain	Narrative			V	GRI 304-3-3
E4	E4-1	AR 1 c	Explanation of how biodiversity strategy interacts with transition plan	Narrative			V	
E4	E4-1	AR 1 d	Disclosure of contribution to impact drivers and possible mitigation actions following mitigation hierarchy and main path dependencies and locked-in assets and resources that are associated with biodiversity and ecosystems change	Narrative			V	GRI 304-3-3
E4	E4-1	AR 1 e	Explanation and quantification of investments and funding supporting the implementation of its transition plan	Narrative			V	
E4	E4-1	AR 1 f	Disclosure of objectives or plans for aligning economic activities (revenues, CapEx)	Narrative			V	
E4	E4-1	AR 1 g	Biodiversity offsets are part of transition plan	Narrative			V	

[10] EU Nature Restoration Plan: https://environment.ec.europa.eu/topics/nature-and-biodiversity/nature-restoration-law_en

[11] ENCORE: Source: https://encore.naturalcapital.finance/en

[12] EXIOBASE: Source: https://www.exiobase.eu/

[13] EU Business for Biodiversity: https://dopa.jrc.ec.europa.eu/kcbd/actions-tracker/#Business%20for%20biodiversity

[14] EU Finance for Biodiversity: https://dopa.jrc.ec.europa.eu/kcbd/dashboard/#Financing%20for%20biodiversity

(Continued)

Table 69 (*Continued*).

ESRS	DR	Paragraph	Name	Data type	Tips and tricks	Related CDR	VSMEs	GRI
E4	E4-1	AR 1 h	Information about how process of implementing and updating transition plan is managed	Narrative			V	
E4	E4-1	AR 1 i	Administrative, management, and supervisory bodies have approved transition plan	Narrative			V	
E4	E4-1	AR 1 j	Indication of metrics and related tools used to measure progress that are integrated in measurement approach (biodiversity and ecosystems)	Narrative			V	
E4	E4-1	AR 1 k	Indication of current challenges and limitations to draft plan in relation to areas of significant impact and actions company is taking to address them (biodiversity and ecosystems)	Narrative			V	
E4	E4-2	22	Policies to manage material impacts, risks, dependencies, and opportunities related to biodiversity and ecosystems [see ESRS 2 – MDR-P]	MDR-P	In relation to each impact, combine TNFD (or similar), Nagoya Protocol,[15] the Convention for Biological Diversity (CBD),[16] IFC Performance Standard 4, 5, 6, and 7,[17] and/or Core Principles from the Accountability			GRI 304-3-3
E4	E4-2	23 a	Disclosure on whether and how biodiversity- and ecosystem-related policies relate to climate change, land-use change (e.g., land artificialisation), freshwater-use change and sea-use change, direct exploitation, invasive alien species, pollution, impacts on the	Narrative				

			state of species (i.e., species population size, species global extinction risk), impacts the extent and condition, and dependencies of ecosystems		Framework,[18] Principle 2 'Respect for Human Rights', materiality assessment, biodiversity assessment methodology and 5.1.1 MDR-P			
E4	E4-2	23 b	Explanation of whether and how biodiversity- and ecosystem-related policy relates to material biodiversity- and ecosystem-related impacts	Narrative				
E4	E4-2	23 c	Explanation of whether and how biodiversity- and ecosystem-related policy relates to material dependencies and material physical and transition risks and opportunities	Narrative				
E4	E4-2	23 d	Explanation of whether and how biodiversity- and ecosystem-related policy supports traceability of products, components, and raw materials with significant actual or potential impacts on biodiversity and ecosystems along value chain	Narrative				

[15] Nagoya Protocol: https://www.cbd.int/abs/

[16] Convention for Biological Diversity (CBD): https://www.cbd.int/convention/

[17] IFC Performance Standard 4, 5, 6, and 7: https://www.ifc.org/wps/wcm/connect/topics_ext_content/ifc_external_corporate_site/sustainability-at- ifc/publications/publications_handbook_pps.

[18] Core Principles from the Accountability Framework, Principle 2 "Respect for Human Rights": https://accountability-framework.org/the- framework/contents/core-principles/

(*Continued*)

Table 69 (*Continued*).

ESRS	DR	Paragraph	Name	Data type	Tips and tricks	Related CDR	VSMEs	GRI
E4	E4-2	23 e	Explanation of whether and how biodiversity- and ecosystem-related policy addresses production, sourcing, or consumption from ecosystems that are managed to maintain or enhance conditions for biodiversity	Narrative				
E4	E4-2	23 f	Explanation of whether and how biodiversity- and ecosystem-related policy addresses social consequences of biodiversity- and ecosystem-related impacts	Narrative				
E4	E4-2	AR 12	Disclosure of how policy refers to production, sourcing, or consumption of raw materials	Narrative			V	
E4	E4-2	AR 12 a	Disclosure of how policy refers to policies limiting procurement from suppliers that cannot demonstrate that they are not contributing to significant conversion of protected areas or key biodiversity areas	Narrative			V	
E4	E4-2	AR 12 b	Disclosure of how policy refers to recognised standards or third-party certifications overseen by regulators	Narrative			V	

E4	E4-2	AR 12 c	Disclosure of how policy addresses raw materials originating from ecosystems that have been managed to maintain or enhance conditions for biodiversity, as demonstrated by regular monitoring and reporting of biodiversity status and gains or losses	Narrative			V	
E4	E4-2	AR 16	Disclosure of how the policy enables avoidance of negative impacts, reduces and minimises unavoidable impacts, restores and rehabilitates degraded ecosystems, and mitigates material biodiversity loss drivers as disclosed in the section, impacts, dependencies, risks, and opportunities	Narrative			V	
E4	E4-2	AR 17 a	Third-party standard of conduct used in policy is objective and achievable based on scientific approach to identifying issues and realistic in assessing how these issues can be addressed under a variety of practical circumstances	Semi-narrative			V	
E4	E4-2	AR 17 b	Third-party standard of conduct used in policy is developed or maintained through process of ongoing consultation with relevant stakeholders with balanced input from all relevant stakeholder groups with no group holding undue authority or veto power over content	Semi-narrative			V	

(Continued)

Table 69 (*Continued*).

ESRS	DR	Paragraph	Name	Data type	Tips and tricks	Related CDR	VSMEs	GRI
E4	E4-2	AR 17 c	Third-party standard of conduct used in policy encourages step-wise approach and continuous improvement in standard and its application of better management practices and requires establishment of meaningful targets and specific milestones to indicate progress against principles and criteria over time	Semi-narrative			V	
E4	E4-2	AR 17 d	Third-party standard of conduct used in policy is verifiable through independent certifying or verifying bodies, which have defined and rigorous assessment procedures that avoid conflicts of interest and are compliant with ISO guidance on accreditation and verification procedures or Article 5(2) of Regulation (EC) No 765/2008	Semi-narrative			V	
E4	E4-2	AR 17 e	Third-party standard of conduct used in policy conforms to ISEAL Code of Good Practice	Semi-narrative			V	
E4	E4-2	24 a	Biodiversity and ecosystem protection policy covering operational sites owned, leased, managed in or near protected area or biodiversity-sensitive area outside protected areas has been adopted	Semi-narrative				

E4	E4-2	24 b	Sustainable land or agriculture practices or policies have been adopted	Semi-narrative		SFDR		
E4	E4-2	24 c	Sustainable oceans or seas practices or policies have been adopted	Semi-narrative		SFDR		
E4	E4-2	24 d	Policies to address deforestation have been adopted	Semi-narrative		SFDR		
E4	E4-3	27	Actions and resources in relation to biodiversity and ecosystems	MDR-A	Combine TNFD (or similar), and other assessment tools with materiality assessment, biodiversity assessment methodology and 5.1.2 MDR-A and 9.3 biodiversity performance score Note: Be thorough with biodiversity offsetting documentation from verifier.			GRI 304-3-3
E4	E4-3	28 a	Disclosure on how the mitigation hierarchy has been applied with regard to biodiversity and ecosystem actions	Narrative			V	GRI 304-3-3
E4	E4-3	28 b	Biodiversity offsets were used in action plan	Semi-narrative				GRI 304-3
E4	E4-3	28 b (i)	Disclosure of aim of biodiversity offset and key performance indicators used	Narrative				GRI 304-3
E4	E4-3	28 b (ii)	Financing effects (direct and indirect costs) of biodiversity offsets	Monetary				GRI 304-3
E4	E4-3	28 b (iii)	Description of biodiversity offsets	Narrative				GRI 304-3
E4	E4-3	28 c	Description of whether and how local and indigenous knowledge and nature-based solutions have been incorporated into biodiversity- and ecosystem-related action	Narrative				
E4	E4-3	AR 20 a	Disclosure of key stakeholders involved and how they are involved, and key stakeholders negatively or positively impacted by action and how they are impacted	Narrative			V	
E4	E4-3	AR 20 b	Explanation of need for appropriate consultations and need to respect decisions of affected communities	Narrative			V	

(Continued)

Table 69 (*Continued*).

ESRS	DR	Paragraph	Name	Data type	Tips and tricks	Related CDR	VSMEs	GRI
E4	E4-3	AR 20 c	Description of whether key action may induce significant negative sustainability impacts for biodiversity and ecosystems	Narrative			V	
E4	E4-3	AR 20 d	Explanation of whether the key action is intended to be a one-time initiative or systematic practice	Narrative			V	
E4	E4-3	AR 20 e	Key action plan is carried out only by undertaking (individual action) using its resources for biodiversity and ecosystems	Semi-narrative			V	GRI 304-3
E4	E4-3	AR 20 e	Key action plan is part of a wider action plan (collective action), of which undertaking is a member for biodiversity and ecosystems	Semi-narrative			V	
E4	E4-3	AR 20 f	Additional information about project, its sponsors, and other participants for biodiversity and ecosystems	Narrative			V	
E4	E4-4	29	Tracking effectiveness of policies and actions through targets	MDR-T	Combine TNFD (or similar), and other assessment tools with materiality assessment, biodiversity assessment methodology and 5.1.4 MDR-T from 9.2 to 9.11 Tabulate by location.			GRI 304-3-3, GRI 304-3
E4	E4-4	32 a	Ecological threshold and allocation of impacts to undertaking were applied when setting target for biodiversity and ecosystems	Semi-narrative				
E4	E4-4	32 a i)	Disclosure of ecological threshold identified and methodology used to identify threshold for biodiversity and ecosystems	Narrative				

E4	E4-4	32 a ii)	Disclosure of how entity-specific threshold was determined for biodiversity and ecosystems	Narrative			
E4	E4-4	32 a iii)	Disclosure of how responsibility for respecting identified ecological threshold is allocated for biodiversity and ecosystems	Narrative			
E4	E4-4	32 b	Target is informed by relevant aspects of EU Biodiversity Strategy for 2030	Semi-narrative			
E4	E4-4	32 c	Disclosure of how the targets relate to the biodiversity and ecosystem impacts, dependencies, risks, and opportunities identified in relation to own operations and upstream and downstream value chains	Narrative			
E4	E4-4	32 d	Disclosure of the geographical scope of the targets	Narrative			
E4	E4-4	32 e	Biodiversity offsets were used in setting target	Semi-narrative			
E4	E4-4	32 f	Layer in mitigation hierarchy to which target can be allocated for biodiversity and ecosystems	Semi-narrative			
E4	E4-4	AR 22	The target addresses shortcomings related to the Substantial Contribution criteria concerning local and indigenous knowledge	Semi-narrative		V	
E4	E4-5	35	Number of sites owned, leased, or managed in or near protected areas or key biodiversity areas that undertaking is negatively affecting	Integer			GRI 304-1, GRI 304-2

(Continued)

Table 69 (*Continued*).

ESRS	DR	Paragraph	Name	Data type	Tips and tricks	Related CDR	VSMEs	GRI
E4	E4-5	35	Area of sites owned, leased, or managed in or near protected areas or key biodiversity areas that undertaking is negatively affecting	Area			V	GRI 304-1, GRI 304-2
E4	E4-5	36	Disclosure of land-use based on lifecycle assessment	Narrative			V	
E4	E4-5	38	Disclosure of metrics considered relevant (land-use change, freshwater-use change, and/or sea-use change)	Narrative				GRI 304-2
E4	E4-5	38 a	Disclosure of conversion over time of land cover	Narrative			V	GRI 304-2
E4	E4-5	38 b	Disclosure of changes over time in management of ecosystem	Narrative			V	GRI304-2
E4	E4-5	38 c	Disclosure of changes in spatial configuration of landscape	Narrative			V	GRI304-2, GRI304-3
E4	E4-5	38 d	Disclosure of changes in ecosystem structural connectivity	Narrative			V	GRI304-2, GRI304-4
E4	E4-5	38 e	Disclosure of functional connectivity	Narrative			V	GRI304-2
E4	E4-5	AR 34 a	Total use of land area	Area			V	
E4	E4-5	AR 34 b	Total sealed area	Area			V	
E4	E4-5	AR 34 c	Nature-oriented area on site	Area			V	
E4	E4-5	AR 34 d	Nature-oriented area off site	Area			V	
E4	E4-5	39	Disclosure of how pathways of introduction and spread of invasive alien species and risks posed by invasive alien species are managed	Narrative	9.4 Invasive species management		V	GRI 304-2

E4	E4-5	AR 32	Number of invasive alien species	Integer			V	
E4	E4-5	AR 32	Area covered by invasive alien species	Area			V	
E4	E4-5	40	Disclosure of metrics considered relevant (state of species)	Narrative			V	GRI304-2
E4	E4-5	40 a	Disclosure of adherence to the recommendations provided by Aligning Accounting Approaches for Nature (Align)	Narrative	If relevant, Align Accounting[19]		V	GRI 304-2
E4	E4-5	40 b	Disclosure of population size, range within specific ecosystems, and extinction risk	Narrative	9.2 Protected area coverage		V	
E4	E4-5	40 c	Disclosure of changes in number of individuals of species within specific area	Narrative			V	GRI 304-2
E4	E4-5	40 d	Information about species at global extinction risk	Narrative			V	
E4	E4-5	40 d (i)	Disclosure of threat status of species and how activities or pressures may affect threat status	Narrative			V	GRI 304-4
E4	E4-5	40 d (ii)	Disclosure of change in relevant habitat for threatened species as proxy for impact on local populations extinction risk	Narrative			V	
E4	E4-5	41 a	Disclosure of ecosystem area coverage	Narrative			V	
E4	E4-5	41 b (i)	Disclosure of quality of ecosystems relative to pre-determined reference state	Narrative			V	
E4	E4-5	41 b (ii)	Disclosure of multiple species within ecosystem	Narrative			V	
E4	E4-5	41 b (iii)	Disclosure of structural components of ecosystem condition	Narrative			V	

(Continued)

[19] https://capitalscoalition.org/project/align/

Table 69 (*Continued*).

ESRS	DR	Paragraph	Name	Data type	Tips and tricks	Related CDR	VSMEs	GRI
E4	E4-6	45 a	Disclosure of quantitative information about potential financial effects of material risks and opportunities arising from biodiversity- and ecosystem-related impacts and dependencies	Narrative	9.11 Ecosystem valuation			
E4	E4-6	45 a	Disclosure of qualitative information about potential financial effects of material risks and opportunities arising from biodiversity- and ecosystem-related impacts and dependencies	Narrative				
E4	E4-6	45 b	Description of effects considered, related impacts, and dependencies (biodiversity and ecosystems)	Narrative				
E4	E4-6	45 c	Disclosure of critical assumptions used in estimates of financial effects of material risks and opportunities arising from biodiversity- and ecosystem-related impacts and dependencies	Narrative				
E4	E4-6	AR 39	Description of related products and services at risk (biodiversity and ecosystems)	Narrative			V	
E4	E4-6	AR 39	Explanation of how time horizons are defined, financial amounts are estimated, and critical assumptions made (biodiversity and ecosystems)	Narrative			V	

Source: CDR (EU) 2023/2772 supplementing Directive 2013/34/EU, *GRI 1, 2, 3: 11, 12, 13, 14, 306, 101, 201, 202, 203, 204, 205, 206, 207,301, 302, 303, 304, 305, 306, 308,401, 402, 403, 404, 405, 406, 407, 408, 409, 410, 411, 413, 414, 415, 416, 417, 418*, and the author.

9.2 Protected Area Coverage (E4-2)

Relevant SDGs: SDG 15: Life on Land.

The assessment and evaluation of the extent of land areas protected and managed for conservation purposes (related to 8.6, Marine Protected Area Coverage) involves the following steps:

1. **Define the assessment scope**: Determine the geographic boundaries and the types of protected areas to be included, such as national parks, nature reserves, or marine protected areas using protected area databases, conservation reports, and governmental records.
2. **Compile data on protected areas**: Gather data on the location, size, and management status of protected areas from official records, governmental agencies, or conservation organisations. Also include donations, contributions, and collaborations with local communities, NGOs, governmental agencies, and stakeholders to address biodiversity conservation, habitat restoration, sustainable land and resource management, and other environmental initiatives, showcasing outcomes and impacts
3. **Calculate coverage percentage**: Calculate the percentage of the total land area that is designated as protected areas within the assessment scope.
4. **Analyse representation**: Analyse the representation of different habitat types or ecosystems within the protected area network. Assess whether key biodiversity hotspots or vulnerable ecosystems are adequately represented and protected.
5. **Identify gaps and prioritise actions**: Identify gaps in protected area coverage and prioritise actions to expand the protected area network, enhance management effectiveness, and strengthen conservation efforts.

9.3 Biodiversity Performance Score (E4-1, E4-2, E4-5)

Relevant SDGs: SDG 15: Life on Land.

The assessment and evaluation of an organisation's overall performance in biodiversity conservation and management involves the following steps:

1. **Define biodiversity criteria**: Using biodiversity performance assessments, sustainability reports, stakeholder consultations identify key indicators or criteria that represent biodiversity conservation efforts, such as habitat protection, species conservation, and ecosystem restoration.

2. **Quantify performance metrics**: Develop quantitative metrics or indicators to assess the organisation's performance in each criterion. For example, the percentage of protected habitat, the number of endangered species supported, or the extent of ecosystem restoration projects. Also include donations, contributions, and collaborations with local communities, NGOs, governmental agencies, and stakeholders to address biodiversity conservation, habitat restoration, sustainable land and resource management, and other environmental initiatives, showcasing outcomes and impacts.
3. **Assign weightings**: Assign weights to each performance metric based on their relative importance in biodiversity conservation. These weights can be determined through stakeholder consultations or expert opinions. Benchmark all weightings against peers and best practice.
4. **Calculate scores**: Evaluate the organisation's performance against each metric and assign scores based on pre-determined benchmarks or targets. Multiply the scores by the corresponding weights and sum them to obtain the biodiversity performance score.
5. **Interpret and communicate results**: Interpret the biodiversity performance score to assess the organisation's biodiversity conservation efforts. Communicate the results to stakeholders, highlighting strengths, areas for improvement, and potential strategies for enhancing biodiversity performance.

9.4 Habitat Restoration Efforts (E4-1, E4-3)

Relevant SDGs: SDG 15: Life on Land.

The assessment and evaluation of an organisation's initiatives to restore and rehabilitate degraded habitats to support biodiversity conservation involves the following steps:

1. **Define restoration goals**: Identify the scope of habitat loss or degradation through habitat restoration project reports, ecological monitoring data, and scientific research. Then define your organisation's desired ecological outcomes and objectives of habitat restoration initiatives. This may include the restoration of specific habitat types, the recovery of target species populations, or the improvement of ecosystem functioning.
2. **Quantify performance metrics**: Develop quantitative metrics or indicators to assess the organisation's performance in each aspect of

restoration and rehabilitation. For example, the percentage of regenerated habitat, the number of endangered species reintroduced, or the extent of ecosystem restoration. Also include donations, contributions, and collaborations with local communities, NGOs, governmental agencies, and stakeholders to address biodiversity conservation, habitat restoration, sustainable land and resource management, and other environmental initiatives, showcasing outcomes and impacts.

3. **Implement restoration actions**: Implement restoration activities according to established plans, which may involve habitat creation, reforestation, wetland restoration, or erosion control measures. Monitor and document the progress and outcomes of each restoration project.
4. **Assess ecological recovery**: Evaluate the success of habitat restoration by assessing indicators of ecological recovery, such as vegetation cover, species diversity, and ecosystem services. Compare the post-restoration conditions with reference sites or historical data.
5. **Monitor long-term sustainability**: Implement long-term monitoring programs to track the sustainability of restored habitats. Assess the persistence of restored habitats, the establishment of self-sustaining ecosystems, and the potential threats to restoration success.
6. **Adaptive management and learning**: Incorporate adaptive management principles into habitat restoration efforts, learning from experiences, and adjusting restoration strategies based on monitoring results. Continuously improve restoration techniques and share lessons learned with relevant stakeholders.

9.5 Species Conservation Efforts (E4-1, E4-3)

Relevant SDGs: SDG 15: Life on Land.

The assessment and evaluation of an organisation's initiatives to conserve and protect endangered or threatened species involves the following steps:

1. **Identify target species**: Using species conservation plans, population surveys, and research publications identify the endangered or threatened species that the organisation aims to protect based on their conservation status and ecological significance.
2. **Quantify performance metrics:** Develop quantitative metrics or indicators to assess the organisation's performance in each aspect of species

conservation. For example, species identification counts, seasonal densities, or population growth. Also include donations, contributions, and collaborations with local communities, NGOs, governmental agencies, and stakeholders to address biodiversity conservation, habitat restoration, sustainable land and resource management, and other environmental initiatives, showcasing outcomes and impacts.

3. **Assess conservation actions**: Evaluate the organisation's efforts to protect target species, such as habitat preservation, captive breeding programs, population monitoring, and anti-poaching measures.
4. **Measure population trends**: Monitor the population trends of target species using appropriate methods, such as population surveys, camera traps, or genetic monitoring. Assess the success of conservation actions based on population stability, recovery, or increase.
5. **Evaluate conservation impact**: Assess the impact of conservation efforts on target species by analysing key indicators, such as habitat occupancy, breeding success, genetic diversity, or range expansion.
6. **Collaborate with stakeholders**: Foster collaborations with relevant stakeholders, including conservation organisations, research institutions, and local communities, to enhance species conservation efforts and share knowledge and resources.

9.6 Invasive Species Management (E4-2)

Relevant SDGs: SDG 15: Life on Land.

The assessment and evaluation of an organisation's efforts to prevent, control, and manage invasive species that pose a threat to biodiversity involves the following steps:

1. **Identify target invasive species**: By using invasive species management plans, monitoring reports, and scientific research, identify the invasive species that are of concern within the organisation's operational or ecological context. These species should be known to cause significant ecological or economic harm.
2. **Assess prevention measures**: Evaluate the effectiveness of measures implemented to prevent the introduction and spread of invasive species. This may include quarantine protocols, monitoring and early detection systems, and risk assessments for new introductions.

3. **Evaluate control and eradication efforts**: Assess the organisation's initiatives to control and eradicate established invasive species populations. This can involve the use of physical removal, biological control methods, or habitat restoration to suppress the invader's impact.

4. **Monitor and track progress**: Implement monitoring programs to track the abundance and distribution of invasive species over time. Regularly assess the success of management actions in reducing invasive species impacts.

5. **Adapt and improve strategies**: Based on monitoring results and ongoing research, adapt management strategies to address emerging invasive species threats. Continuously learn from experiences and share best practices within the organisation and with relevant stakeholders.

9.7 Ecological Connectivity (E4-1, E4-3)

Relevant SDGs: SDG 15: Life on Land.

The assessment and evaluation of the functional connectivity of habitats to facilitate species movement and ecological processes involves the following steps:

1. **Identify and map target habitats**: Begin with reviewing habitat connectivity studies, landscape ecology research, and geographic information systems (GIS) data to identify the key habitats or ecological corridors that need to be connected to support species movement and genetic exchange that are related to your organisation in some way. Then map the target habitats and identify potential barriers, such as roads, infrastructure, or fragmented landscapes, that hinder connectivity.

2. **Quantify connectivity metrics**: Use spatial analysis tools to measure connectivity metrics, such as patch size, isolation, proximity, and landscape permeability. These metrics can be calculated using indices such as the Integral Index of Connectivity[20] (aka Landscape Connectivity Index) or the Circuitscape model.[21] Also include donations, contributions, and collaborations with local communities, NGOs, governmental

[20] Pascual-Hortal, L., Saura, S. Comparison and development of new graph-based landscape connectivity indices: towards the priorization of habitat patches and corridors for conservation. Landscape Ecol 21, 959–967 (2006). https://doi.org/10.1007/s10980-006-0013-z

[21] For more information, see https://circuitscape.org/

agencies, and stakeholders to address biodiversity conservation, habitat restoration, sustainable land and resource management, and other environmental initiatives, showcasing outcomes and impacts.

3. **Evaluate functional connectivity**: Assess the functional connectivity by considering species-specific movement patterns, dispersal abilities, and habitat preferences. This can involve modelling species distribution and movement using species distribution models or movement ecology models.

4. **Identify conservation actions**: Based on the connectivity assessment, identify priority areas for habitat restoration, creation of wildlife corridors, or mitigation measures to improve ecological connectivity.

9.8 Sustainable Agriculture Practices (E4-1, E4-2, E5-1)

Relevant SDGs: SDG 2: Zero Hunger, SDG 15: Life on Land.

The assessment and evaluation of sustainable agricultural practices that promote biodiversity conservation and minimise negative impacts on ecosystems involves the following steps:

1. **Identify key sustainability criteria**: By using sustainable agriculture guidelines, certification schemes, farm assessments, and related research, define the criteria that represent sustainable agriculture practices, such as soil health, water management, biodiversity conservation, and pesticide use reduction in your organisation's operational locations.

2. **Quantify performance indicators**: Develop quantitative indicators to assess the organisation's performance in each sustainability criterion. For example, the adoption of organic farming practices, the use of cover crops, or the implementation of agroforestry systems. Also include donations, contributions, and collaborations with local communities, NGOs, governmental agencies, and stakeholders to address biodiversity conservation, habitat restoration, sustainable land and resource management, and other environmental initiatives, showcasing outcomes and impacts.

3. **Assess compliance and adoption**: Evaluate the extent to which the organisation complies with sustainable agriculture standards and has adopted the identified practices. This can involve on-site inspections, interviews with farmers, and verification of documentation.

4. **Calculate performance scores**: Assign scores to the organisation's performance against each sustainability indicator, reflecting the level of compliance or adoption. Multiply the scores by pre-determined weights and sum them to obtain the sustainable agriculture performance score.

5. **Identify improvement opportunities**: Identify areas where the organisation can improve sustainable agriculture practices. Provide recommendations and support for implementing best practices, promoting knowledge sharing, and fostering collaborations with farmers and stakeholders.

9.9 Sustainable Fisheries Practices (E4-2)

Relevant SDGs: SDG 14: Life Below Water.

The assessment and evaluation of an organisation's practices and policies to ensure the sustainability of fisheries and minimise negative impacts on marine biodiversity involves the following steps:

1. **Define sustainability criteria**: Using fishery management plans, catch data, and scientific assessments establish criteria for sustainable fisheries, including aspects such as fishing quotas, gear selectivity, bycatch reduction, and compliance with fishing regulations.

2. **Assess compliance and practices**: Evaluate the organisation's adherence to sustainable fishing practices and regulatory requirements. This can involve analysing catch data, monitoring fishing vessel activities, and assessing gear technology.

3. **Measure fish stock status**: Monitor the status of fish stocks targeted by the organisation through stock assessments and data on population abundance, biomass, and recruitment rates. Assess whether fishing efforts are within sustainable limits.

4. **Evaluate ecosystem impacts**: Consider the broader ecosystem impacts of fishing activities, such as impacts on non-target species, habitats, and the overall marine food web. Assess the organisation's efforts to minimise negative ecosystem impacts.

5. **Promote sustainable practices**: Identify opportunities for improving sustainability in fisheries, such as promoting ecosystem-based fisheries management, supporting marine protected areas, and engaging with fishers and communities to adopt responsible fishing practices.

9.10 Conservation of Critical Ecosystems (E4-3)

Relevant SDGs: SDG 15: Life on Land.

The assessment and evaluation of an organisation's initiatives to conserve and protect critical ecosystems with high biodiversity value involves the following steps:

1. **Identify critical ecosystems**: By using ecosystem assessments, biodiversity surveys, conservation reports, and relevant research, identify ecosystems of high biodiversity value that are vulnerable to degradation or loss, such as tropical rainforests, coral reefs, or wetlands. Within each ecosystem, identify key indicators related to biodiversity value, for example, extent and density of rare or endangered species, conservation or regeneration impacts, etc.
2. **Quantify performance indicators**: Develop quantitative indicators to assess the organisation's performance in each ecosystem criterion. Also include donations, contributions, and collaborations with local communities, NGOs, governmental agencies, and stakeholders to address biodiversity conservation, habitat restoration, sustainable land and resource management, and other environmental initiatives, showcasing outcomes and impacts.
3. **Assess conservation efforts**: Evaluate the organisation's efforts to protect and manage critical ecosystems, including measures to prevent habitat loss, mitigate degradation, and restore ecosystem functions.
4. **Monitor ecosystem health**: Monitor key indicators of ecosystem health, such as species richness, habitat condition, and ecological processes. This can involve field surveys, remote sensing, or data from monitoring programs.
5. **Evaluate effectiveness**: Assess the effectiveness of conservation actions by comparing the status and trends of critical ecosystems over time, considering factors such as deforestation rates, coral reef bleaching events, or wetland area loss.
6. **Engage stakeholders**: Engage with local communities, indigenous groups, and other stakeholders to involve them in conservation efforts, build capacity, and support sustainable livelihoods linked to the conservation of critical ecosystems.

9.11 Ecosystem Services Valuation (E4-6)

Relevant SDGs: SDG 15: Life on Land.

The assessment and evaluation of the economic value of ecosystem services provided by biodiversity-rich habitats involves the following steps:

1. **Identify ecosystem services**: Using an appropriate ecosystem services framework that encompasses natural capital conservation, as well as ecosystem services assessments, economic valuation studies, and environmental impact assessments, identify the range of ecosystem services provided by biodiversity-rich habitats, such as carbon sequestration, water purification, pollination, or ecotourism.
2. **Quantify service values**: Apply economic valuation methods to estimate the economic value of ecosystem services. This can involve market-based approaches (e.g., estimating the value of ecotourism revenue) or non-market valuation techniques (e.g., contingent valuation surveys).
3. **Assess potential impacts**: Evaluate the potential impacts of your organisation's activities on the provision of ecosystem services. Consider the direct and indirect effects on the quantity and quality of services.
4. **Monitor changes in value**: Monitor changes in the value of ecosystem services over time to assess the organisation's impact on the provision of services. This can include measuring changes in visitor numbers, changes in water quality, or changes in the market value of ecosystem-related products.
5. **Promote sustainable use**: Identify opportunities for sustainable use and management of biodiversity-rich habitats to maintain the provision of ecosystem services. Implement measures to minimise negative impacts and enhance the value of ecosystem services.
6. **Estimate financial impacts**: Estimate potential costs and benefits of your organisation's activities on the provision of ecosystem services.

10

E-4 Circular Economy

In this chapter: Assessment of circular economy impacts and actions to prevent and mitigate negative impacts. Lifecycle Assessments and adaption to a sustainable economy. Risks and opportunities related to circular economy impacts and their financial effects. Common indicators and metrics.

Resource use, including the depletion of non-renewable resources and the regenerative production of renewable resources, has significant implications for the sustainability and resilience of our economies and ecosystems. As global resource consumption continues to rise, it is crucial for organisations to assess their impacts and take action to minimise negative consequences. The transition towards a circular economy, characterised by reducing waste, maximising resource efficiency, and maintaining the value of products and materials throughout their lifetime, is pivotal in ensuring long-term environmental and economic sustainability. This methodology provides a framework for evaluating the material positive and negative impacts, disclosing actions taken to prevent or mitigate negative impacts, and outlining plans to align strategies and business models with circular economy principles.

The assessment of resource use and circular economy impacts encompasses a range of considerations, such as the environmental implications of resource depletion, waste generation, and the decoupling of economic growth from material consumption. By adhering to this methodology, undertakings can comprehensively evaluate their resource use practices, identify areas for improvement, and implement measures to minimise negative impacts. Furthermore, this assessment enables undertakings to identify material risks and opportunities associated with resource dependencies and manage them effectively.

The disclosure requirements outlined in this methodology align with various global initiatives and frameworks, including the EU Biodiversity

Strategy, the Post-2020 Global Biodiversity Framework, and the European Green Deal. By adhering to these requirements, undertakings demonstrate their commitment to sustainable resource management, biodiversity conservation, and the transition to a circular economy. The financial implications of material risks and opportunities arising from resource use and circular economy impacts are also considered, ensuring that undertakings have a holistic understanding of the short-, medium-, and long-term financial effects.

- None of the data points are applicable to your ESG report if the topic is deemed not material, with the exception of IRO1 that are to be disclosed irrespective of the outcome of your materiality assessment,
- Not all DPs are to be reported if the undertaking has not adopted the respective policies, implemented the respective actions, or set the respective targets in relation to a sustainability matter that has been assessed to be material.
- Some data points may be phased-in for the first three years, Annex 2: List of Phased-in Disclosure Requirements.

TIPS AND TRICKS

An excellent example of combining TCFD and TNFD approaches: Forico Future Fibre, An Illustrative Example of Integrated TCFD + TNFD Disclosures, September 2023. forico.com.au/news/world-first-illustrative-example-of-nature-and-climate-related-financial-disclosures

ESRS E5: CIRCULAR ECONOMY		
GOVERNANCE & STRATEGY	**IMPACTS, RISKS, & OPPORTUNITIES**	**METRICS & TARGETS**
INCLUDES CROSS-CUTTING DISCLOSURES: • IRO-1: Description of *processes to identify and assess* material resource use and circular economy-related impacts, risks, and opportunities	• E5-1: *Policies* related to resource use and circular economy • E5-2: *Action plans* and *resources* in relation to resource use and circular economy policies	• E5-3: *Targets and tracking effectiveness* related to resource use and circular economy • E5-4: *Resource inflows* • E5-5: *Resource outflows* including waste and end of life management • E5-6: *Potential financial effects* from resource use and circular economy-related risks and opportunities

Before diving into detailed metrics and targets, begin with describing your materiality assessment and organisational governance initiatives that directly interact with resource use and circular economy goals, most importantly, the processes and methodologies to identify and assess impacts, risks, and opportunities related to resource use and circular economy issues (see **Chapter 3: Single and Double Materiality Assessment and 10.1 Circular Economy Assessment Methodology**).

Table 70 Resource use and circular economy assessment indicators relating to policies, impacts, risks, and opportunities.

ESRS	DR	Paragraph	Name	Data type	Tips and tricks	Related CDR	VSMEs	GRI
E5	IRO-1	11a	Disclosure of methodologies, assumptions, and tools used in the screening in order to identify actual and potential impacts, risks, and opportunities in own operations and upstream and downstream value chain	Narrative	To create policy, combine 5.1.1 MDR-P with stakeholder engagement, materiality assessment, and 10.1 circular economy assessment and 10.3 Reduction of Waste Generation by Reuse, Re-manufacture, and Recycle See also, EU Circular Economy[1].			
E5	**IRO-1**	**11 b**	**Information about the process for conducting consultations regarding resource and circular economy**	**Narrative**				
E5	IRO-1	AR 7 a	Disclosure of business units associated with resource use and circular economy material impacts, risks, and opportunities	Narrative				GRI 306-3-3
E5	IRO-1	AR 7 b	Disclosure of material resources used	Narrative				
E5	IRO-1	AR 7 c	Disclosure of material impacts and risks of staying in business as usual	Narrative				
E5	IRO-1	AR 7 d	Disclosure of material opportunities related to circular economy	Narrative				
E5	IRO-1	AR 7 e	Disclosure of material impacts and risks of transition to circular economy	Narrative				
E5	IRO-1	AR 7 f	Disclosure of stages of value chain where resource use, risks, and negative impacts are concentrated	Narrative				

ESRS 2	E5	62	Disclosures to be reported in case the undertaking has not adopted policies and/or actions and/or targets					
E5	E5-1	14	Policies to manage its material impacts, risks, and opportunities related to resource use and circular economy	MDR-P				GRI 301-3-3, GRI 306-3-3
E5	E5-1	15a	Disclosure of whether and how policy addresses transitioning away from extraction of virgin resources, including relative increases in use of secondary (recycled) resources	Narrative				
E5	E5-1	15b	Disclosure of whether and how policy addresses sustainable sourcing and use of renewable resources	Narrative				
E5	E5-1	AR 9 a	Description of whether and how policy addresses waste hierarchy for prevention, preparing for reuse, recycling, other recovery, disposal	Narrative			V	
E5	E5-1	AR 9 b	Description of whether and how policy addresses prioritisation of strategies to avoid or minimise waste over waste treatment strategies	Narrative			V	

Source: CDR (EU) 2023/2772 supplementing Directive 2013/34/EU, *GRI 1, 2, 3: 11, 12, 13, 14, 306, 101, 201, 202, 203, 204, 205, 206, 207,301, 302, 303, 304, 305, 306, 308,401, 402, 403, 404, 405, 406, 407, 408, 409, 410, 411, 413, 414, 415, 416, 417, 418*, and the author.

[1] EU Circular Economy: https://eur-lex.europa.eu/resource.html?uri=cellar:9903b325-6388-11ea-b735-01aa75ed71a1.0017.02/DOC_1&format=PDF

10.1 Circular Economy Assessment Methodology

Comprehensive resource use and circularity assessments are highly complex and require significant organisational commitment, so I will present a methodology overview that aims to provide an understanding of your organisation's relationship with resource consumption and the transition towards a circular economy.[2] This methodology and subsequent indicators will enable you, in general terms, to evaluate your organisation's material positive and negative impacts, disclose actions taken to prevent or mitigate negative impacts, and outline plans and capacity to adapt its strategy and business models in line with circular economy principles. The methodology consists of the following components:

1. **Assessing resource use impacts:**

- Identify the types of resources used by your organisation, including non-renewable resources and renewable resources, and track them to source through your value chain. During your stakeholder engagement, you will have begun mapping your supplier networks and their input flows, which should highlight potential hotspots or flag impacts that could be addressed through collaborative efforts.
- During your materiality assessment, evaluate the material positive and negative impacts associated with resource use, considering factors such as resource depletion, environmental degradation, and waste generation.
- Quantify the impacts in terms of resource consumption, waste generation and management, and the associated environmental and social consequences.

2. **Develop actions and targets to prevent or mitigate negative impacts:**

- Document and disclose the measures taken by your organisation to prevent or mitigate actual or potential material negative impacts arising from resource use and the circular economy. Apply consistent methodologies to evaluating resource thresholds and relate them to targets and actions. For example, implementing resource-efficient practices, adopting circular business models, enabling recycling and reuse, and reducing waste generation.

---

[2] For further information regarding Circular Transition Indicators, see: *The Circular Transition Indicators v4.0*, by the World Business Council for Sustainable Development, 2023.

- Provide evidence of the outcomes and effectiveness of these actions, highlighting resource savings, waste reduction, and improvements in resource efficiency.

3. **Adaptation strategy and business models:**

- Beginning with addressing the most significant resource use impacts, outline your organisation's plans and capacity to adapt its strategy and business models in line with circular economy principles, for example measures to minimise waste, integrate extended producer responsibility (EPR) management and tracking, maintain the value of products, materials, and resources, and enhance their efficient use in production and consumption.
- Introduce or adapt policies that clearly state your organisation's commitment to circular economy frameworks and principles, including waste reduction targets, closed-loop systems, and sustainable resource management practices.

4. **Managing material risks and opportunities:**

- Identify and assess the nature, type, and extent of your organisation's material risks and opportunities related to its impacts and dependencies on resource use and the circular economy.
- Implement strategies and practices to manage these risks and seize opportunities, considering aspects such as resource price volatility, regulatory changes, and stakeholder expectations.
- Provide transparency on how your organisation monitors, evaluates, and manages resource-related risks and opportunities.

5. **Financial effects:**

- Evaluate the short-, medium-, and long-term financial effects of material risks and opportunities arising from your organisation's impacts and dependencies on resource use and the circular economy.
- Consider the potential costs associated with negative impacts, such as resource scarcity, price fluctuations, and regulatory compliance, as well as the potential benefits of positive impacts, such as cost savings, resource efficiency gains, and market differentiation.

Table 71 Indictors and metrics relating to resource use and circular economy, and potential financial effects.

ESRS	DR	Paragraph	Name	Data type	Tips and tricks	Related CDR	VSMEs	GRI
E5	E5-2	19	Actions and resources in relation to resource use and circular economy	MDR-A	For actions, combine TNFD (or similar), materiality assessment, 10.1 circular economy assessment and 10.3 Reduction of Waste Generation by Reuse, Re-manufacture, and Recycle assessment methodology and 5.1.2 MDR-A			GRI 301-3-3, GRI 306-3-3, GRI 306-2
E5	E5-2	20 a	Description of higher levels of resource efficiency in use of technical and biological materials and water	Narrative			V	
E5	E5-2	20b	Description of higher rates of use of secondary raw materials	Narrative			V	
E5	E5-2	20c	Description of application of circular design	Narrative			V	
E5	E5-2	20d	Description of application of circular business practices	Narrative			V	
E5	E5-2	20e	Description of actions taken to prevent waste generation	Narrative			V	GRI 306-2
E5	E5-2	20f	Description of optimisation of waste management	Narrative			V	GRI 306-2
E5	E5-2	AR 11	Information about collective action on development of collaborations or initiatives increasing circularity of products and materials	Narrative			V	
E5	E5-2	AR 12 a	Description of contribution to circular economy	Narrative			V	
E5	E5-2	AR 12 b	Description of other stakeholders involved in collective action resource use and circular economy	Narrative			V	
E5	E5-2	AR 12 c	Description of organisation of project resource use and circular economy	Narrative			V	

E5	E5-3	23	Tracking effectiveness of policies and actions through targets	MDR-T	For targets, combine TNFD (or similar), materiality assessment, 10.1 circular economy assessment and 10.3 Reduction of Waste Generation by Reuse, Re-manufacture, and Recycle assessment methodology and 5.1.4 MDR-T			GRI 301-3-3, GRI 306-3-3
E5	E5-3	24	Disclosure of how target relates to resource use and circular economy	Narrative				
E5	E5-3	24 a	Disclosure of how target relates to increase of circular design	Narrative				
E5	E5-3	24 b	Disclosure of how target relates to increase of circular material use rate	Narrative				
E5	E5-3	24 c	Disclosure of how target relates to minimisation of primary raw material	Narrative				
E5	E5-3	24 d	Disclosure of how target relates to reversal of depletion of stock of renewable resources	Narrative				
E5	E5-3	24 e	Target relates to waste management	Semi-narrative				
E5	E5-3	24 e	Disclosure of how target relates to waste management	Narrative				
E5	E5-3	24 f	Disclosure of how target relates to other matters related to resource use or circular economy	Narrative				
E5	E5-3	25	Layer in waste hierarchy to which target relates	Semi-narrative				
E5	E5-3	26 a	Disclosure of ecological threshold identified and methodology used to identify ecological threshold relating to resource use and circular economy	Narrative			V	
E5	E5-3	26 b	Disclosure of how ecological entity-specific threshold was determined for resource use and circular economy	Narrative			V	
E5	E5-3	26 c	Disclosure of how responsibility for respecting identified ecological threshold is allocated	Narrative			V	
E5	E5-3	27	The targets being set and presented are mandatory if required by legislation	Semi-narrative				

(Continued)

Table 71 (*Continued*).

ESRS	DR	Paragraph	Name	Data type	Tips and tricks	Related CDR	VSMEs	GRI
E5	E5-4	30	Disclosure of information on material resource inflows	Narrative	10.2 Lifecycle assessment supported with material indicators 10.3 to 10.10 and cross-referenced with circularity analysis and with pollution indicators, e.g., for hazardous waste and non-hazardous waste see 7.2			GRI 306-1
E5	E5-4	31a	Overall total weight of products and technical and biological materials used during the reporting period	Mass				GRI 301-1
E5	E5-4	31b	Percentage of biological materials (and biofuels used for non-energy purposes)	Per cent				
E5	E5-4	31c	The absolute weight of secondary reused or recycled components, secondary intermediary products, and secondary materials used to manufacture the organisation's products and services (including packaging)	Mass			V	GRI 301-2
E5	E5-4	31c	Percentage of secondary reused or recycled components, secondary intermediary products, and secondary materials	Per cent				
E5	E5-4	32	Description of methodologies used to calculate data and key assumptions used	Narrative				
E5	E5-4	AR 22	Description of materials that are sourced from by-products or waste stream	Narrative			V	
E5	E5-4	AR 25	Description of how double counting was avoided and of choices made	Narrative				
E5	E5-5	35	Description of the key products and materials that come out of the organisation's production process	Narrative				
E5	E5-5	36a	Disclosure of the expected durability of the products placed on the market, in relation to the industry average for each product group	Table/per cent				

E5	E5-5	36b	Disclosure of the reparability of products	Narrative				
E5	E5-5	36c	The rates of recyclable content in products	Per cent			V	
E5	E5-5	36c	The rates of recyclable content in products packaging	Per cent			V	
E5	E5-5	40	Description of methodologies used to calculate data (resource outflows)	Narrative				GRI 306-2
E5	E5-5	37 a	Total waste generated	Mass			V	GRI 306-3
E5	E5-5	37 b	Hazardous waste diverted from disposal	Mass			V	GRI 306-4
E5	E5-5	37b(i)	Hazardous waste diverted from disposal due to preparation for reuse	Mass			V	
E5	E5-5	37b(ii)	Hazardous waste diverted from disposal due to recycling	Mass			V	
E5	E5-5	37b(ii)	Hazardous waste diverted from disposal due to other recovery operations	Mass				
E5	E5-5	37b	Non-hazardous waste diverted from disposal	Mass			V	
E5	E5-5	37b(i)	Non-hazardous waste diverted from disposal due to preparation for reuse	Mass			V	
E5	E5-5	37b(ii)	Non-hazardous waste diverted from disposal due to recycling	Mass			V	
E5	E5-5	37b(ii)	Non-hazardous waste diverted from disposal due to other recovery operations	Mass				
E5	E5-5	37 c	Hazardous waste directed to disposal	Mass				GRI 306-5
E5	E5-5	37 c(i)	Hazardous waste directed to disposal by incineration	Mass				
E5	E5-5	37 c(ii)	Hazardous waste directed to disposal by landfilling	Mass				
E5	E5-5	37 c(ii)	Hazardous waste directed to disposal by other disposal operations	Mass				

(Continued)

Table 71 (*Continued*).

ESRS	DR	Paragraph	Name	Data type	Tips and tricks	Related CDR	VSMEs	GRI
E5	E5-5	37 c	Non-hazardous waste directed to disposal	Mass				
E5	E5-5	37 c(i)	Non-hazardous waste directed to disposal by incineration	Mass				
E5	E5-5	37 c(ii)	Non-hazardous waste directed to disposal by landfilling	Mass				
E5	E5-5	37 c(ii)	Non-hazardous waste directed to disposal by other disposal operations	Mass				
E5	E5-5	37 d	Non-recycled waste	Mass		SFDR		
E5	E5-5	37 d	Percentage of non-recycled waste	Per cent		SFDR		
E5	E5-5	38	Disclosure of composition of waste	Narrative			V	GRI 306-3, GRI 306-4, GRI 306-5
E5	E5-5	38 a	Disclosure of waste streams relevant to organisation's sector or activities	Narrative				GRI 306-3, GRI 306-5
E5	E5-5	38 b	Disclosure of materials that are present in waste	Narrative				GRI 306-3
E5	E5-5	39	Total amount of hazardous waste	Mass		SFDR		GRI 306-3
E5	E5-5	39	Total amount of radioactive waste	Mass		SFDR		GRI 306-3
E5	E5-5	40	Description of methodologies used to calculate data (waste generated)	Narrative				GRI 306-2, GRI 306-3, GRI 306-4, GRI 306-5

E5	E5-5	AR 28	Disclosure of its engagement in product end-of-life waste management	Narrative			V	GRI 306-2
E5	E5-6	43 a	Disclosure of quantitative information about potential financial effects of material risks and opportunities arising from resource use and circular economy-related impacts	Narrative				GRI 306-2
E5	E5-6	43 a	Disclosure of qualitative information of potential financial effects of material risks and opportunities arising from resource use and circular economy-related impacts	Narrative				
E5	E5-6	43 b	Description of effects considered and related impacts for resource use and circular economy	Narrative				
E5	E5-6	43 c	Disclosure of critical assumptions used in estimates of financial effects of material risks and opportunities arising from resource use and circular economy-related impacts	Narrative				
E5	E5-6	AR 35	Description of related products and services at risk for resource use and circular economy	Narrative			V	
E5	E5-6	AR 35	Explanation of how time horizons are defined, financial amounts are estimated, and critical assumptions made for resource use and circular economy	Narrative			V	

Source: CDR (EU) 2023/2772 supplementing Directive 2013/34/EU, *GRI 1, 2, 3: 11, 12, 13, 14, 306, 101, 201, 202, 203, 204, 205, 206, 207,301, 302, 303, 304, 305, 306, 308,401, 402, 403, 404, 405, 406, 407, 408, 409, 410, 411, 413, 414, 415, 416, 417, 418*, and the author.

10.2 Lifecycle Assessment (LCA) and Circularity Indicators (E5-1, E5-4, E5-5, E5-6)

Relevant SDGs: SDG 12: Responsible Consumption and Production.

Lifecycle assessments (LCA),[3] for example ISO 14040:2006, are highly complex and require significant organisational commitment to undertake. Following is a brief summary of the LCA process and tracking material flows as indicators of activity or practice in the context of ESG reporting. Completion of the LCA is beyond the scope of this book and expert assistance is recommended; these indicators[4] will therefore focus on specific terms relevant to LCAs, the tracking of circular material inflow and outflow, and their financial impacts.[5,6]

1. **Application of LCA**: The organisation determines and discloses the extent to which it applies LCA to all of its goods and services. An LCA study involves a thorough inventory of the energy and materials that are required across the entire value chain of a product, process, or service, and calculates the corresponding emissions to the environment. LCA thus assesses cumulative potential environmental impacts from resource extraction to end-of-life management, commonly referred to as cradle to grave. The aim is to document and improve the overall environmental profile of the product or service by serving as a holistic baseline upon which carbon footprints can be accurately compared. LCA data and studies are normally the responsibility of product/service engineering and design functions, who typically hold information related to circularity practices and material flow tracking.

2. **Circular material inflow**: As part of the LCA, an organisation assesses and discloses whether the resources entering the company, such as materials, parts, or products, are renewable or non-virgin. This indicates

---

[3] Your LCA approach must be based on EU laws covering the use of the Environmental Footprint methods to measure and communicate the life cycle environmental performance of products and organisations. For further information, see: Commission Recommendation C92021) 9332 final of 16.12.2021 and Annexes 1–4.

[4] For further information, see: United Nations Environment Programme (UNEP), 2016, *Global Guidance for life cycle impact assessment indicators*, ISBN: 978-92-807-3630-4.

[5] For further information regarding social impacts relating to life cycle assessments, see: United Nations Environment Programme (UNEP), 2020, *Guidelines for Social Life Cycle Assessment of Products and Organizations.*

[6] For further information regarding land-use relating to life cycle assessments, see: *Land-use related environmental indicators for Life Cycle Assessment.* JRC Technical report.

the degree to which the organisation relies on sustainable and circular material inputs. Relevant ESG criteria must be applied to material inflows through stakeholder engagement and materiality assessments.

3. **Circular material outflow**: The organisation tracks and discloses the material flows leaving the company, including materials, parts, products, by-products, use of product, and waste streams. The focus is on whether these material flows are recovered or recycled, renewable, reusable, or somehow contributing to circularity and waste minimisation.

4. **Disclosure**: The organisation discloses the results of its LCA studies and provides information on the tracking of product material flows, highlighting the degree of circularity achieved. Financial impacts must be related to risks and opportunities for promoting circularity both internally and through the value chain.

10.3 Reduction of Waste Generation by Reuse, Re-manufacture, and Recycle (ESRS 2 IRO-1, E5-1, E5-2, E5-3)

Relevant SDGs: SDG 12: Responsible Consumption and Production.

The measurement and evaluation methodology for the reduction of waste generation by reused, re-manufactured, and recycled materials involves tracking the total waste generated during a reporting period. Total waste includes mineral, non-mineral, and hazardous waste treated by any waste treatment technology, excluding waste treated on-site or off-site through closed-loop recycling, reuse, or re-manufacturing processes. Waste should be measured in kilograms (kg) or metric tonnes (mt) and reported based on weight rather than volume (litres, m^3). The waste generation data should also be normalised by dividing it by the amount of net value added (NVA) generated in the same reporting period.

1. **Waste generation calculation**: Measure the total amount of waste generated during the reporting period by weighing or metering the waste at each specific business site. Exclude waste treated through closed-loop recycling, reuse, or re-manufacturing processes. Possible sources of information include:

    a. Weighing or metering waste at each specific business site.

    b. Bills from waste management companies: If it is difficult to directly measure waste generated, organisations can calculate the amount of waste generated based on bills received from waste management companies. These bills typically include the type of waste and the amount of waste managed in kg or mt.

c. Accounts payable: When facility managers or general services administrators are not present, waste management cost calculations can be found in the accounts payable.

d. Waste disposal contractors: In countries where waste treatment is required by law and waste disposal contractors are involved, information on waste managed can be obtained from the bills provided by the waste management company.

e. Sales of waste materials: If the waste generated by an entity can be sold as a raw material, relevant information can be found on invoices issued by your organisation selling waste materials (accounts receivable).

f. Operational reporting: When recycled, reused, or re-manufactured materials are returned to the processes of the reporting entity (closed-loop processes), the related figures should be collected at each business site and reported through operational reporting.

2. **Normalisation**: Divide the total waste generated by the NVA (expressed in currency, such as EUR, USD, and GBP) generated in the same reporting period. This normalisation allows for comparing waste generation relative to the economic output.

3. **Calculation of difference**: Compute the difference in waste generation between the current reporting period (t) and the previous reporting period ($t - 1$) to monitor progress and improvements in waste reduction efforts.

4. **Measurement of reused, re-manufactured, and recycled waste**: Measure the amount of waste that is reused, re-manufactured, or recycled during the reporting period. Ideally, distinguish among the three options (reuse, recycling, re-manufacturing) if possible. The measurement should be in kilograms (kg) or metric tonnes (mt).

10.4 Design for Sustainability (E5-1)

Relevant SDGs: SDG 12: Responsible Consumption and Production.

The measurement and evaluation of your organisation's integration of sustainability principles in product design, including considerations for recyclability, durability, and resource efficiency, involves the following steps:

1. **Define design criteria**: Identify the key design criteria relevant to sustainability, such as recyclability, material efficiency, or eco-friendly materials.
2. **Assess design practices**: Evaluate the organisation's design practices and processes, considering factors such as the use of lifecycle thinking, consideration of environmental impacts, or incorporation of sustainable design principles.
3. **Quantify sustainable design features**: Quantify the presence of sustainable design features in products, such as the percentage of recycled or recyclable materials used, energy efficiency improvements, or use of environmentally friendly manufacturing processes.
4. **Monitor and improve**: Continuously monitor the integration of sustainable design practices over time. Set targets for increased adoption of sustainable design principles and implement strategies such as design guidelines, employee training, or collaboration with suppliers to drive sustainable design innovation.

10.5 Supply Chain Transparency (E5-1)

Relevant SDGs: SDG 12: Responsible Consumption and Production.

The measurement and evaluation of your organisation's transparency and traceability of its supply chain, including the visibility of suppliers and their adherence to sustainability standards, involves the following steps:

1. **Identify supply chain tiers**: Identify the key tiers or levels within your organisation's supply chain (see Chapter 2: Stakeholder Engagement), including direct suppliers, sub-tier suppliers, and raw material sources.
2. **Assess supplier transparency**: Evaluate the transparency of suppliers, including their disclosure of relevant sustainability information, adherence to sustainability standards, or participation in certification programs.
3. **Quantify supply chain transparency**: Quantify the percentage of suppliers that demonstrate transparency and compliance with sustainability standards within each supply chain tier.

4. The formula for quantifying supply chain transparency is as follows:

   Supply Chain Transparency Percentage = (Number of Transparent and Compliant Suppliers/Total Number of Suppliers) × 100

5. **Calculate the supply chain transparency index:** Calculate the supply chain transparency index as the weighted average of the transparency percentages for each supply chain tier, reflecting the relative importance of each tier.

6. **Monitor and improve:** Continuously monitor the supply chain transparency index over time. Engage with suppliers to improve transparency and promote sustainable practices. Establish mechanisms for monitoring supplier performance and conducting regular audits.

10.6 Eco-Design Integration (E5-1)

Relevant SDGs: SDG 12: Responsible Consumption and Production.

The measurement and evaluation of the integration of eco-design principles in the organisation's product development process, focusing on reducing environmental impacts throughout the product lifecycle, involves the following steps:

1. **Identify eco-design criteria**: Identify the key eco-design criteria relevant to the organisation's products, such as energy efficiency, material selection, or recyclability.

2. **Assess eco-design practices**: Evaluate the organisation's practices and processes related to eco-design, including the consideration of environmental impacts, use of lifecycle assessment methodologies, or integration of sustainable design principles.

3. **Quantify eco-design features**: Quantify the presence of eco-design features in products, such as energy-efficient design improvements, use of environmentally friendly materials, or reduction of hazardous substances.

4. **Monitor and improve**: Continuously monitor the integration of eco-design practices over time. Set targets for increased adoption of eco-design principles and implement strategies such as design guidelines, training programs, or collaboration with suppliers to drive eco-design innovation.

10.7 Closed-Loop Supply Chain Integration (E5-1)

Relevant SDGs: SDG 12: Responsible Consumption and Production.

The measurement and evaluation of your organisation's integration of closed-loop supply chain practices, which involve the recovery, recycling, or reuse of products and materials, involves the following steps:

1. **Identify closed-loop practices:** Identify the key closed-loop practices relevant to the organisation's operations, such as product take-back programs, material recovery, or re-manufacturing.
2. **Assess implementation:** Evaluate the organisation's implementation of closed-loop practices, including the availability and effectiveness of take-back programs, the percentage of products or materials recycled or reused, or the integration of recycled content in products.
3. **Quantify closed-loop performance:** Measure the quantity or value of products or materials that are successfully recovered, recycled, or reused within the closed-loop supply chain. This can involve tracking the volume of returned products, recycling rates, or the incorporation of recycled content.
4. The formula for quantifying closed-loop performance is as follows:

   Closed-loop Performance = (Quantity or Value of Recovered, Recycled, or Reused Products/Total Quantity or Value of Products) × 100
5. **Monitor and optimise:** Continuously monitor the closed-loop supply chain performance over time. Identify areas for improvement, such as increasing collection rates, improving sorting and recycling processes, or exploring opportunities for product redesign to enhance recyclability.

10.8 Product Life Extension (E5-1, E5-3)

Relevant SDGs: SDG 12: Responsible Consumption and Production.

The measurement and evaluation of your organisation's efforts to extend the lifespan of products through repair, refurbishment, or upgrades, involves the following steps:

1. **Identify product categories**: Identify the key product categories relevant to the organisation's operations, such as electronics, appliances, or furniture.

2. **Assess life extension initiatives**: Evaluate the organisation's initiatives to extend the lifespan of products, including repair services, refurbishment programs, or upgrade options.

3. **Quantify impact**: Measure the number or percentage of products that undergo life extension measures, such as repairs, refurbishments, or upgrades. Track the increased lifespan of products resulting from these interventions. The formula for evaluating product life extension can be expressed as follows:

   Product Life Extension = (Extended Product Life/Original Product Life) × 100

4. **Monitor and improve**: Continuously monitor the product life extension rate over time to assess the effectiveness of life extension initiatives. Implement strategies to increase the rate, such as promoting repair services, providing spare parts, or designing products for durability and modularity.

10.9 Material Recovery and Recycling Efficiency (aka Material Circularity Rate) (E5-1, E5-3, E5-6)

Relevant SDGs: SDG 12: Responsible Consumption and Production.

The measurement and evaluation of the efficiency of material recovery and recycling processes within your organisation's operations involves the following steps:

Define material categories: Identify the key material categories relevant to the organisation's operations, such as plastics, metals, or paper.

1. **Quantify material inputs**: Measure the total amount of materials consumed by the organisation for each material category, using appropriate units (e.g., weight, volume).

2. **Assess material recovery and recycling**: Determine the amount of materials recovered and recycled by the organisation for each material category. This can involve tracking recycling rates, material recovery facilities, or recycling partnerships.

3. **Calculate the recycling efficiency**: Calculate the material recovery and recycling efficiency as the ratio of recovered and recycled materials to the total materials consumed:

   Material Recovery and Recycling Efficiency = (Recovered or Recycled Material/Total Material Input) × 100

4. **Set targets and improve**: Set targets to increase the material recovery and recycling efficiency over time. Implement strategies such as optimising sorting and recycling processes, increasing collection rates, or exploring opportunities for material substitution to enhance recyclability.

10.10 Resource Efficiency Ratio (E5-2, E5-3)

Relevant SDGs: SDG 12: Responsible Consumption and Production.

The measurement and evaluation of the efficiency of resource use within your organisation's operations, comparing the output (e.g., products or services) to the input of resources (e.g., raw materials, energy), involves the following steps:

1. **Define resource inputs and outputs**: Identify the key resources used by the organisation, such as raw materials, water, or energy, and the corresponding output in terms of products or services.
2. **Quantify resource inputs**: Measure the total amount of resources consumed by the organisation for each resource category, using appropriate units (e.g., weight, volume, or energy units).
3. **Measure output**: Determine the quantity or value of the organisation's products or services generated during a specific time period.
4. **Calculate the resource efficiency ratio**: Calculate the resource efficiency ratio as the ratio of output to input resources. This can be expressed as a percentage or as a unitless ratio:

   Resource Efficiency Ratio = (Output/Input) × 100

5. **Set targets and improve**: Set targets to increase the resource efficiency ratio over time. Implement strategies such as process optimi sation, material substitution, energy efficiency measures, or product redesign to reduce resource inputs while maintaining or increasing output.

11

S1: Own Workforce

In this chapter: Assessment of own workforce impacts and actions to prevent and mitigate negative impacts. Management of workforce dependencies and their financial effects. Risks and opportunities related to own workforce impacts and their financial effects. Common indicators and metrics.

The workforce is a vital component to your organisation, and its well-being and positive engagement are crucial for overall sustainability and success. Understanding and managing the material impacts on the workforce is essential for fostering a positive and inclusive work environment. This methodology provides a framework for assessing both the positive and negative impacts that the undertaking has on its own workforce, recognising the importance of fair employment practices, occupational health and safety, diversity and inclusion, and employee well-being.

Your organisation is responsible for implementing actions to prevent, mitigate, or remediate any negative impacts on your workforce. This includes measures to address issues such as workplace safety, employee rights, fair compensation, professional development, work–life balance, and employee engagement. By disclosing these actions and their outcomes, your organisation demonstrates commitment to creating a supportive and inclusive work environment that promotes the well-being and satisfaction of their employees.

In summary, this methodology provides a robust framework for assessing the material impacts, risks, and opportunities related to your organisation's workforce, identified as Own Workforce. By adopting this approach, your organisation can foster a positive work environment, promote fair employment practices, and maximise the potential of their staff. Additionally, the disclosure of actions taken and the financial effects enhances transparency and demonstrates your organisation's commitment to the well-being and success of its own workforce. Remember that there are more details on the support website https://esgreportpro.com along with examples of some key metrics.

In the following chapter:

- None of the data points are applicable to your ESG report if the topic is deemed not material, with the exception of IRO1 that are to be disclosed irrespective of the outcome of your materiality assessment.
- Not all DPs are to be reported if the undertaking has not adopted the respective policies, implemented the respective actions, or set the respective targets in relation to a sustainability matter that has been assessed to be material.
- Organisations with less than 750 employees during the financial year of their reporting may omit all data points in this chapter for the first one year of preparation, unless deemed to be material.
- Some data points may be phased-in for the first three years, Annex 2: List of phased-in Disclosure Requirements.

ESRS S1: OWN WORKFORCE		
GOVERNANCE & STRATEGY	**IMPACTS, RISKS, & OPPORTUNITIES**	**METRICS & TARGETS**
INCLUDES CROSS-CUTTING DISCLOSURES: • SBM-2: Views and interests of own workforce *stakeholders* • SBM-3: *Material* impacts, risks, and opportunities and their interaction with strategy and business model(s) • Information on *violations of UNGC principles and OECD guidelines* • *Lack of processes and compliance mechanisms* to monitor compliance with UN Global Compact principles and OECD Guidelines for Multinational Enterprises	• S1-1: *policies* related to own workforce • S1-2: Processes for *engaging* with own workers and workers' representatives about impacts • S1-3: Processes to *remediate negative impacts* and channels for own workers to raise concerns • S1-4: *Taking action* on material impacts on own workforce, and approaches to *mitigating* material risks and pursuing material opportunities related to own workforce, and effectiveness of those actions • E5-2: *Action plans* and *resources* in relation to resource use and circular economy policies	• S1-5: *Targets* and *tracking effectiveness* of policies, actions, and targets related to managing material negative impacts, advancing positive impacts, and managing material risk and opportunities • S1-6: *Characteristics* of the own employees • S1-7: Characteristics of *non-employee workers* in the organisation's • S1-8: *Collective bargaining* coverage and social dialogue • S1-9: *Diversity indicators* • S1-10: Adequate *wages* • S1-11: Social *protection* • S1-12: Persons with *disabilities* • S1-13: *Training and skills development* indicators • S1-14: *Health and safety* indicators • S1-15: *Work–life balance* indicators • S1-14: *Health and safety* indicators • S1-16: *Compensation* indicators (pay gap and total compensation) • S1-17: *Incidents*, *complaints*, and severe *human rights* impacts and incidents

Before diving into detailed metrics and targets, begin with describing your materiality assessment and organisational governance initiatives that directly interact with own workforce goals, most importantly, the processes and methodologies to identify and assess impacts, risks, and opportunities related to own workforce issues (see **Chapter 3: Single and Double Materiality Assessment and 11.1 Own Workforce Assessment Methodology**).

Table 72 Own workforce assessment indicators relating to policies, strategies, and business models.

ESRS	DR	Paragraph	Name	Data type	Tips and tricks	Related CDR	VSMEs	GRI
S1	SBM-3	14	All people in its own workforce who can be materially impacted by undertaking are included in scope of disclosure under MDR and General Disclosures	Semi-narrative	Stakeholder engagement assessment with materiality assessment, 5.1.1 MDR-P			
S1	SBM-3	14 a	Description of types of employees and non-employees in its own workforce subject to material impacts	Narrative				
S1	SBM-3	14 b	Material negative impacts occurrence (own workforce)	Semi-narrative				
S1	SBM-3	14 c	Description of activities that result in positive impacts and types of employees and non-employees in its own workforce that are positively affected or could be positively affected	Narrative				
S1	SBM-3	14 d	Description of material risks and opportunities arising from impacts and dependencies on own workforce	Narrative				
S1	SBM-3	14 e	Description of material impacts on workers that may arise from transition plans for reducing negative impacts on environment and achieving greener and climate-neutral operations	Narrative				
S1	SBM-3	14 f (i)	Information about type of operations at significant risk of incidents of forced labour or compulsory labour	Narrative		SFDR		GRI 409-1
S1	SBM-3	14 f (ii)	Information about countries or geographic areas with operations considered at significant risk of incidents of forced labour or compulsory labour	Narrative		SFDR		GRI 409-1

S1	SBM-3	14 g (i)	Information about type of operations at significant risk of incidents of child labour	Narrative		SFDR		GRI 408-1
S1	SBM-3	14 g (ii)	Information about countries or geographic areas with operations considered at significant risk of incidents of child labour	Narrative		SFDR		GRI 408-1
S1	SBM-3	15	Disclosure of how understanding of people in its own workforce/value chain workers with particular characteristics, working in particular contexts, or undertaking particular activities may be at greater risk of harm has been developed	Narrative				
S1	SBM-3	16	Disclosure of which material risks and opportunities arising from impacts and dependencies on people in its own workforce relate to specific groups of people	Narrative				
ESRS 2	S1	62	Disclosures to be reported in case the undertaking has not adopted policies and/or actions and/or targets	Narrative				
S1	S1-1	19	Policies to manage material impacts, risks, and opportunities related to its own workforce	MDR-P	Stakeholder engagement assessment with materiality assessment, human resource policy audit, 5.1.1 MDR-P and indictors where appropriate, especially training and development, see 11.13, 11.14, 11.16			GRI 2-23, GRI 401-3-3, GRI 402-3-3
S1	S1-1	19	Policies to manage material impacts, risks, and opportunities related to own workforce, including for specific groups within workforce or all own workforce	Semi-narrative				GRI 2-23
S1	S1-1	AR10	Disclosure of explanations of significant changes to policies adopted during reporting year	Narrative			V	GRI 2-23
S1	S1-1	20	Description of relevant human rights policy commitments relevant to own workforce	Narrative		SFDR	V	GRI 2-23
S1	S1-1	20a	Disclosure of general approach in relation to respect for human rights including labour rights, of people in its own workforce	Narrative		SFDR	V	GRI 2-23

(Continued)

Table 72 (*Continued*).

ESRS	DR	Paragraph	Name	Data type	Tips and tricks	Related CDR	VSMEs	GRI
S1	S1-1	20b	Disclosure of general approach in relation to engagement with people in its own workforce	Narrative		SFDR	V	GRI 2-23, GRI 2-29
S1	S1-1	20c	Disclosure of general approach in relation to measures to provide and/or enable remedy for human rights impacts	Narrative		SFDR	V	GRI 2-23, GRI 2-25, GRI 401-3-3, GRI 402-3-3, GRI 403-3-3, GRI 404-3-3, GRI 405-3-3, GRI 406-3-3, GRI 407-3-3, GRI 408-3-3, GRI 409-3-3
S1	S1-1	21	Disclosure of whether and how policies are aligned with relevant internationally recognised instruments	Narrative		SFDR	V	GRI 2-23
S1	S1-1	22	Policies explicitly address trafficking in human beings, forced labour, or compulsory labour and child labour	Semi-narrative		SFDR		GRI 408-3-3, GRI 408-1, GRI 409-3-3, GRI 409-1
S1	S1-1	23	Workplace accident prevention policy or management system is in place	Semi-narrative		SFDR		GRI 403-1

S1	S1-1	24a	Specific policies aimed at elimination of discrimination are in place	Semi-narrative				GRI 405-3-3, GRI 406-3-3
S1	S1-1	24b	Grounds for discrimination are specifically covered in policy	Semi-narrative				
S1	S1-1	24c	Disclosure of specific policy commitments related to inclusion and/or positive action for people from groups at particular risk of vulnerability in own workforce	Narrative				
S1	S1-1	24d	Disclosure of whether and how policies are implemented through specific procedures to ensure discrimination is prevented, mitigated, and acted upon once detected, as well as to advance diversity and inclusion	Narrative				GRI 406-3-3
S1	S1-1	AR 14	Disclosure on an illustration of the types of communication of its policies to those individuals, group of individuals, or organisations for whom they are relevant	Narrative			V	GRI 2-23
S1	S1-1	AR 17 a)	Policies and procedures that make qualifications, skills, and experience the basis for the recruitment, placement, training, and advancement are in place or planned	Semi-narrative			V	
S1	S1-1	AR 17 b)	Has or planning to assign responsibility at top management level for equal treatment and opportunities in employment, issue clear company-wide policies and procedures to guide equal employment practices, and link advancement to desired performance in this area	Semi-narrative			V	

(Continued)

Table 72 (*Continued*).

ESRS	DR	Paragraph	Name	Data type	Tips and tricks	Related CDR	VSMEs	GRI
S1	S1-1	AR 17 c)	Staff training on non-discrimination policies and practices are planned or in place	Semi-narrative			V	
S1	S1-1	AR 17 d)	Adjustments to the physical environment to ensure health and safety for workers, customers, and other visitors with disabilities are planned or in place	Semi-narrative			V	
S1	S1-1	AR 17 e)	Plans to evaluate or has evaluated whether a there is a risk that job requirements have been defined in a way that would systematically disadvantage certain groups	Semi-narrative			V	
S1	S1-1	AR 17 f)	Planning to keep or keeping an up-to-date records on recruitment, training, and promotion that provide a transparent view of opportunities for employees and their progression	Semi-narrative			V	
S1	S1-1	AR 17 g)	Has put in place or plans to put in place grievance procedures to address complaints, handle appeals, and provide recourse for employees when discrimination is identified, and is alert to formal structures and informal cultural issues that can prevent employees from raising concerns and grievances	Semi-narrative			V	
S1	S1-1	AR 17 h)	Have or plans to have programs to promote access to skills development	Semi-narrative			V	GRI 404-2

Source: CDR (EU) 2023/2772 supplementing Directive 2013/34/EU, *GRI 1, 2, 3: 11, 12, 13, 14, 306, 101, 201, 202, 203, 204, 205, 206, 207,301, 302, 303, 304, 305, 306, 308,401, 402, 403, 404, 405, 406, 407, 408, 409, 410, 411, 413, 414, 415, 416, 417, 418*, and the author.

11.1 Own Workforce Methodology

The methodology for assessing the impacts, risks, and opportunities related to your organisation's own workforce (which includes contracted labour, consultants, and other workforce members who received payment for their services by your organisation) consists of several key components. The following components should be combined with specific metrics detailed in this chapter and provide a structured approach to understanding and disclosing the material aspects of the workforce's influence on the undertaking and vice versa. The methodology includes the following:

1. **Collection of secure data:**

   Own workplace reporting requires substantial data resources, most of which is sensitive and subject to privacy regulations. As part of your initial reporting scope, determine the most appropriate solutions for sourcing and managing data. Strip all personal details and aggregate into generic dataset formats, for example, gender and location, demographic information cross-referenced to position (for example, by gender and/or age: up to 30 years/30–50 years/50+ years and/or seniority: Senior management/Middle management /Line staff):

Table 73 Employee headcount by gender.

Gender	# of employees (head count)
Male	
Female	
Other	
Not reported	
Total employees	

Source: CDR (EU) 2023/2772 supplementing Directive 2013/34/EU.

Table 74 Employee headcount by location with at least 50 employees.

Country	# of employees (head count)
Country 1	
Country 2	
Country 3	
…	
Total for all countries	

Source: CDR (EU) 2023/2772 supplementing Directive 2013/34/EU.

2. **Identification of material impacts:**

- Through stakeholder engagement (Chapter 2) and your materiality assessment (Chapter 3), identify and assess the material positive and negative impacts that your organisation has on its own workforce. Consider various aspects such as fair employment practices, occupational health and safety, diversity and inclusion, employee well-being, and work–life balance.
- Evaluate the extent and significance of these impacts to determine their materiality.

3. **Actions and outcomes:**

- Disclose the actions taken by your organisation to prevent, mitigate, or remediate any negative impacts on the workforce. Provide a comprehensive account of the measures implemented to ensure workplace safety, employee rights, fair compensation, professional development, and employee engagement.
- Tabulate social and human rights matters, their related policies, actions, and targets. Report on the outcomes and results of these actions to demonstrate the effectiveness of the undertaken initiatives.

4. **Analysis of material risks and opportunities:**

- Evaluate the nature, type, and extent of the material risks associated with your organisation's impacts on its own workforce. Identify potential opportunities for improvement and enhancement in areas such as employee satisfaction, productivity, talent retention, and organisational performance.
- Conduct a thorough analysis of these risks and opportunities, considering both short-term and long-term implications.

5. **Management of workforce dependencies:**

- Recognise the dependencies of your organisation on its own workforce and understand their influence on business operations and performance. Develop strategies and plans to adapt your organisation's strategy and business model in alignment with the workforce's needs and expectations.

- Implement measures to manage workforce dependencies effectively, ensuring your organisation's ability to respond to changes and challenges in a dynamic work environment.

6. **Financial effects:**

- Assess the financial effects of material risks and opportunities arising from your organisation's impacts and dependencies on its own workforce. Consider short-term, medium-term, and long-term time horizons to capture the financial implications over different planning periods.
- Quantify the potential costs, savings, investments, and returns associated with managing workforce-related risks and capitalising on opportunities.

7. **Continuous monitoring:**

- Worker engagement and consultations with management should form a regular and meaningful dialogue that drives and oversees a regular monitoring program. Both engagement and consultations can be assessed and evaluated using the following indicators, which should be fully embedded from policy level downwards.

11.1.1 Worker engagement index (S1-2)

Relevant SDGs: SDG 8 (Decent Work and Economic Growth), SDG 10 (Reduced Inequalities), and SDG 16 (Peace, Justice, and Strong Institutions).

Worker engagement index measures the level of worker engagement and participation in decision-making processes within the organisation and includes the following steps:

1. **Design worker engagement surveys:** Develop surveys to gather feedback from workers on their perception of engagement opportunities, communication channels, involvement in decision-making, and overall satisfaction with their ability to contribute to the organisation.
2. **Analyse survey responses:** Evaluate survey responses to assess the level of worker engagement and identify areas for improvement in engaging with workers and workers' representatives.

11.1.2 Frequency of worker-management communication (S1-2)

Relevant SDGs: SDG 8 (Decent Work and Economic Growth), SDG 10 (Reduced Inequalities), SDG 16 (Peace, Justice, and Strong Institutions).

Frequency of worker-management communication measures the regularity and effectiveness of communication channels between workers and management.

1. **Document communication channels:** Identify and document the various communication channels established by the organisation for workers to engage with management, such as regular meetings, suggestion boxes, online platforms, or feedback sessions.
2. **Assess communication frequency:** Track the frequency of communication events or meetings between workers and management through records or meeting minutes.
3. **Worker feedback analysis:** Analyse the feedback received from workers through various communication channels to assess the effectiveness of the communication process and the extent to which worker inputs are considered in decision-making.
4. To calculate the average frequency of worker-management communication, using the following formula:

   Average Frequency = Total Number of Communication Events/Time Period

where:

'Total Number of Communication Events' refers to the overall count of communication events or interactions between workers and management during a specific time period.

'Time Period' represents the duration over which the communication events were observed or recorded. It could be measured in days, weeks, months, or any other suitable unit of time.

Table 75 Social and human rights matters, policies, actions, and targets template.

Social and human rights matters	Non-exhaustive list of factors to consider in materiality assessment	Examples of policies	Examples of actions	Examples of targets
Working conditions				
Secure employment	Percentage of employees with temporary contracts, ratio of non-employee workers to employees, social protection	No-layoff policy, limits on renewing temporary contracts, employer provision of social protection where state provision is lacking	Offer permanent contracts to employees with temporary contracts, implement plans for social protection where state provision is lacking	Increasing the percentage of workforce with employment contracts (especially permanent contracts) and social protection
Working time	Percentage of employees with part time or zero-hour contracts, employee satisfaction with working time	Limitations on overtime, long and split shifts, and night and weekend work, adequate lead time for scheduling	Shift work rotation, extend advance notice of scheduling, reduction of excessive overtime	Increasing the percentage of workforce with flexible working time arrangements
Adequate wages	EU, national, or local legal definitions of adequate wages, fair wages, and minimum wages	Policy to pay all own workers an adequate wage	Negotiation of fair wages in collective bargaining agreements, verification that employment agencies pay a fair wage	Paying all own workers an adequate wage
Social dialogue/existence of work councils/ information, consultation, and participation rights of workers	Extent of workplace, cross-border, and board-level representation through trade unions and/or works councils	Policy to encourage institutions for social dialogue, regular information and consultation with worker representatives, consultation before final decisions on employment-related issues are reached	Expansion of sustainability issues dealt with in social dialogue, increase in number of meetings, increase in resources for works councils	Extending social dialogue to more establishments and/or countries

(*Continued*)

Table 75 (*Continued*).

Social and human rights matters	Non-exhaustive list of factors to consider in materiality assessment	Examples of policies	Examples of actions	Examples of targets
Freedom of association/ collective bargaining including the rate of workers covered by collective agreements	Percentage of own workforce covered by collective bargaining agreements, work stoppages	Non-interference in trade union formation and recruitment (including trade union access to undertakings), bargaining in good faith, recognition of adequate time off for duties, facilities and dismissal protection for workers' representatives, no discrimination of trade union members and workers' representatives	Expansion of sustainability issues dealt with in collective bargaining, increase in resources for workers' representatives	Increasing the percentage of own workers covered by collective bargaining, negotiating collective bargaining agreements over sustainability issues
Work–life balance	Family-related leave, flexible working hours, access to childcare	Provision of family leave, flexible working time, access to day care facilities for all workers	Expand family leave eligibility and flexible working time arrangements, increasing provision of day care	Extending work-life measures to a greater percentage of own workers
Health and safety	Coverage by H&S system, rate of fatalities, non-fatal accidents, work-related ill health, work days lost	Coverage of all own workers in H&S management system	Increase health and safety training, investment in safer equipment	Reducing the rate of injuries and work time lost due to injuries
Equal treatment and opportunities for all				
Gender equality and equal pay for work of equal value	Percentage of women in top management and workforce, male-female wage gap	Policy for gender equality and equal pay for equal work	Targeted recruitment and promotion of women, reduction in pay gap through negotiation of collective bargaining agreements	Increasing the percentage of women in the workforce and top management, reducing the male–female wage gap

Training and skills development	Amount and distribution of training, Percentage of employees with regular performance, and development reviews	Policy for enhancing skills and career prospects for employees	Skills audits, training to fill skills gaps	Increasing the percentage of employees receiving training and regular skills development reviews
The employment and inclusion of people with disabilities	Percentage employment and accessibility measures for employees with disabilities	Policy for making workplaces accessible to persons with disabilities	Increasing accessibility measures	Increasing the percentage of persons with disabilities in own workforce
Measures against violence and harassment in the workplace	Prevalence of violence and harassment	Zero tolerance policy for violence and harassment at workplace	Improving complaint mechanisms, increasing sanctions against violence and harassment, providing training for prevention to management	Extending measures to all workplaces
Diversity	Representation of women and/or ethnic groups or minorities in own workforce. Age distribution in own workforce. Percentage of persons with disabilities within the own workforce	Policy for inclusiveness (i.e., ethnic diversity or minority groups) and affirmative action	Training on diversity and inclusion (including ethnicity considerations), targeted recruitment of underrepresented groups	Increasing the percentage of underrepresented groups in own workforce and top management
Other work-related rights				
Child labour	Type of operations and geographical areas at risk of child labour	Policy for identifying where child labour occurs, identifying where exposure of young workers to hazardous work occurs, and preventing risk of exposure	Age verification measures, partnerships with organisations to eliminate child labour, measures against worst forms of child labour	Extending measures for preventing exposure of young persons to hazardous work to a greater percentage of operations

(Continued)

Table 75 (*Continued*).

Social and human rights matters	Non-exhaustive list of factors to consider in materiality assessment	Examples of policies	Examples of actions	Examples of targets
Forced labour	Type of operations and geographical areas at risk of forced labour	Policy for identifying where forced labour occurs and reducing risk of forced labour	Measures ensuring free consent to employment without threat of penalty, contracts in understandable language, freedom to terminate employment without penalty, disciplinary measures should not obligate labour, free consent to overtime, freedom of movement (including to exit workplace), fair treatment for migrant workers, monitoring employment agencies	Extending measures for preventing child labour to a greater number of operations
Adequate housing	Should be included in analysis but defined in CSRD/ESRS and dependent on location factors	Policy for minimum housing quality standards	Not defined	Not defined
Water and sanitation	Should be included in analysis but defined in CSRD/ESRS and dependent on location factors	Policy for minimum hygiene standards	Not defined	Not defined
Privacy	Should be included in analysis but defined in CSRD/ESRS and dependent on location factors	Policy for workforce privacy of information	Not defined	Not defined

Source: CDR (EU) 2023/2772 supplementing Directive 2013/34/EU.

Table 76 Own workforce assessment indicators relating stakeholder engagement, impacts, risks, and opportunities.

ESRS	DR	Paragraph	Name	Data type	Tips and tricks	Related CDR	VSMEs	GRI
S1	S1-2	27	Disclosure of whether and how perspectives of own workforce inform decisions or activities aimed at managing actual and potential impacts	Narrative	To define processes for engaging with own workers and workers' representatives about impacts, refer back to Chapter 2: Stakeholder Engagement combined with own workforce assessment methodology and 11.1.1, 11.1.2 as well as cross-reference with Business Conduct: 15.1, 15.2, 15.9, and 15.10 See also: – UN Guiding Principles on Business & Human Rights[1] – UN Global Compact Principles[2] – OECD Guidelines for Multinational Enterprises[3] – ILO Declaration of Fundamental Rights and Principles at Work[4]			GRI 3-3, GRI 401-3-3, GRI 402-3-3, GRI 403-3-3, GRI 404-3-3, GRI 405-3-3, GRI 406-3-3, GRI 407-3-3, GRI 408-3-3, GRI 409-3-3
S1	S1-2	27a	Engagement occurs with own workforce or their representatives	Semi-narrative				
S1	S1-2	27b	Disclosure of stage at which engagement occurs, type of engagement and frequency of engagement	Narrative				
S1	S1-2	27c	Disclosure of function and most senior role within undertaking that has operational responsibility for ensuring that engagement happens and that results inform organisation's approach	Narrative				
S1	S1-2	27d	Disclosure of Global Framework Agreement or other agreements related to respect of human rights of workers	Narrative				
S1	S1-2	27e	Disclosure of how effectiveness of engagement with its own workforce is assessed	Narrative				GRI 2-29

[1] https://www.ohchr.org/sites/default/files/documents/publications/guidingprinciplesbusinesshr_en.pdf
[2] https://unglobalcompact.org/what-is-gc/mission/principles
[3] https://mneguidelines.oecd.org/mneguidelines/
[4] https://www.ilo.org/ilo-declaration-fundamental-principles-and-rights-work

Table 76 (*Continued*).

ESRS	DR	Paragraph	Name	Data type	Tips and tricks	Related CDR	VSMEs	GRI
S1	S1-2	28	Disclosure of steps taken to gain insight into perspectives of people in its own workforce that may be particularly vulnerable to impacts and/or marginalised	Narrative				GRI 2-29
S1	S1-2	29	Statement in case the undertaking has not adopted a general process to engage with its own workforce	Narrative				
S1	S1-2	29	Disclosure of timeframe for adoption of general process to engage with its own workforce in case the undertaking has not adopted a general process for engagement	Narrative			V	
S1	S1-2	AR 25 a	Disclosure of how undertaking engages with at-risk or persons in vulnerable situations	Narrative			V	
S1	S1-2	AR 25 b	Disclosure of how potential barriers to engagement with people in its workforce are taken into account	Narrative			V	
S1	S1-2	AR 25 c	Disclosure of how people in its workforce are provided with information that is understandable and accessible through appropriate communication channels	Narrative			V	

S1	S1-2	AR 25 d	Disclosure of any conflicting interests that have arisen among different workers and how these conflicting interests have been resolved	Narrative			V	
S1	S1-2	AR 25 e	Disclosure of how undertaking seeks to respect human rights of all stakeholders engaged	Narrative			V	
S1	S1-2	AR 26	Information about effectiveness of processes for engaging with its own workforce from previous reporting periods	Narrative			V	
S1	S1-3	32a	Disclosure of general approach to and processes for providing or contributing to remedy where undertaking has caused or contributed to a material negative impact on people in its own workforce	Narrative	Processes to remediate negative impacts and channels for own workers to raise concerns combine your Stakeholder Engagement combined with own workforce assessment methodology, 5.1.2 MDR-A and 15.9 Access to remedy			GRI 2-25
S1	S1-3	32b	Disclosure of specific channels in place for its own workforce to raise concerns or needs directly with undertaking and have them addressed	Narrative				GRI 2-25, GRI 403-2
S1	S1-3	AR 29	Third-party mechanisms are accessible to all own workforce	Semi-narrative			V	
S1	S1-3	AR 30	Disclosure of how own workforce and their workers' representatives are able to access channels at a level of undertaking they are employed by or contracted to work for	Narrative			V	
S1	S1-3	32c	Grievance or complaints handling mechanisms related to employee matters exist	Semi-narrative		SFDR		

(Continued)

Table 76 (*Continued*).

ESRS	DR	Paragraph	Name	Data type	Tips and tricks	Related CDR	VSMEs	GRI
S1	S1-3	32d	Disclosure of processes through which undertaking supports or requires availability of channels	Narrative				GRI 2-26
S1	S1-3	32e	Disclosure of how issues raised and addressed are tracked and monitored and how effectiveness of channels is ensured	Narrative				GRI 2-25
S1	S1-3	33	Disclosure of whether and how it is assessed that its own workforce is aware of and trust structures or processes as way to raise their concerns or needs and have them addressed	Narrative				GRI 2-25, GRI 403-3
S1	S1-3	33	Policies regarding protection against retaliation for individuals that use channels to raise concerns or needs are in place	Semi-narrative				
S1	S1-3	34	Disclosure of timeframe for channel or processes for raising concerns to be in place if it has not adopted a channel	Narrative			V	

Source: CDR (EU) 2023/2772 supplementing Directive 2013/34/EU, *GRI 1, 2, 3: 11, 12, 13, 14, 306, 101, 201, 202, 203, 204, 205, 206, 207,301, 302, 303, 304, 305, 306, 308,401, 402, 403, 404, 405, 406, 407, 408, 409, 410, 411, 413, 414, 415, 416, 417, 418*, and the author.

Table 77 Own workforce assessment indicators relating to actions and targets.

ESRS	DR	Paragraph	Name	Data type	Tips and tricks	Related CDR	VSMEs	GRI
S1	S1-4	37	Action plans and resources to manage its material impacts, risks, and opportunities related to its own workforce	MDR-A	Disclose all planned or underway actions to prevent, mitigate, or remediate material negative impacts by reviewing stakeholder engagement with own workforce assessment methodology, 5.1.2 MDR-A with 11.1.1 Worker Engagement Index, and 11.1.2 Frequency of Worker-Management Communication			GRI 401-3-3
S1	S1-4	38a	Description of action taken, planned or underway to prevent or mitigate negative impacts on own workforce	Narrative				GRI 402-3-3, GRI 403-3-3, GRI 403-9, GRI 403-10, GRI 404-3-3, GRI 405-3-3, GRI 406-3-3, GRI 407-3-3, GRI 408-3-3, GRI 409-3-3
S1	S1-4	38b	Disclosure on whether and how action has been taken to provide or enable remedy in relation to actual material impact	Narrative				GRI 402-3-3, GRI 403-3-3, GRI 403-9, GRI 403-10, GRI 404-3-3, GRI 405-3-3, GRI 406-3-3, GRI 407-3-3, GRI 408-3-3, GRI 409-3-3
S1	S1-4	38c	Description of additional initiatives or actions with primary purpose of delivering positive impacts for own workforce	Narrative				GRI 402-3-3, GRI 403-3-3, GRI 403-9, GRI 403-10, GRI 404-3-3, GRI 405-3-3, GRI 406-3-3, GRI 407-3-3, GRI 408-3-3, GRI 409-3-3
S1	S1-4	38d	Description of how effectiveness of actions and initiatives in delivering outcomes for own workforce is tracked and assessed	Narrative				GRI 402-3-3, GRI 403-3-3, GRI 403-9, GRI 403-10, GRI 404-3-3, GRI 405-3-3, GRI 406-3-3, GRI 407-3-3, GRI 408-3-3, GRI 409-3-3

(Continued)

Table 77 (*Continued*).

ESRS	DR	Paragraph	Name	Data type	Tips and tricks	Related CDR	VSMEs	GRI
S1	S1-4	39	Description of process through which it identifies what action is needed and appropriate in response to particular actual or potential negative impact on own workforce	Narrative				GRI 3-3, GRI 401-3-3, GRI 402-3-3, GRI 403-3-3, GRI 403-9, GRI 403-10, GRI 404-3-3, GRI 405-3-3, GRI 406-3-3, GRI 407-3-3, GRI 408-3-3, GRI 409-3-3
S1	S1-4	40a	Description of what action is planned or underway to mitigate material risks arising from impacts and dependencies on own workforce and how effectiveness is tracked	Narrative				GRI 3-3
S1	S1-4	40b	Description of what action is planned or underway to pursue material opportunities in relation to own workforce	Narrative				
S1	S1-4	41	Disclosure of whether and how it is ensured that own practices do not cause or contribute to material negative impacts on own workforce	Narrative				
S1	S1-4	43	Disclosure of resources are allocated to the management of material impacts	Narrative				
S1	S1-4	AR 33 a	Disclosure of general and specific approaches to addressing material negative impacts	Narrative			V	

S1	S1-4	AR 33 b	Disclosure of initiatives aimed at contributing to additional material positive impacts	Narrative			V	
S1	S1-4	AR 33 c	Disclosure of how far undertaking has progressed in efforts during reporting period	Narrative			V	
S1	S1-4	AR 33 d	Disclosure of aims for continued improvement	Narrative			V	
S1	S1-4	AR 35	Disclosure of how undertaking seeks to use leverage with relevant business relationships to manage material negative impacts affecting own workforce	Narrative			V	GRI 2-24
S1	S1-4	AR 36	Disclosure of how the initiative, and its own involvement, is aiming to address the material impact concerned	Narrative			V	
S1	S1-4	AR 40 a	Disclosure of whether and how workers and workers' representatives play a role in decisions regarding design and implementation of programmes or processes whose primary aim is to deliver positive impacts for workers	Narrative			V	GRI 3-3, GRI 401-3-3, GRI 402-3-3, GRI 403-3-3, GRI 404-3-3, GRI 405-3-3 GRI 406-3-3, GRI 407-3-3, GRI 408-3-3, GRI 409-3-3
S1	S1-4	AR 40 b	Information about intended or achieved positive outcomes of programmes or processes for own workforce	Narrative			V	

(Continued)

Table 77 (*Continued*).

ESRS	DR	Paragraph	Name	Data type	Tips and tricks	Related CDR	VSMEs	GRI
S1	S1-4	AR 41	Initiatives or processes whose primary aim is to deliver positive impacts for own workforce are designed also to support achievement of one or more of Sustainable Development Goals	Semi-narrative			V	GRI 203-2
S1	S1-4	AR 43	Information about measures taken to mitigate negative impacts on workers that arise from transition to greener, climate-neutral economy	Narrative			V	
S1	S1-4	AR 48	Description of internal functions that are involved in managing impacts and types of action taken by internal functions to address negative and advance positive impacts	Narrative			V	
S1	S1-5	46	Targets set to manage material impacts, risks, and opportunities related to own workforce	MDR-T	Disclose all planned or underway targets and specific metrics to prevent, mitigate, or remediate material negative impacts by reviewing stakeholder engagement with own workforce assessment			GRI 401-3-3
S1	S1-5	47a	Disclosure of whether and how own workforce or workforce's representatives were engaged directly in setting targets	Narrative				

S1	S1-5	47b	Disclosure of whether and how own workforce or workforce's representatives were engaged directly in tracking performance against targets	Narrative	methodology, 5.1.3 MDR-M, 5.1.4 MDR-T, and 11.1.1 Worker Engagement Index, and 11.1.2 Frequency of Worker-Management Communication			GRI 3-3, GRI 401-3-3, GRI 402-3-3, GRI 403-3-3, GRI 404-3-3, GRI 405-3-3, GRI 406-3-3, GRI 407-3-3, GRI 408-3-3, GRI 409-3-3
S1	S1-5	47c	Disclosure of whether and how own workforce or workforce's representatives were engaged directly in identifying lessons or improvements as result of undertaking performance	Narrative				GRI 3-3, GRI 401-3-3, GRI 402-3-3, GRI 403-3-3, GRI 404-3-3, GRI 405-3-3, GRI 406-3-3, GRI 407-3-3, GRI 408-3-3, GRI 409-3-3
S1	S1-5	AR 49 a	Disclosure of intended outcomes to be achieved in lives of people in its own workforce	Narrative			V	
S1	S1-5	AR 49 b	Information about stability over time of target in terms of definitions and methodologies to enable comparability	Semi-narrative			V	
S1	S1-5	AR 49 c	Disclosure of references to standards or commitments which targets are based on	Narrative			V	

Source: CDR (EU) 2023/2772 supplementing Directive 2013/34/EU, *GRI 1, 2, 3: 11, 12, 13, 14, 306, 101, 201, 202, 203, 204, 205, 206, 207,301, 302, 303, 304, 305, 306, 308,401, 402, 403, 404, 405, 406, 407, 408, 409, 410, 411, 413, 414, 415, 416, 417, 418*, and the author.

Table 78 Indicators and metrics related to own workforce characteristics.

ESRS	DR	Paragraph	Name	Data type	Tips and tricks	Related CDR	VSMEs	GRI
S1	S1-6	50a	Characteristics of organisation's employees – number of employees by gender table	Table	Describe characteristics of the employees from stakeholder engagement combined with own workforce assessment methodology and relevant indicators – tabulate key data – see 11.2 Note that if totals are included with breakdowns, then only two tables (by gender and by region) are required. Remember to cross-reference with financial reports.		V	GRI 2-7, GRI 405-1
S1	S1-6	50a	Number of employees by head count	Table/integer				GRI 2-7, GRI 405-1
S1	S1-6	50a	Average number of employees by head count	Table/integer				GRI 2-7, GRI 405-1
S1	S1-6	50a	Gender of employees in countries with 50 or more employees	Table			V	GRI 2-7, GRI 405-1
S1	S1-6	50a	Number of employees in countries with 50 or more employees	Table/integer			V	GRI 2-7, GRI 405-1
S1	S1-6	50a	Average number of employees in countries with 50 or more employees	Table/integer				GRI 2-7, GRI 405-1
S1	S1-6	50b	Table relating to contract type and gender of employees by the total number by head count or full-time equivalent (FTE) of: total employees, permanent employees, temporary employees, non-guaranteed hours employees, full-time employees, part-time employees	Table			V	GRI 2-7
S1	S1-6	51	Table relating to contract type and region of employees by the total number by head count or full-time equivalent (FTE) of: total employees, permanent employees, temporary employees, non-guaranteed hours employees, full-time employees, part-time employees	Table			V	GRI 2-7
S1	S1-6	50b	Average number of employees by head count or full-time equivalent	Table/Decimal				GRI 2-7
S1	S1-6	50c	Number of employee turnover	Integer				GRI 401-1
S1	S1-6	50c	Percentage of employee turnover	Per cent				GRI 401-1

S1	S1-6	50d	Description of methodologies and assumptions used to compile data concerning employees	Narrative				GRI 2-7
S1	S1-6	50 d (i)	Employee numbers are reported in head count or full-time equivalent	Semi-narrative				GRI 2-7
S1	S1-6	50 d (ii)	Employees numbers are reported at the end of reporting period/average/other methodology	Semi-narrative				GRI 2-7
S1	S1-6	50e	Disclosure of contextual information necessary to understand data concerning employees	Narrative				GRI 2-7
S1	S1-6	52	Further detailed breakdown by gender and by region table	Table			V	GRI 2-7
S1	S1-6	52 a	Number of full-time employees by head count or full-time equivalent	Table, decimal/ integer			V	GRI 2-7
S1	S1-6	52 b	Number of part-time employees by head count or full-time equivalent	Table, decimal/ integer			V	GRI 2-7
S1	S1-6	50f	Disclosure of cross-reference head count of the total number of employees and breakdowns by gender and by country for countries in which the undertaking has 50 or more employee and related to most representative number in financial statements	Narrative				
S1	S1-7	55 a	Number of non-employees in own workforce	Decimal	Non-employees included in own workforce assessment methodology with 11.2 and 11.3. If you cannot report exact figures, then estimate to the nearest 10 (for organisations with less than 1000 non-employees) and the nearest 100 for organisations with more than 1000 non-employees, and explain plus identify information from actual data versus estimates.			GRI 2-8
S1	S1-7	55 a	Number of non-employees in own workforce – self-employed people	Decimal				GRI 2-8
S1	S1-7	55 a	Number of non-employees in own workforce – people provided by undertakings primarily engaged in employment activities	Decimal				GRI 2-8
S1	S1-7	56	Disclosure of the most common types of non-employees (for example, self-employed people, people provided by undertakings primarily engaged in employment activities, and other types relevant to the undertaking), their relationship with the undertaking, and the type of work that they perform	Narrative			V	GRI 2-8

(Continued)

Table 78 (*Continued*).

ESRS	DR	Paragraph	Name	Data type	Tips and tricks	Related CDR	VSMEs	GRI
S1	S1-7	55 b	Description of methodologies and assumptions used to compile data (non-employees)	Narrative				GRI 2-8
S1	S1-7	55 b (i)	Non-employee numbers are reported in head count/ full-time equivalent	Semi-narrative				GRI 2-8
S1	S1-7	55 b (ii)	Non-employee numbers are reported at the end of reporting period/average/other methodology	Semi-narrative				GRI 2-8
S1	S1-7	55c	Disclosure of contextual information necessary to understand data (non-employee workers)	Narrative				GRI 2-8
S1	S1-7	57	Description of basis of preparation of non-employees estimated number	Narrative				
S1	S1-8	60 a	Percentage of total employees covered by collective bargaining agreements	Per cent	Collective bargaining and social dialogue – combine stakeholder engagement with own workforce assessment methodology and 11.4 Union density and collective bargaining coverage		V	GRI 2-30
S1	S1-8	60 b	Percentage of its employees covered by collective bargaining agreements are within coverage rate by country (in the EEA)	Table/per cent				
S1	S1-8	60 c	Percentage of own employees covered by collective bargaining agreements (outside EEA) by region	Table/per cent			V	
S1	S1-8	61	Working conditions and terms of employment for employees not covered by collective bargaining agreements are determined based on collective bargaining agreements that cover other employees or based on collective bargaining agreements from other undertakings	Semi-narrative			V	GRI 2-30

S1	S1-8	62	Description of extent to which working conditions and terms of employment of non-employees in own workforce are determined or influenced by collective bargaining agreements	Narrative			V	
S1	S1-8	63a	Percentage of employees in country (EEA) covered by workers' representatives	Per cent				
S1	S1-8	63b	Disclosure of existence of any agreement with employees for representation by European Works Council[5] (EWC), Societas Europaea (SE) Works Council, or Societas Cooperativa Europaea (SCE) Works Council	Narrative				
S1	S1-8	AR 70	Own workforce in region (non-EEA) covered by collective bargaining agreements by coverage rate and by region	Table/ semi-narrative				
S1	S1-9	66a	Number of employees (head count) at top management level	Integer	Diversity metrics – use stakeholder engagement combined with own workforce assessment methodology and 11.5 Proportion of women in managerial positions and 11.6 Number and percentage of women board members			GRI 405-1
S1	S1-9	66a	Percentage of employees at top management level	Per cent				GRI 405-1
S1	S1-9	66b	Number of employees (head count) under 30 years old	Integer				GRI 405-1
S1	S1-9	66b	Percentage of employees under 30 years old	Per cent				GRI 405-1
S1	S1-9	66b	Number of employees (head count) between 30 and 50 years old	Integer				GRI 405-1
S1	S1-9	66b	Percentage of employees between 30 and 50 years old	Per cent				GRI 405-1
S1	S1-9	66b	Number of employees (head count) over 50 years old	Integer				GRI 405-1
S1	S1-9	66b	Percentage of employees over 50 years old	Per cent				GRI 405-1

(Continued)

[5] https://ec.europa.eu/social/main.jsp?catId=707&intPageId=211&langId=en

Table 78 (*Continued*).

ESRS	DR	Paragraph	Name	Data type	Tips and tricks	Related CDR	VSMEs	GRI
S1	S1-9	AR 71	Disclosure of own definition of top management used	Narrative				
S1	S1-10	69	All employees are paid adequate wage, in line with applicable benchmarks	Semi-narrative	Adequate wages – use stakeholder engagement combined with own workforce assessment methodology and 11.7 Living wage gap and 11.18 Gender pay gap: Equality of remuneration			GRI 202-1
S1	S1-10	70	Adequate wages by country table	Table/numerical				GRI 202-1
S1	S1-10	70	Percentage of employees paid below the applicable adequate wage benchmark	Table/per cent				GRI 202-1
S1	S1-10	71	Percentage of non-employees paid below adequate wage	Table/per cent			V	GRI 202-1
S1	S1-11	74 a	All employees in own workforce are covered by social protection, through public programs or through benefits offered, against loss of income due to sickness	Semi-narrative	Social protection – use stakeholder engagement combined with own workforce assessment methodology and: 11.2 Contingent and subcontracted workers 11.3 Long-term work contracts 11.8 Social Security Coverage Rate 11.9 Social Protection Expenditure			
S1	S1-11	74 b	All employees in own workforce are covered by social protection, through public programs, or through benefits offered, against loss of income due to unemployment starting from when own worker is working for undertaking	Semi-narrative				
S1	S1-11	74 c	All employees in own workforce are covered by social protection, through public programs, or through benefits offered, against loss of income due to employment injury and acquired disability	Semi-narrative				
S1	S1-11	74 d	All employees in own workforce are covered by social protection, through public programs, or through benefits offered, against loss of income due to parental leave	Semi-narrative				

S1	S1-11	74 e	All employees in own workforce are covered by social protection, through public programs, or through benefits offered, against loss of income due to retirement	Semi-narrative				
S1	S1-11	75	Social protection employees by country [table] by types of events and type of employees	Table				GRI 401-2
S1	S1-11	75	Disclosure of types of employees who are not covered by social protection, through public programs or through benefits offered, against loss of income due to sickness	Narrative			V	GRI 401-2
S1	S1-11	75	Disclosure of types of employees who are not covered by social protection, through public programs or through benefits offered, against loss of income due to unemployment starting from when own worker is working for undertaking	Narrative				GRI 401-2
S1	S1-11	75	Disclosure of types of employees who are not covered by social protection, through public programs or through benefits offered, against loss of income due to employment injury and acquired disability	Narrative				GRI 401-2
S1	S1-11	75	Disclosure of types of employees who are not covered by social protection, through public programs or through benefits offered, against loss of income due to maternity leave	Narrative				GRI 401-2
S1	S1-11	75	Disclosure of types of employees who are not covered by social protection, through public programs or through benefits offered, against loss of income due to retirement	Narrative				GRI 401-2

(Continued)

Table 78 (*Continued*).

ESRS	DR	Paragraph	Name	Data type	Tips and tricks	Related CDR	VSMEs	GRI
S1	S1-12	79	Percentage of persons with disabilities among employees subject to legal restrictions on collection of data	Per cent	Persons with disabilities – use stakeholder engagement combined with own workforce assessment methodology and: 11.10 Hiring of Vulnerable Groups 11.11 Disability Inclusion in Workforce 11.12 Accessibility of Facilities			GRI 405-1
S1	S1-12	80	Percentage of employees with disabilities in own workforce breakdown by gender	Table/numerical			V	
S1	S1-12	AR 76	Disclosure of contextual information necessary to understand data and how data has been compiled for persons with disabilities	Narrative				
S1	S1-13	83 a	Training and skills development indicators gender	Table	Training and skills development – use stakeholder engagement combined with own workforce assessment methodology with 11.13 Average Hours of Training per Year per Employee and 11.14 Expenditure on Employee Training per Year per Employee			GRI 404-3
S1	S1-13	83 a	Percentage of employees who participated in regular performance and career development reviews	Table/per cent				GRI 404-3
S1	S1-13	83 b	Average number of training hours by gender	Table			V	GRI 404-1
S1	S1-13	83 b	Average number of training hours per person for employees	Table/decimal			V	GRI 404-1
S1	S1-13	84	Percentage of employees who participated in regular performance and career development reviews by employee category	Table/per cent			V	GRI 404-1
S1	S1-13	84	Average number of employees who participated in regular performance and career development reviews by employee category	Table/decimal			V	GRI 404-1, GRI 404-3

S1	S1-13	85	Percentage of non-employees who participated in regular performance and career development reviews	Table/per cent			V	
S1	S1-14	88 a	Percentage of people in its own workforce who are covered by health and safety management system based on legal requirements and/or recognised standards or guidelines	Per cent	Health and safety metrics – use stakeholder engagement combined with own workforce assessment methodology with 11.15, Frequency/incident rates of occupational injuries and 11.16 Expenditures on employee health and safety as a proportion of revenue			GRI 403-8
S1	S1-14	88b	Number of fatalities in own workforce as a result of work-related injuries and work-related ill health	Integer			V	GRI 403-9
S1	S1-14	88b	Number of fatalities as a result of work-related injuries and work-related ill health of other workers working on organisation's sites	Integer			V	GRI 403-10
S1	S1-14	88c	Number of recordable work-related accidents for own workforce	Integer		SFDR	V	GRI 403-9
S1	S1-14	88c	Rate of recordable work-related accidents for own workforce	Per cent				
S1	S1-14	88d	Number of cases of recordable work-related ill health of employees	Integer			V	GRI 403-10
S1	S1-14	88e	Number of days lost to work-related injuries and fatalities from work-related accidents, work-related ill health and fatalities from ill health related to employees	Integer		SFDR		
S1	S1-14	89	Number of cases of recordable work-related ill health of non-employees	Integer			V	
S1	S1-14	89	Number of days lost to work-related injuries and fatalities from work-related accidents, work-related ill health and fatalities from ill health related to non-employees	Integer			V	

(*Continued*)

Table 78 (*Continued*).

ESRS	DR	Paragraph	Name	Data type	Tips and tricks	Related CDR	VSMEs	GRI
S1	S1-14	90	Percentage of own workforce who are covered by health and safety management system based on legal requirements and/or recognised standards or guidelines and which has been internally audited and/ or audited or certified by external party	Per cent			V	GRI 403-8
S1	S1-14	AR 81	Description of underlying standards for internal audit or external certification of health and safety management systems	Narrative			V	
S1	S1-14	AR 82	Number of fatalities in own workforce as a result of work-related injuries	Integer			V	GRI 403-9, GRI 403-10
S1	S1-14	AR 82	Number of fatalities in own workforce as a result of work-related ill health	Integer			V	GRI 403-9, GRI 403-10
S1	S1-14	AR 82	Number of fatalities as a result of work-related injuries of other workers working on organisation's sites	Integer			V	GRI 403-9, GRI 403-10
S1	S1-14	AR 82	Number of fatalities as a result of work-related ill health of other workers working on the organisation's sites	Integer			V	GRI 403-9, GRI 403-10
S1	S1-14	AR 94	Number of cases of recordable work-related ill health detected among former own workforce	Integer			V	
S1	S1-14	AR 81	Description of underlying standards for internal audit or external certification of health and safety management system	Narrative			V	

S1	S1-15	93 a	Percentage of employees entitled to take family-related leave	Per cent	Work–life balance metrics – use materiality assessment and HR records		V	GRI 401-3
S1	S1-15	93 b	Percentage of entitled employees who took family-related leave	Per cent			V	GRI 401-3
S1	S1-15	93 b	Percentage of entitled employees who took family-related leave by gender table	Table/numerical			V	GRI 401-3
S1	S1-15	94	All employees are entitled to family-related leaves through social policy and/or collective bargaining agreements	Semi-narrative				
S1	S1-16	97 a	Gender pay gap [table]	Table	Remuneration metrics – use stakeholder engagement combined with own workforce assessment methodology with 11.17 Employee turnover rate 11.18 Gender pay gap: Equality of remuneration 11.19 Employee wages and benefits as a proportion of revenue, with breakdown by employment type and gender, 11.20 CEO-to-worker pay ratio 11.21 Distribution of surplus/ profits	SFDR/ BENCH	V	GRI 405-2
S1	S1-16	97 a	Gender pay gap	Table/per cent		SFDR/ BENCH	V	GRI 405-2
S1	S1-16	97 b	Annual total remuneration ratio	Per cent		SFDR		GRI 2-21, GRI 405-2
S1	S1-16	97 c	Disclosure of contextual information necessary to understand data, how data has been compiled, and other changes to underlying data that are to be considered	Narrative				GRI 2-21, GRI 405-2
S1	S1-16	98	Gender pay gap breakdown by employee category and/or country/segment [table]	Table/numerical			V	GRI 405-2
S1	S1-16	98	Gender pay gap breakdown by ordinary basic salary and complementary/variable components	Table/numerical			V	GRI 405-2
S1	S1-16	99	Remuneration ratio adjusted for purchasing power differences between countries	Per cent			V	
S1	S1-16	99	Description of methodology used for calculation of remuneration ratio adjusted for purchasing power differences between countries	Narrative			V	

(Continued)

Table 78 (*Continued*).

ESRS	DR	Paragraph	Name	Data type	Tips and tricks	Related CDR	VSMEs	GRI
S1	S1-17	103 a	Number of incidents of discrimination [table]	Table	Incidents, Complaints and Severe Human Rights Impacts – All incidents of discrimination and human rights offenses must be fully documented in HR and reported to the Board, link with Governance and Business Conduct and: 11.22 Fines paid or payable due to settlements 11.23 Harassment and discrimination at the workplace 15.9 Access to remedy Cross-reference any occurrence with: UN Guiding Principles on Business & Human Rights UN Global Compact Principles Global Framework Agreement	SFDR		
S1	S1-17	103 a	Number of incidents of discrimination	Table/integer				GRI 406-1
S1	S1-17	103 b	Number of complaints filed through channels for people in own workforce to raise concerns	Integer				
S1	S1-17	103 b	Number of complaints filed to National Contact Points for OECD Multinational Enterprises	Integer				
S1	S1-17	103 c	Amount of material fines, penalties, and compensation for damages as a result of violations regarding social and human rights factors	Monetary				GRI 2-27
S1	S1-17	103 c	Information about reconciliation of material fines, penalties, and compensation for damages as a result of violations regarding social and human rights factors with most relevant amount presented in financial statements	Narrative				GRI 2-27
S1	S1-17	103 d	Disclosure of contextual information necessary to understand data and how data has been compiled (work-related grievances, incidents, and complaints related to social and human rights matters)	Narrative				GRI 2-27

S1	S1-17	104 a	Number of severe human rights issues and incidents connected to own workforce	Integer	OECD Guidelines for Multinational Enterprises ILO Declaration of Fundamental Rights and Principles at Work	SFDR/ BENCH	V	
S1	S1-17	104 a	Number of severe human rights issues and incidents connected to own workforce that are cases of non-respect of UN Guiding Principles and OECD Guidelines for Multinational Enterprises	Integer		SFDR/ BENCH	V	
S1	S1-17	104 a	No severe human rights issues and incidents connected to own workforce have occurred	Semi-narrative		SFDR/ BENCH	V	
S1	S1-17	104 b	Amount of material fines, penalties, and compensation for severe human rights issues and incidents connected to own workforce	Monetary				GRI 2-27
S1	S1-17	104 b	Information about reconciliation of amount of material fines, penalties, and compensation for severe human rights issues and incidents connected to own workforce with most relevant amount presented in financial statements	Narrative				GRI 2-27
S1	S1-17	AR 103	Disclosure of the status of incidents and/or complaints and actions taken	Narrative			V	GRI 406-1
S1	S1-17	AR 106	Number of severe human rights cases where undertaking played role securing remedy for those affected	Integer			V	

Source: CDR (EU) 2023/2772 supplementing Directive 2013/34/EU, *GRI 1, 2, 3: 11, 12, 13, 14, 306, 101, 201, 202, 203, 204, 205, 206, 207,301, 302, 303, 304, 305, 306, 308,401, 402, 403, 404, 405, 406, 407, 408, 409, 410, 411, 413, 414, 415, 416, 417, 418*, and the author.

11.2 Contingent and Subcontracted Workers (S1-7)

Relevant SDGs: SDG 1 (No Poverty), SDG 3 (Good Health and Well-being), SDG 8 (Decent Work and Economic Growth), SDG 10 (Reduced Inequalities).

Measures the use of contingent and/or subcontracted workers by an organisation. It includes the total number of such workers, the proportion under contract with suppliers vs. independent workers, compliance with ethical recruitment practices and international standards, and a time series comparison of the proportion of contingent and/or subcontracted workers with changes in company growth. Calculate and tabulate:

1. Total number of contingent and/or subcontracted workers: Identify the total number of workers in actual headcount and as a percentage of the total workforce.

Table 79 Employee headcount by employment type.

	Reporting period:				
	Female	**Male**	**Other**	**Not disclosed**	**Total**
Number of employees (head count/FTE)					
Permanent employees (head count/FTE)					
Temporary employees (head count/FTE)					
Non-guaranteed hours employees (head count/FTE)					
Full-time equivalent employees (head count/FTE)					
Part-time equivalent employees (head count/FTE)					

Source: CDR (EU) 2023/2772 supplementing Directive 2013/34/EU.

Table 80 Employee headcount by employment and region type.

	Reporting period:		
	Region A	**Region B**	**Total**
Number of employees (head count/FTE)			
Permanent employees (head count/FTE)			
Temporary employees (head count/FTE)			
Non-guaranteed hours employees (head count/FTE)			
Full-time equivalent employees (head count/FTE)			
Part-time equivalent employees (head count/FTE)			

Source: CDR (EU) 2023/2772 supplementing Directive 2013/34/EU.

2. Proportion of workers under contract with suppliers vs. independent workers: Provide information on this proportion. The following formula can be referred to:

   Proportion of Workers under Contract with Suppliers = (Number of Workers under Contract with Suppliers/Total Number of Contingent and/or Subcontracted Workers) × 100

   Proportion of Independent Workers = (Number of Independent Workers/Total Number of Contingent and/or Subcontracted Workers) × 100

3. Documented compliance with ethical recruitment practices and formal international standards or protocols: Indicate whether or not the organisation abides by these practices/standards.

4. Time series comparison: Provide a comparison of changes in company growth with changes in the proportion of contingent and/or subcontracted workers over the most recent five years.

11.3 Long-Term Work Contracts (S1-6, S1-11)

Relevant SDGs: SDG 1 (No Poverty), SDG 3 (Good Health and Well-being), SDG 8 (Decent Work and Economic Growth), SDG 10 (Reduced Inequalities)

This indicator measures the age of the organisation and the percentage of employees falling into different categories of contract length. The categories include 0–6 months, 6–12 months, 12–24 months, and more than 24 months. The following formula can be referred to:

Percentage of Employees in a Contract Length Category = (Number of Employees in the Category/Total Number of Employees) × 100

11.4 Union Density and Collective Bargaining Coverage (S1-8)

Relevant SDGs: SDG 1 (No Poverty), SDG 3 (Good Health and Well-being), SDG 8 (Decent Work and Economic Growth), SDG 10 (Reduced Inequalities)

This indicator assesses an organisation's commitment to workers' rights to unionise and engage in collective bargaining. It aims to identify whether the organisation discourages or forbids worker participation in trade unions, the legality of union membership, and the percentage of workers covered by collective bargaining agreements. The indicator also assesses the percentage of employees versus subcontracted or contingent workers in an organisation and its suppliers.

The organisation is required to report on several aspects, including whether they discourage, obstruct, or forbid worker participation in trade unions, legality of union membership, percentage of own workers and suppliers' workers belonging to a trade union, percentage of own workers and suppliers' workers covered by collective bargaining agreements, and the percentage of employees versus subcontracted or contingent workers in the organisation and its suppliers. The information is reported annually on a country-by-country basis over the most recent five-year period.

1. Determine the total number of employees in the reporting entity at the end of the reporting period. This can be expressed as headcount or full-time equivalent (FTE), but the chosen approach should be consistently applied between periods.
2. Identify the employees who are covered by collective bargaining agreements, expressing this number in terms of headcount or FTE, consistent with the denominator.
3. Calculate the percentage of employees covered by collective bargaining agreements by dividing the number of employees covered by collective agreements by the total workforce and multiplying by 100. The following formulae can be referred to:

- Percentage of Employees Belonging to a Trade Union = (Number of Employees Belonging to a Trade Union/Total Number of Employees) × 100
- Percentage of Workers Covered by Collective Bargaining Agreements = (Number of Workers Covered by Collective Bargaining Agreements/Total Number of Workers) × 100

Narrative information is crucial to provide context, as there may be instances where collective agreements are not permitted by regulators, not requested by employees, or not reached among relevant stakeholders.

Results should be tabulated:

Table 81 Collective bargaining coverage presentation.

	Collective bargaining coverage			Social dialogue
Coverage rate %	Own employees EEA (countries >50 FTE)	Non-employee workers EEA (countries > 50 FTE)	Own employees non-EEA (regions > 50 FTE)	Workplace representation EEA only (countries >50 FTE)
0%–19%				Country A
20%–39%		Country A	Region A	
40%–59%	Country A			Country B
60%–79%	Country B		Region B	
80%–100%		Country B		

Source: CDR (EU) 2023/2772 supplementing Directive 2013/34/EU.

11.5 Proportion of Women in Managerial Positions (S1-9)

Relevant SDGs: SDG 1 (No Poverty), SDG 3 (Good Health and Well-being), SDG 8 (Decent Work and Economic Growth), SDG 10 (Reduced Inequalities).

This indicator measures the percentage of women holding managerial positions within an organisation. It provides insight into the gender balance of leadership roles and can indicate progress toward gender equality in the workplace. The indicator is calculated by identifying the total number of managers in an organisation at the end of the reporting period, expressed as headcount or full-time equivalent (FTE), and determining the proportion of those managers who are women. The proportion of women in managerial positions (PWM) can be calculated using the following equation:

$$PWM = (TWM/TNM) \times 100$$

where:

PWM = percentage of women managers in the organisation.

TWM = total number of women managers in the organisation (headcount or FTE).

TNM = total number of all managers in the organisation (headcount or FTE).

The information should be reported annually for the past five years, and the same approach should be applied consistently between periods. PWM scores of ≥0.4 are considered sustainable, while PWM scores of <0.4 are considered unsustainable.

11.6 Number and Percentage of Women Board Members (S1-9)

Relevant SDGs: SDG 1 (No Poverty), SDG 3 (Good Health and Well-being), SDG 8 (Decent Work and Economic Growth), SDG 10 (Reduced Inequalities).

This indicator measures the representation of women on the board of directors or governance function of an organisation over a five-year trend, which is calculated as follows:

$$RWBt = WBt/40$$

where:

RWB = ratio of women members on a board (as a percentage) to the sustainability norm of no less than 40%.

WB = actual percentage of women on the board of directors or governance function of an organisation.

40 = normative minimum percentage of women on the board of directors or governance function of an organisation.

t = most recent year.

And where: RWB scores of ≥1.0 are sustainable, and RWB scores of <1.0 are unsustainable and then tabulated.

Table 82 Ratio of women on the board (RWB) template.

Year	T1	T2	T3	T4	T5
RWB					

Source: CDR (EU) 2023/2772 supplementing Directive 2013/34/EU.

11.7 Living Wage Gap (S1-10)

Relevant SDGs: SDG 1 (No Poverty), SDG 3 (Good Health and Well-being), SDG 8 (Decent Work and Economic Growth), SDG 10 (Reduced Inequalities).

The living wage gap measures the difference between the actual wages paid to employees and the normative living wages that represent the 50th percentile of wage norms for individuals in average households in specific countries. It aims to assess the extent to which employees are paid enough to meet basic needs and maintain a decent standard of living.

The actual wages and benefits paid to individual employees are calculated as annual compensation, including regular pay and in-kind benefits that reduce the amount of cash income needed for a decent living standard. The living wage gap is calculated by subtracting the actual wage from the normative living wage, and the sum of all individual gaps greater than zero is reported for the most recent five years. The living wage gap score of ≤0 is considered sustainable, while scores of >0 are unsustainable.

The formula for calculating the living wage gap based on the provided methodology is as follows:

Living Wage Gap = Max(0, Normative Living Wage – Actual Wage)

where:

Normative living wage refers to the wage level required for a decent living standard, which encompasses both regular pay and in-kind benefits that contribute to meeting essential needs.

Actual wage represents the annual compensation received by an individual employee, including regular pay and in-kind benefits that reduce the amount of cash income needed for a decent living standard.

11.8 Social Security Coverage Rate (S1-11)

Relevant SDGs: SDG 1 (No Poverty), SDG 3 (Good Health and Well-being), SDG 8 (Decent Work and Economic Growth), SDG 10 (Reduced Inequalities).

Social security coverage rate measures the percentage of the population covered by social security programs, such as pensions, healthcare, disability benefits, and unemployment benefits.

1. Determine the total population: Obtain the total population of the country or region under consideration.
2. Calculate the number of individuals covered by social security: Determine the number of individuals who are enrolled and eligible for social security benefits.
3. Calculate the social security coverage rate:

   Social Security Coverage Rate = (Number of Individuals Covered/Total Population) × 100

11.9 Social Protection Expenditure (S1-12)

Relevant SDGs: SDG 1 (No Poverty), SDG 3 (Good Health and Well-being), SDG 8 (Decent Work and Economic Growth), SDG 10 (Reduced Inequalities).

Social protection expenditure refers to the financial resources allocated by governments or organisations to provide social security benefits and support to individuals and families.

Identify social protection expenditure: Sum up the total expenditure allocated to social security programs, including pensions, healthcare, unemployment benefits, and other social support initiatives.

To calculate the social protection expenditure, the following formula can be used:

Social Protection Expenditure = Direct Expenditure + Indirect Expenditure

where:

'Direct Expenditure' refers to the amount of financial resources allocated specifically to social protection programs and initiatives. This includes funds spent on social assistance programs such as cash transfers, pensions, unemployment benefits, disability benefits, and other forms of direct support provided to individuals or households.

'Indirect Expenditure' represents the financial resources used to support social protection indirectly. This can include expenditures on administrative costs, program management, infrastructure, staff salaries, and other related expenses necessary for the functioning and implementation of social protection programs.

11.10 Hiring of Vulnerable Groups (S1-12)

Relevant SDGs: SDG 8 (Decent Work and Economic Growth), SDG 10 (Reduced Inequalities).

This indicator measures your organisation's commitment to social inclusivity by identifying vulnerable groups in society and disclosing the percentage of total permanent employees who belong to at least one of the identified groups. You need to identify vulnerable groups in society and calculate the percentage of employees who belong to at least one of these groups. The equation for this calculation is:

$$PVP = (EVP/TNE) \times 100$$

where:

PVP is the percentage of employees in vulnerable populations.
EVP is the number of employees in vulnerable populations.
TNE is the total number of employees.

11.11 Disability Inclusion in Workforce (S1-12)

Relevant SDGs: SDG 8 (Decent Work and Economic Growth), SDG 10 (Reduced Inequalities).

Disability inclusion in workforce measures the representation and participation of people with disabilities within an organisation's workforce.

1. Identify the total workforce: Count the total number of employees working within the organisation.
2. Determine the number of employees with disabilities: Identify the number of employees who self-disclose as individuals with disabilities.
3. Calculate the disability inclusion rate:

   Disability Inclusion Rate = (Number of Employees with Disabilities/ Total Workforce) × 100

11.12 Accessibility of Facilities (S1-12)

Relevant SDGs: SDG 10 (Reduced Inequalities), SDG 11 (Sustainable Cities and Communities).

Accessibility of facilities measures the extent to which an organisation's physical spaces and facilities are designed and adapted to accommodate the needs of people with disabilities.

1. Assess facility accessibility: Conduct an audit or assessment of the organisation's facilities to evaluate their compliance with accessibility standards and guidelines. This includes considering factors such as wheelchair accessibility, ramps, elevators, signage, parking spaces, and restroom facilities.
2. Assign accessibility scores: Assign scores or ratings to different aspects of facility accessibility based on compliance with accessibility standards. This can be done on a scale or through a categorical assessment (e.g., fully accessible, partially accessible, not accessible).
3. Calculate accessibility score: Aggregate the scores or ratings across different aspects of facility accessibility to obtain an overall accessibility score or rating.

Note: The methodology and sources of information for these indicators may vary depending on the specific organisation and its practices. It is important to consider local regulations, industry standards, and the specific needs of people with disabilities when conducting assessments and gathering data.

11.13 Average Hours of Training per Year per Employee (S1-13)

Relevant SDGs: SDG 3 (Good Health and Well-being), SDG 8 (Decent Work and Economic Growth).

The Average Hours of Training per Year per Employee indicator measures the average amount of training received by employees in a reporting period. It encompasses internal and external training programs, supported education, sabbatical periods with guaranteed return to employment, and training on specific topics such as health and safety. The indicator is calculated using the following equations:

Average training hours per employee = Total number of training hours provided to employees/Total number of employees

Average training hours per employee category = Total number of training hours provided to each category of employees/Total number of employees in the category

where:

The denominator can be expressed as either headcount or full-time equivalent (FTE) and should be consistently applied within and between periods.

Note that training expenses are usually recorded as a separate line item in the profit and loss statement under operating costs or general expenses. Accounts payable may contain a dedicated account for recording training costs, such as employee training expenses. Additionally, reference management accounting systems, internal management reports, and balanced scorecard reports as training indicators are often included as key performance indicators for organisational learning and growth.

11.14 Expenditure on Employee Training per Year per Employee (S1-13)

Relevant SDGs: SDG 3 (Good Health and Well-being), SDG 8 (Decent Work and Economic Growth).

The Expenditure on Employee Training per Year per Employee indicator measures the amount of financial investment in employee training and development. It considers both direct and indirect costs associated with training programs, including course fees, trainers' fees, training facilities, training equipment, and related travel expenses. The indicator can be calculated using the following equations:

Average training expenditures per employee = Total amount of training expenses/Total number of employees

Average training expenditures per employee category = Total amount of training expenses for each category of employees/Total number of employees in the category

where:

The denominator for these equations can be expressed as either headcount or full-time equivalent (FTE), and the chosen approach should be consistently applied within and between periods. It is recommended to provide a breakdown of the data by employment categories.

Training expenses should include the costs associated with course fees, trainers' fees, training facilities, training equipment, and travel expenses. Data sources may include financial records, accounts payable, and budgeting systems. Additionally, multinational organisations are encouraged to disclose training expenditures and hours of training by country and, if applicable, by gender. This promotes transparency and allows for a more comprehensive understanding of training investments across different contexts.

11.15 Frequency/Incident Rates of Occupational Injuries (S1-14)

Relevant SDGs: SDG 3 (Good Health and Well-being), SDG 8 (Decent Work and Economic Growth).

This indicator measures the rate of occupational injuries, accidents, and diseases in the workplace, taking into account the number of lost workdays as a result of these incidents.

The frequency rate is calculated as the number of new injury cases divided by the total number of hours worked by workers in the reporting period. The incident rate is calculated as the total number of lost days expressed in terms of the number of hours divided by the total number of hours worked by workers in the reporting period. Organisations must specify whether 'days' mean calendar days or scheduled workdays and when the lost-days count begins. Reporting on mental health and stress is encouraged, and multinational organisations should disclose this indicator by gender. Any occupational injuries, illnesses, and deaths resulting from an occupational accident are deemed unsustainable.

The formulas for calculating the frequency rate and incident rate of occupational injuries based on the provided methodology are as follows:

1. Frequency rate:

   Frequency Rate = (Number of New Injury Cases/Total Number of Hours Worked) × 1,000,000

   where:

   'Number of New Injury Cases' represents the total number of new injury cases that occurred in the reporting period.

   'Total Number of Hours Worked' refers to the cumulative number of hours worked by all workers within the organisation during the reporting period.

2. Incident rate:

   Incident Rate = (Total Number of Lost Days/Total Number of Hours Worked) × 1000

   where:

   'Total Number of Lost Days' represents the cumulative number of lost workdays due to occupational injuries, accidents, or diseases in the reporting period.

   Total Number of Hours Worked refers to the cumulative number of hours worked by all workers within the organisation during the reporting period.

Organisations must establish arrangements in accordance with national laws or regulations to record occupational accidents, occupational diseases, commuting accidents, dangerous occurrences, and incidents. A person authorised to prepare and keep records of all these occurrences should be identified, and organisations must prepare appropriate records for inspection purposes and as information for workers' representatives and health services.

11.16 Expenditures on Employee Health and Safety as a Proportion of Revenue (S1-14)

Relevant SDGs: SDG 3 (Good Health and Well-being), SDG 8 (Decent Work and Economic Growth).

The Expenditures on Employee Health and Safety as a Proportion of Revenue indicator measures the percentage of an entity's total revenue allocated to expenses related to occupational safety and health. It encompasses costs for insurance programs, direct financing of healthcare activities, and expenditures on working environment issues related to occupational safety and health. This indicator also includes reporting on mental health and stress, acknowledging their significance, particularly in the services sector. The calculation considers both operating costs and capital expenditures related to health and safety initiatives.

1. Total all the expenses incurred during the reporting period for occupational safety and health-related insurance programs, healthcare activities directly financed by your organisation, and costs associated with the working environment.
2. Divide the total expenditure on health and safety by the total revenue generated during the reporting period:

   Expenditures on Employee Health and Safety as a Proportion of Revenue = (Total Expenditures on Health and Safety/Total Revenue) × 100

The result is expressed as a percentage, representing the proportion of revenue allocated to employee health and safety expenditures.

Organisations may also maintain specific registers, such as a register of medical visits, as required by the legislation of the country of operation. For capital expenditures related to health and safety, relevant information can be sourced from capital budget reports. Operating costs and immediate expenses can be found in the organisation's profit and loss (P&L) statement and are normally categorised as production overheads or selling expenses. Additionally, multinational organisations are encouraged to disclose health and safety expenditures by country, similar to 11.15.

11.17 Employee Turnover Rate (S1-15, G1-1)

Relevant SDGs: SDG 3 (Good Health and Well-being), SDG 8 (Decent Work and Economic Growth).

This indicator measures the rate at which employees leave your organisation within a given year. You should determine and disclose the number of employees who left the entity during the year, divided by the average number of employees during that same year:

Employee turnover rate = (Number of employees who left during the year/Average number of employees during the year) × 100

11.18 Gender Pay Gap: Equality of Remuneration (S1-16)

Relevant SDGs: SDG 3 (Good Health and Well-being), SDG 5 (Gender Equality), SDG 8 (Decent Work and Economic Growth).

This indicator measures the gender pay gap at an organisation, both overall and by occupational level. Gender pay gaps can be calculated in terms of women's pay as a percentage of men's, and in a way that includes not only base salary or wages, but also compensation associated with incentives and rewards (but no overtime). All such calculations should not be adjusted in ways that take differences in other factors into account, such as hours worked, age, experience, or education. All calculations shall also include both full- and part-time employees, with all averages to be expressed in terms of the median rather than the mean. The formula for calculating the gender pay gap based on the provided methodology is as follows:

Gender Pay Gap = ((Median Men's Pay − Median Women's Pay)/Median Men's Pay) × 100

where:

'Median Men's Pay' refers to the median earnings of male employees within the organisation, considering base salary, wages, and compensation associated with incentives and rewards (excluding overtime).

'Median Women's Pay' represents the median earnings of female employees within the organisation, considering the same elements as mentioned above.

11.19 Employee Wages and Benefits as a Proportion of Revenue, with Breakdown by Employment Type and Gender (S1-10, S1-16)

Relevant SDGs: SDG 3 (Good Health and Well-being), SDG 5 (Gender Equality), SDG 8 (Decent Work and Economic Growth).

The Employee Wages and Benefits as a Proportion of Revenue indicator measures the proportion of an entity's total revenue that is allocated to employee wages and benefits. It provides insights into the company's commitment to compensating its workforce and can be further analysed by employment type and gender.

To calculate this indicator, follow these steps:

1. Compute the total payroll, which includes employee salaries and payments made to government institutions on behalf of employees (e.g., contributions, pensions, employment taxes, levies, and employment

funds). Exclude training costs, costs of protective equipment, or other expenses directly related to the employee's job function.

2. Divide the total amount of employee wages and benefits by the total revenue generated during the reporting period.

Employee Wages and Benefits as a Proportion of Revenue = (Total Employee Wages and Benefits/Total Revenue) × 100

Additionally, it is recommended to provide a breakdown and cross reference of employee wages and benefits by the following categories to match other data requirements:

- Employees and supervised workers.
- Type of employment contract (permanent or temporary).
- Type of employment (full-time or part-time).
- Age group: under 30 years, 30–50 years, over 50 years.
- Country and region.
- Gender.

11.20 CEO-to-Worker Pay Ratio (S1-16)

Relevant SDGs: SDG 3 (Good Health and Well-being), SDG 5 (Gender Equality), SDG 8 (Decent Work and Economic Growth).

This indicator measures the ratio between the total compensation of the CEO and the median compensation of the lowest-paid quartile of workers, expressed as a score. A score of ≤1.0 is considered sustainable, while a score of >1.0 is unsustainable. The CEO's total compensation includes salary, bonuses, benefits, restricted stock grants, long-term incentive pay-outs, and options realised and/or granted. Worker compensation is defined as wages and all benefits (excluding overtime). The equation to calculate the CEO-to-worker pay ratio is:

CWP = (CC/WC)/30where:
CWP is the CEO:worker pay ratio.
CC is the CEO compensation.
WC is the median compensation of the lowest-paid quartile of workers.
t is a specific year.
30 is the normative maximum CEO:worker pay ratio in any year.

11.21 Distribution of Surplus/Profits (S1-16)

Relevant SDGs: SDG 3 (Good Health and Well-being), SDG 5 (Gender Equality), SDG 8 (Decent Work and Economic Growth).

This indicator measures the distribution of a company's surplus or profits among various stakeholders, including members, workers, ESOPs, shareholders, and reinvestment in the organisation or other purposes or programs. It includes six sub-indicators.

The total surplus/profits distributed (TSP) is calculated as the sum of the amounts distributed to members/workers/producers (MWP), ESOPs, shareholders (SHR), reinvestment in the organisation (RIO), and other purposes or programs (OTH). The percentage of surplus/profits distributed to each stakeholder group is then calculated by dividing their respective amounts by the total surplus/profits distributed (TSP). The formula for the distribution of surplus/profits is:

- Distribution of surplus/profits = $TSPt = MWPt + ESOt + SHRt + RIOt + OTHt$
- Percentage of surplus/profits distributed to MWP = $MWPt/TSPt$
- Percentage of surplus/profits distributed to ESO = $ESOt/TSPt$
- Percentage of surplus/profits distributed to SHR = $SHRt/TSPt$
- Percentage of surplus/profits distributed to RIO = $RIOt/TSPt$
- Percentage of surplus/profits distributed to OTH = $OTHt/TSPt$

11.22 Fines Paid or Payable Due to Settlements (S1-17, G1-4)

Relevant SDGs: SDG 3 (Good Health and Well-being), SDG 5 (Gender Equality), SDG 8 (Decent Work and Economic Growth)

All fines paid or payable by your organisation due to unlawful behaviours, including corruption, must be summed up and disclosed on an annual basis. Any fines paid or payable due to settlements, attributable to unlawful behaviours of any type, by your organisation in any one year will be deemed unsustainable.

Your organisation must total all fines paid or payable by your organisation due to unlawful behaviours on an annual basis and disclose them.

11.23 Harassment and Discrimination at the Workplace (S1-17, G1-1)

Relevant SDGs: SDG 3 (Good Health and Well-being), SDG 5 (Gender Equality), SDG 8 (Decent Work and Economic Growth).

This indicator measures whether your organisation has policies, training courses, or mechanisms in place to address harassment and discrimination at the workplace. It also assesses whether there are any incidents of harassment and discrimination, and if so, whether they have been reported to designated units or senior management, and whether there are safeguards in place to prevent retaliation and protect the confidentiality of the complainant.

Your organisation should disclose whether it has policies, training courses, or mechanisms to address harassment and discrimination at the workplace. It should also disclose whether any incidents of harassment and discrimination have occurred and, if so, whether they have been reported to the designated unit, senior management, or board of directors. Your organisation should also disclose whether safeguards are in place to prevent retaliation and protect the confidentiality of the complainant.

12

S2 Value Chain Workers

In this chapter: Assessment of value chain workers' impacts and actions to prevent and mitigate negative impacts. Management of value chain workers' dependencies and their financial effects. Risks and opportunities related to value chain workers' impacts and their financial effects. Common indicators and metrics.

In today's interconnected global economy, the value chain extends beyond the boundaries of an undertaking, encompassing various stakeholders, suppliers, and business relationships. Recognising the significance of workers within the value chain is crucial to ensure their well-being, rights, and equitable treatment throughout all stages of production and distribution. Impacts on value chain workers are either caused or contributed to by your organisation, and they encompass a range of positive and negative actual or potential effects.

Accordingly, the objective of this methodology is to establish a framework for disclosing information that sheds light on the impacts and actions related to value chain workers. Additionally, the methodology highlights associated risks and opportunities related to your organisation's impacts and dependencies on workers, as well as the financial implications across different time horizons. In summary, it is imperative for organisations to follow a systematic and comprehensive approach in identifying and managing the material impacts on workers. By prioritising fair working conditions, ensuring equal treatment and opportunities for all workers, and upholding fundamental labour rights, organisations can strive to create a positive and sustainable impact on the workers within its value chain. These efforts contribute to a responsible and ethical approach to value chain management, aligning with the principles of sustainability and social responsibility.

In the following chapter:

- None of the data points are applicable to your ESG report if the topic is deemed not material, with the exception of IRO1 that are to be disclosed irrespective of the outcome of your materiality assessment.
- Not all DPs are to be reported if the undertaking has not adopted the respective policies, implemented the respective actions, or set the respective targets in relation to a sustainability matter that has been assessed to be material.
- Organisations with less than 750 employees during the financial year of their reporting may omit all data points in this chapter for the first two years of preparation, unless deemed to be material.

ESRS S2: VALUE CHAIN WORKERS		
GOVERNANCE & STRATEGY	**IMPACTS, RISKS, & OPPORTUNITIES**	**METRICS & TARGETS**
INCLUDES CROSS-CUTTING DISCLOSURES: • SBM-2: Views and interests of *stakeholders* • SBM-3: *Material* impacts, risks, and opportunities and their interaction with strategy and business model(s) • Information on **violations of UNGC principles and OECD guidelines, monitoring processes and compliance**	• S2-1: *Policies* related to value chain workers • S2-2: Processes for *engaging* with value chain workers about impacts • S2-3: Processes to *remediate negative impacts* and channels for value chain workers to raise concerns • S2-4: **Taking action** on material impacts on value chain workers, and approaches to **mitigating** material risks and pursuing material opportunities related to value chain workers, and effectiveness of those actions	• S2-5: *Targets* and *tracking effectiveness* of policies, actions, and targets related to managing material negative impacts, advancing positive impacts, and managing material risk and opportunities

Before diving into detailed metrics and targets, begin with describing your materiality assessment and organisational governance initiatives that directly interact with value chain workers goals, most importantly, the processes and methodologies to identify and assess impacts, risks, and opportunities related to value chain workers issues (see **Chapter 3: Single and Double Materiality Assessment and 12.1 Value Chain Workers Assessment Methodology**).

Table 83 Value chain workers assessment indicators relating to policies and stakeholder engagement.

ESRS	DR	Paragraph	Name	Data type	Tips and tricks	Related CDR	VSMEs	GRI 2-25
S2	SBM-3	11	All value chain workers who can be materially impacted by undertaking are included in the scope of disclosure under ESRS 2	Semi-narrative	To what extent do material impacts, risks, and opportunities associated with value chain workers interact with strategy and business model – use stakeholder engagement, materiality assessment and 12.1 Value chain worker assessment methodology			
S2	SBM-3	11 a)	Description of types of value chain workers subject to material impacts	Narrative				
S2	SBM-3	11 a i)-v)	Type of value chain workers subject to material impacts by own operations or through value chain	Semi-narrative				
S2	SBM-3	11 b	Disclosure of geographies or commodities for which there is significant risk of child labour, or of forced or compulsory labour, among workers in undertaking's value chain	Narrative		SFDR		GRI 409-1
S2	SBM-3	11 c	Material negative impacts occurrence on value chain workers	Semi-narrative				GRI 401-3-3, GRI 402-3-3, GRI 403-3-3, GRI 404-3-3, GRI 405-3-3, GRI 406-3-3, GRI 407-3-3, GRI 408-3-3, GRI 409-3-3
S2	SBM-3	11 d	Description of activities that result in positive impacts and types of value chain workers that are positively affected or could be positively affected	Narrative				
S2	SBM-3	11 e	Description of material risks and opportunities arising from impacts and dependencies on value chain workers	Narrative				

S2	SBM-3	12	Disclosure of whether and how the undertaking has developed an understanding of how workers with particular characteristics, those working in particular contexts, or those undertaking particular activities may be at greater risk of harm	Narrative				GRI2-29
S2	SBM-3	13	Disclosure of which material risks and opportunities arising from impacts and dependencies on value chain workers are impacts on specific groups	Narrative				
ESRS 2	S2		Disclosures to be reported in case the undertaking has not adopted policies and/or actions and/or targets	Narrative				
S2	S2-1	16	Policies to manage material impacts, risks, and opportunities related to value chain workers, including for specific groups of value chain workers or all value chain workers	MDR-P	Policies related to value chain workers – use stakeholder engagement assessment with materiality assessment, 5.1.1 MDR-P, 15.9 Access to remedy and 12.2 Supplier compliance with labour laws and regulations		V	GRI 2-23, GRI 401-3-3, GRI 402-3-3, GRI 403-3-3, GRI 404-3-3, GRI 405-3-3, GRI 406-3-3, GRI 407-3-3, GRI 408-3-3, GRI 409-3-3
S2	S2-1	17	Description of relevant human rights policy commitments relevant to value chain workers	Narrative		SFDR		GRI 2-23
S2	S2-1	17a	Disclosure of general approach in relation to respect for human rights relevant to value chain workers	Narrative		SFDR		GRI 2-23
S2	S2-1	17b	Disclosure of general approach in relation to engagement with value chain workers	Narrative	12.3 Labour rights and freedom of association in the supply chain	SFDR		GRI 2-23, GRI 2-29

(*Continued*)

Table 83 (*Continued*).

ESRS	DR	Paragraph	Name	Data type	Tips and tricks	Related CDR	VSMEs	GRI 2-25
S2	S2-1	17 c	Disclosure of general approach in relation to measures to provide and. or enable remedy for human rights impacts	Narrative	12.4 Percentage of workers in the supply chain covered by collective bargaining agreements See also: – UN Guiding Principles on Business & Human Rights[1] – UN Global Compact Principles[2] – OECD Guidelines for Multinational Enterprises[3] – ILO Declaration of Fundamental Rights and Principles at Work[4]	SFDR		GRI 2-23, GRI 2-25, GRI 401-3-3, GRI 402-3-3, GRI 404-3-3, GRI 405-3-3, GRI 406-3-3, GRI 407-3-3, GRI 408-3-3, GRI 409-3-3
S2	S2-1	18	Sustainability matters addressed by policy	Semi-narrative				GRI 408-3-3, GRI 408-1, GRI 409-3-3, GRI 409-1
S2	S2-1	18	Supplier Code of Conduct regarding value chain workers					
S2	S2-1	19	Description of how policies are aligned with relevant internationally recognised instruments					
S2	S2-1	19	Disclosure of extent and indication of nature of cases of non-respect of the UN Guiding Principles on Business and Human Rights, ILO Declaration on Fundamental Principles and Rights at Work or OECD Guidelines for Multinational Enterprises that involve value chain workers	Narrative				GRI 2-23
S2	S2-1	AR 12	Disclosure of explanations of significant changes to policies adopted during reporting year	Narrative			V	
S2	S2-1	AR 16	Disclosure on an illustration of the types of communication of its policies to those individuals, group of individuals, or entities for whom they are relevant	Narrative			V	GRI 2-23

S2	S2-2	22	Disclosure of how perspectives of value chain workers inform decisions or activities aimed at managing actual and potential impacts	Narrative	Process for engaging with value chain workers about impacts, related processes to mediate them, and channels to raise concerns – use stakeholder engagement with value chain workers assessment and material indicators, and specifically related to SBM-3: Impacts, Risks and Opportunities, plus: 12.5 Supplier diversity and inclusion metrics 12.6 Supplier employee wages and benefits compared to living wage standards			GRI 3-3, GRI 2-29, GRI 401-3-3, GRI 402-3-3, GRI 403-3-3, GRI 404-3-3, GRI 405-3-3, GRI 406-3-3, GRI 407-3-3, GRI 408-3-3, GRI 409-3-3
S2	S2-2	22 a	Engagement occurs with value chain workers or their legitimate representatives directly, or with credible proxies	Semi-narrative				GRI 3-3, GRI 2-29, GRI 402-3-3, GRI 403-3-3, GRI 404-3-3, GRI 405-3-3, GRI 406-3-3, GRI 407-3-3, GRI 408-3-3, GRI 409-3-3
S2	S2-2	22 b	Disclosure of stage at which engagement occurs, type of engagement, and frequency of engagement	Narrative				GRI 3-3, GRI 2-29, GRI 402-3-3, GRI 403-3-3, GRI 404-3-3, GRI 405-3-3, GRI 406-3-3, GRI 407-3-3, GRI 408-3-3, GRI 409-3-3
S2	S2-2	22 c	Disclosure of function and most senior role within undertaking that has operational responsibility for ensuring that engagement happens and that results inform undertakings approach	Narrative				GRI 3-3, GRI 402-3-3, GRI 403-3-3, GRI 404-3-3, GRI 405-3-3, GRI 406-3-3, GRI 407-3-3, GRI 408-3-3, GRI 409-3-3
S2	S2-2	22 d	Disclosure of Global Framework Agreement[5] or other agreements related to respect of human rights of workers	Narrative				GRI 3-3, GRI 402-3-3, GRI 403-3-3, GRI 404-3-3, GRI 405-3-3, GRI 406-3-3, GRI 407-3-3, GRI 408-3-3, GRI 409-3-3

(*Continued*)

[1] https://www.ohchr.org/sites/default/files/documents/publications/guidingprinciplesbusinesshr_en.pdf
[2] https://unglobalcompact.org/what-is-gc/mission/principles
[3] https://mneguidelines.oecd.org/mneguidelines/
[4] https://www.ilo.org/ilo-declaration-fundamental-principles-and-rights-work
[5] https://www.industriall-union.org/global-framework-agreements

Table 83 (*Continued*).

ESRS	DR	Paragraph	Name	Data type	Tips and tricks	Related CDR	VSMEs	GRI 2-25
S2	S2-2	22 e	Disclosure of how effectiveness of engagement with value chain workers is assessed	Narrative				GRI 3-3, GRI 2-29, GRI 402-3-3, GRI 403-3-3, GRI 404-3-3, GRI 405-3-3, GRI 406-3-3, GRI 407-3-3, GRI 408-3-3, GRI 409-3-3
S2	S2-2	23	Disclosure of steps taken to gain insight into perspectives of value chain workers/consumers and end-users that may be particularly vulnerable to impacts and/or marginalised	Narrative				GRI 2-29
S2	S2-2	24	Statement in case the undertaking has not adopted a general process to engage with its own workforce	Narrative				
S2	S2-2	24	Disclosure of timeframe for adoption of general process to engage with value chain workers in case the undertaking has not adopted a general process for engagement	Narrative			V	
S2	S2-2	AR 17	Type of role or function handling with engagement	Semi-narrative			V	
S2	S2-3	27 a	Disclosure of general approach to and processes for providing or contributing to remedy where undertaking has identified that it connected with a material negative impact on value chain workers	Narrative				GRI 2-25

S2	S2-3	27 b	Disclosure of specific channels in place for value chain workers to raise concerns or needs directly with undertaking and have them addressed	Narrative			GRI 2-25
S2	S2-3	27 c	Disclosure of processes through which undertaking supports or requires availability of channels	Narrative			
S2	S2-3	27 d	Disclosure of how issues raised and addressed are tracked and monitored and how effectiveness of channels is ensured	Narrative			GRI 2-26
S2	S2-3	28	Disclosure of how it is assessed that value chain workers are aware of and trust structures or processes as a way to raise their concerns or needs and have them addressed	Narrative			GRI 2-25
S2	S2-3	29	Disclosure of timeframe for channel or processes for raising concerns to be in place in case the undertaking has not adopted a general process for engagement	Narrative		V	
S2	S2-3	AR 25	Grievances are treated confidentially and with respect to rights of privacy and data protection	Semi-narrative		V	
S2	S2-3	AR 25	Value chain workers are allowed to use anonymously channels to raise concerns or needs	Semi-narrative		V	

Source: CDR (EU) 2023/2772 supplementing Directive 2013/34/EU, *GRI 1, 2, 3: 11, 12, 13, 14, 306, 101, 201, 202, 203, 204, 205, 206, 207,301, 302, 303, 304, 305, 306, 308,401, 402, 403, 404, 405, 406, 407, 408, 409, 410, 411, 413, 414, 415, 416, 417, 418*, and the author.

12.1 Value Chain Workers Assessment Methodology

Due to a lack or proximity and direct authority, value chain workers can be difficult to engage with, so the focus of disclosures is how your organisation reaches out to them and harnesses their inputs, as well as indicators for critical issues.

The methodology for assessing the impacts, risks, and opportunities related to workers in the value chain begins with a comprehensive analysis of your organisation's operations, products, services, and business relationships to identify key partners and potential issues. Through stakeholder engagement with value chain workers and their representatives, you can then understand and disclose material impacts, and the associated risks and opportunities. The methodology comprises the following components:

1. Identification of material impacts:

- Identify and evaluate the material positive and negative impacts on workers in the value chain caused or contributed to by your organisation's operations and relationships. Consider aspects such as labour rights, working conditions, occupational health and safety, fair wages, human rights, and social well-being.
- Assess the extent and significance of these impacts to determine their materiality.

2. Actions and outcomes:

- Disclose the actions taken by your organisation to prevent, mitigate, or remediate actual or potential negative impacts on workers in the value chain. Provide a transparent account of the measures implemented to ensure fair labour practices, supply chain transparency, responsible sourcing, and supplier engagement.
- Report on the outcomes and results of these actions to demonstrate the effectiveness of your organisation's initiatives and their impact on improving workers' conditions.

3. Analysis of material risks and opportunities:

- Analyse the nature, type, and extent of the material risks associated with the organisation's impacts and dependencies on workers in the value chain. Identify potential opportunities for enhancing workers' rights, welfare, and empowerment throughout the value chain.

- Conduct a thorough assessment of these risks and opportunities, considering short-term and long-term implications for your organisation and the value chain workers.

4. Management of value chain dependencies:

- Recognise the dependencies of your organisation on value chain workers and the influence of these dependencies on your organisation's operations and reputation. Develop strategies and initiatives to manage and improve workers' conditions, human rights, and social sustainability within the value chain.
- Collaborate with suppliers, contractors, and other business partners to foster responsible and ethical practices across the value chain.

5. Financial effects:

- Assess the financial effects of material risks and opportunities arising from your organisation's impacts and dependencies on workers in the value chain. Consider short-term, medium-term, and long-term time horizons to capture the financial implications over different planning periods.
- Quantify the potential costs, investments, savings, and returns associated with managing value chain-related risks and capitalising on opportunities.

The goal of this methodology is to gain a holistic understanding of your organisation's impacts on value chain workers, take appropriate actions to mitigate risks and enhance worker well-being, and make informed decisions that promote fair and responsible practices throughout the value chain. In combination with the following indicators and metrics, this methodology facilitates transparent reporting, enabling stakeholders to assess the organisation's commitment to workers' rights and social sustainability within its value chain.

Table 84 Value chain workers assessment indicators relating to actions and targets.

ESRS	DR	Paragraph	Name	Data type	Tips and tricks	Related CDR	VSMEs	GRI 2-25
S2	S2-4	MDR-A	Action plans and resources to manage its material impacts, risks, and opportunities related to value chain workers	MDR-A	Disclose all planned or underway actions to prevent, mitigate, or remediate material negative impacts by reviewing stakeholder engagement with own workforce assessment methodology, 5.1.2 MDR-A, and 12.10 Worker empowerment programs or initiatives in the supply chain and where appropriate cross-reference with: 11.1.1 Worker engagement index and 11.1.2 Frequency of worker-management communication			GRI 401-3-3
S2	S2-4	32 a	Description of action planned or underway to prevent, mitigate, or remediate material negative impacts on value chain workers	Narrative				GRI 2-25, GRI 401-3-3, GRI 402-3-3, GRI 403-3-3, GRI 403-7, GRI 404-3-3, GRI 405-3-3, GRI 406-3-3, GRI 407-3-3, GRI 408-3-3, GRI 409-3-3
S2	S2-4	32 b	Description of whether and how action to provide or enable remedy in relation to an actual material impact	Narrative				GRI 2-25, GRI 401-3-3, GRI 402-3-3, GRI 403-3-3, GRI 403-7, GRI 404-3-3, GRI 405-3-3, GRI 406-3-3, GRI 407-3-3, GRI 408-3-3, GRI 409-3-3
S2	S2-4	32 c	Description of additional initiatives or processes with primary purpose of delivering positive impacts for value chain workers	Narrative				GRI 2-25, GRI 401-3-3, GRI 402-3-3, GRI 403-3-3, GRI 403-7, GRI 404-3-3, GRI 405-3-3, GRI 406-3-3, GRI 407-3-3, GRI 408-3-3, GRI 409-3-3
S2	S2-4	32 d	Description of how effectiveness of actions or initiatives in delivering outcomes for value chain workers is tracked and assessed	Narrative				GRI 2-25, GRI 401-3-3, GRI 402-3-3, GRI 403-3-3, GRI 403-7, GRI 404-3-3, GRI 405-3-3, GRI 406-3-3, GRI 407-3-3, GRI 408-3-3, GRI 409-3-3

S2	S2-4	33 a	Description of approach to identifying what action is needed and appropriate in response to particular actual or potential material negative impact on value chain workers	Narrative				GRI 2-25, GRI 3-3, GRI 401-3-3, GRI 402-3-3, GRI 403-3-3, GRI 403-7, GRI 404-3-3, GRI 405-3-3, GRI 406-3-3, GRI 407-3-3, GRI 408-3-3, GRI 409-3-3
S2	S2-4	33 b	Description of approach to taking action in relation to specific material impacts on value chain workers	Narrative				GRI 2-25, GRI 3-3, GRI 401-3-3, GRI 402-3-3, GRI 403-3-3, GRI 403-7, GRI 404-3-3, GRI 405-3-3, GRI 406-3-3, GRI 407-3-3, GRI 408-3-3, GRI 409-3-3
S2	S2-4	33 c	Description of approach to ensuring that processes to provide or enable remedy in the event of material negative impacts on value chain workers are available and effective in their implementation and outcomes	Narrative				GRI 3-3
S2	S2-4	34 a	Description of what action is planned or underway to mitigate material risks arising from impacts and dependencies on value chain workers and how effectiveness is tracked	Narrative				
S2	S2-4	34 b	Description of what action is planned or underway to pursue material opportunities in relation to value chain workers	Narrative				
S2	S2-4	35	Disclosure of how it is ensured that own practices do not cause or contribute to material negative impacts on value chain workers	Narrative				

(*Continued*)

Table 84 (*Continued*).

ESRS	DR	Paragraph	Name	Data type	Tips and tricks	Related CDR	VSMEs	GRI 2-25
S2	S2-4	36	Disclosure of severe human rights issues and incidents connected to upstream and downstream value chain	Narrative		SFDR		GRI 402-3-3, GRI 401-3-3, GRI 402-3-3, GRI 403-3-3, GRI 404-3-3, GRI 405-3-3, GRI 406-3-3, GRI 407-3-3, GRI 408-3-3, GRI 409-3-3
S2	S2-4	38	Disclosure of resources allocated to management of material impacts	Narrative				
S2	S2-4	AR 28 a	Disclosure of general and specific approaches to addressing material negative impacts	Narrative			V	
S2	S2-4	AR 28 b	Disclosure of initiatives aimed at contributing to additional material positive impacts	Narrative			V	
S2	S2-4	AR 28 c	Disclosure of how far undertaking has progressed in efforts during the reporting period	Narrative			V	
S2	S2-4	AR 28 d	Disclosure of aims for continued improvement	Narrative			V	
S2	S2-4	AR 30	Disclosure of how undertaking seeks to use leverage with relevant business relationships to manage material negative impacts affecting value chain workers	Narrative			V	GRI 2-24
S2	S2-4	AR 31	Disclosure of how participation in industry or multi-stakeholder initiative and undertaking's own involvement is aiming to address material impacts	Narrative			V	GRI 2-25, GRI 3-3

S2	S2-4	AR 36 a)	Disclosure of how value chain workers and legitimate representatives or their credible proxies play a role in decisions regarding design and implementation of programmes or processes	Narrative			V	GRI 3-3, GRI 401-3-3, GRI 402-3-3, GRI 406-3-3, GRI 407-3-3, GRI 408-3-3, GRI 409-3-3
S2	S2-4	AR 36 b)	Information about intended or achieved positive outcomes of programmes or processes for value chain workers	Narrative			V	
S2	S2-4	AR 37	Initiatives or processes whose primary aim is to deliver positive impacts for value chain workers are designed also to support achievement of one or more of Sustainable Development Goals	Semi-narrative			V	GRI 203-2
S2	S2-4	AR 43	Information about whether and how process(es) to manage material risks related to value chain workers are integrated into existing risk management process(es)	Narrative				
S2	S2-4	AR 44	Description of internal functions that are involved in managing impacts and types of action taken by internal functions to address negative and advance positive impacts	Narrative			V	
S2	S2-5	41	Targets set to manage material impacts, risks, and opportunities related to value chain workers [see ESRS 2–MDR-T]	MDR-T	Disclose all planned or underway targets and specific metrics to prevent, mitigate, or remediate material negative impacts by reviewing stakeholder engagement with own workforce assessment			
S2	S2-5	42 a	Disclosure of how value chain workers, their legitimate representatives, or credible proxies were engaged directly in setting targets	Narrative				

(Continued)

Table 84 (*Continued*).

ESRS	DR	Paragraph	Name	Data type	Tips and tricks	Related CDR	VSMEs	GRI 2-25
S2	S2-5	42 b	Disclosure of how value chain workers, their legitimate representatives, or credible proxies were engaged directly in tracking performance against targets	Narrative	methodology, 5.1.3 MDR-M, 5.1.4 MDR-T, and 12.2 to 12.12			GRI 3-3, GRI 402-1, GRI 401-3-3, GRI 402-3-3, GRI 403-3-3, GRI 404-3-3, GRI 405-3-3, GRI 406-3-3, GRI 407-3-3, GRI 408-3-3, GRI 409-3-3
S2	S2-5	42 c	Disclosure of how value chain workers, their legitimate representatives, or credible proxies were engaged directly in identifying lessons or improvements as result of undertaking's performance	Narrative				GRI 3-3, GRI 402-1, GRI 401-3-3, GRI 40-3-3, GRI 403-3-3, GRI 404-3-3, GRI 405-3-3, GRI 406-3-3, GRI 407-3-3, GRI 408-3-3, GRI 409-3-3
S2	S2-5	AR 45 a	Disclosure of intended outcomes to be achieved in lives of value chain workers	Narrative			V	
S2	S2-5	AR 45 b	Information about stability over time of target in terms of definitions and methodologies to enable comparability	Narrative			V	
S2	S2-5	AR 45 c	Disclosure of references to standards or commitments on which target is based	Narrative			V	

Source: CDR (EU) 2023/2772 supplementing Directive 2013/34/EU, *GRI 1, 2, 3: 11, 12, 13, 14, 306, 101, 201, 202, 203, 204, 205, 206, 207,301, 302, 303, 304, 305, 306, 308,401, 402, 403, 404, 405, 406, 407, 408, 409, 410, 411, 413, 414, 415, 416, 417, 418*, and the author.

12.2 Supplier Compliance with Labour Laws and Regulations (S2-1)

Relevant SDG: SDG 8 (Decent Work and Economic Growth).

This indicator assesses the extent to which suppliers in your organisation's value chain comply with labour laws and regulations. It focuses on adherence to local and international labour standards:

1. Conduct regular supplier audits to evaluate compliance with labour laws and regulations.
2. Develop a scoring system or checklist based on relevant labour standards, such as minimum wage, working hours, child labour, and occupational health and safety.
3. Assign scores for each compliance criterion during the audit.
4. Calculate the compliance score for each supplier by summing the scores obtained across all compliance criteria.
5. Express compliance as a percentage by dividing the compliance score by the maximum possible score and multiplying by 100.

12.3 Labour Rights and Freedom of Association in the Supply Chain (S2-1)

Relevant SDG: SDG 8 (Decent Work and Economic Growth).

This indicator evaluates the respect for labour rights and freedom of association within your organisation's supply chain. It assesses the extent to which workers can exercise their right to join or form trade unions and engage in collective bargaining.

Methodology:

1. Conduct assessments and audits to evaluate the presence and effectiveness of trade unions and collective bargaining agreements within the supply chain.
2. Develop a scoring system or checklist based on internationally recognised labour rights frameworks, such as ILO conventions.
3. Evaluate the presence of trade unions, workers' freedom to associate, and collective bargaining agreements.
4. Assign scores for each criterion during the assessment.

5. Calculate the overall labour rights and freedom of association score for each supplier by summing the scores obtained across all criteria.
6. Express the labour rights and freedom of association score as a percentage by dividing the obtained score by the maximum possible score and multiplying by 100.

12.4 Percentage of Workers in the Supply Chain Covered by Collective Bargaining Agreements (S2-1)

Relevant SDG: SDG 8 (Decent Work and Economic Growth).

This indicator measures the proportion of workers in your organisation's supply chain who are covered by collective bargaining agreements. It assesses the extent to which workers have negotiated terms and conditions of employment.

Methodology:

1. Identify the total number of workers in the supply chain (TW) and the number of workers covered by collective bargaining agreements (TCB).
2. Calculate the percentage of workers covered by collective bargaining agreements using the formula:

   Percentage of workers covered = (TCB/TW) × 100

12.5 Supplier Diversity and Inclusion Metrics (S2-1, S2-5)

Relevant SDG: SDG 5 (Gender Equality) and SDG 10 (Reduced Inequalities).

This indicator assesses your organisation's efforts to promote diversity and inclusion within its supply chain. It evaluates the representation of underrepresented groups among the organisation's suppliers.

1. Collect demographic data on suppliers' ownership, management, and workforce composition, including gender and minority representation.
2. Calculate diversity metrics using the following formulas:

   Percentage of women-owned suppliers = (Number of women-owned suppliers/Total number of suppliers) × 100

   Percentage of minority-owned suppliers = (Number of minority-owned suppliers/Total number of suppliers) × 100

Percentage of diverse employees within supplier organisations = (Number of diverse employees/Total number of employees) × 100

12.6 Supplier Employee Wages and Benefits Compared to Living Wage Standards (S2-1, S2-5)

Relevant SDG: SDG 1 (No Poverty) and SDG 8 (Decent Work and Economic Growth).

This indicator assesses whether supplier employees receive wages and benefits that meet or exceed the living wage standards in their respective regions. It focuses on ensuring fair compensation for workers.

1. Gather data on supplier employee wages and benefits through payroll records, surveys, or supplier self-disclosures.
2. Determine the living wage standards specific to each region or industry.
3. Compare the supplier employee wages with the living wage benchmarks using the following formula:

   Compliance percentage = (Number of employees earning above living wage/Total number of employees) × 100.

12.7 Supplier Training and Capacity Building Programs (S2-2)

Relevant SDG: SDG 8 (Decent Work and Economic Growth).

This indicator assesses the presence and effectiveness of training and capacity building programs implemented by suppliers to enhance workers' skills, knowledge, and overall well-being. It focuses on initiatives that support professional development, improve working conditions, and promote worker empowerment.

1. Collect data on supplier training programs, including the number of training sessions, topics covered, number of participants, and outcomes.
2. Evaluate the relevance and impact of training initiatives through pre- and post-training assessments, performance evaluations, or feedback mechanisms.
3. Calculate the training effectiveness score for each supplier by considering factors such as knowledge gain, skill improvement, and behavioural changes.

4. Express the training effectiveness score as a percentage by dividing the obtained score by the maximum possible score and multiplying by 100.

12.8 Health and Safety Incidents Reported by Suppliers (S2-3)

Relevant SDG: SDG 3 (Good Health and Well-being).

This indicator measures the number and severity of health and safety incidents reported by suppliers in your organisation's value chain. It focuses on ensuring safe and healthy working conditions for supplier employees.

1. Collect incident reports from suppliers, including data on the number, type, and severity of health and safety incidents.
2. Calculate the incident rate per employee or per hour worked using the formula:

   Incident rate = (Total number of incidents/Total number of employees or hours worked) × 1000.
3. Analyse incident trends over time and compare rates across suppliers.

12.9 Supplier Employee Satisfaction and Well-being Indicators (S2-3)

Relevant SDG: SDG 3 (Good Health and Well-being) and SDG 8 (Decent Work and Economic Growth).

This indicator assesses the satisfaction and well-being of employees within the organisation's supply chain. It evaluates factors such as job satisfaction, work–life balance, mental health, and overall employee well-being.

1. Conduct employee satisfaction surveys within the supply chain, focusing on factors related to job satisfaction, work environment, work–life balance, and overall well-being.
2. Use a standardised survey instrument or develop a custom survey based on industry best practices.
3. Analyse survey responses quantitatively to calculate satisfaction scores for each supplier.
4. Consider additional qualitative feedback through focus groups or interviews to gain deeper insights into employee well-being.

12.10 Worker Empowerment Programs or Initiatives in the Supply Chain (S2-4)

Relevant SDG: SDG 8 (Decent Work and Economic Growth).

This indicator evaluates the presence and effectiveness of programs or initiatives implemented in the supply chain to empower workers, promote their rights, and enhance their well-being.

1. Identify and assess the existence and impact of worker empowerment programs or initiatives within the supply chain.
2. Develop a scoring system to evaluate the presence and effectiveness of initiatives, considering factors such as training programs, grievance mechanisms, worker representation, and participation in decision-making.
3. Assign scores to each criterion during the assessment.
4. Calculate the overall worker empowerment score for each supplier by summing the scores obtained across all criteria.
5. Express the worker empowerment score as a percentage by dividing the obtained score by the maximum possible score and multiplying by 100.

12.11 Disability Inclusion in Value Chain Workforce (S2-5)

Relevant SDGs: SDG 8 (Decent Work and Economic Growth) and SDG 10 (Reduced Inequalities).

Disability Inclusion in Workforce measures the representation and participation of people with disabilities within an organisation's workforce.

1. Identify the total workforce: Count the total number of employees working within the organisation.
2. Determine the number of employees with disabilities: Identify the number of employees who self-disclose as individuals with disabilities.
3. Calculate the disability inclusion rate: Divide the number of employees with disabilities by the total workforce and multiply the result by 100.

   Disability Inclusion Rate = (Number of Employees with Disabilities/ Total Workforce) × 100

12.12 Supplier Turnover Rates (S2-5)

Relevant SDG: SDG 8 (Decent Work and Economic Growth)

This indicator measures the rate of employee turnover within suppliers in your organisation's value chain. It assesses the stability and retention of the workforce, which can be indicative of factors such as job satisfaction, working conditions, and employee engagement.

1. Calculate the supplier turnover rate by dividing the number of employees who left the supplier during a specific period (L) by the average number of employees during that period (E), and multiplying by 100:

   Turnover rate = $(L/E) \times 100$

2. Analyse turnover trends over time and compare rates across suppliers.
3. Conduct exit interviews or surveys to gather information on the reasons for turnover.

13

S3 Affected Communities

In this chapter: Assessment of affected communities, impacts and actions to prevent and mitigate negative impacts. Management of affected communities' dependencies and their financial effects. Risks and opportunities related to communities, and their financial effects. Common indicators and metrics.

Understanding and addressing the impacts of organisations on communities is of utmost importance for fostering sustainable development and ensuring social responsibility. Organisations play a significant role in shaping the economic, social, and cultural fabric of communities. Their actions can have both positive and negative effects on the well-being, rights, and livelihoods of the people living in those communities. Therefore, it is crucial for organisations to proactively consider and manage their impacts on communities to promote equitable outcomes, protect human rights, and contribute to the overall welfare of society. By recognising the significance of community impacts, organisations can strive towards creating positive change, fostering meaningful engagement, and upholding the rights and well-being of the communities they interact with.

In the following chapter:

- None of the data points are applicable to your ESG report if the topic is deemed not material, with the exception of IRO1 that are to be disclosed irrespective of the outcome of your materiality assessment.
- Not all DPs are to be reported if the undertaking has not adopted the respective policies, implemented the respective actions, or set the respective targets in relation to a sustainability matter that has been assessed to be material.
- Organisations with less than 750 employees during the financial year of their reporting may omit all data points in this chapter for the first two years of preparation, unless deemed to be material.

ESRS S3: AFFECTED COMMUNIITES		
GOVERNANCE & STRATEGY	**IMPACTS, RISKS, & OPPORTUNITIES**	**METRICS & TARGETS**
• INCLUDES CROSS-CUTTING DISCLOSURES: • SBM-2: Views and interests of *stakeholders* • SBM-3: *Material* impacts, risks, and opportunities and their interaction with strategy and business model(s) • Information on *violations of UNGC principles and OECD guidelines, monitoring processes, and compliance*	• S3-1: *Policies* related to affected communities • S3-2: Processes for *engaging* with affected communities about impacts • S3-3: Processes to *remediate negative impacts* and channels for affected communities to raise concerns • S3-4: *Taking action* on material impacts on affected communities, and approaches to *mitigating* material risks and pursuing material opportunities related to affected communities, and effectiveness of those actions	• S3-5: *Targets* and *tracking effectiveness* of policies, actions, and targets related to managing material negative impacts, advancing positive impacts, and managing material risk and opportunities

Before diving into detailed metrics and targets, begin with describing your materiality assessment and organisational governance initiatives that directly interact with affected communities' goals, most importantly, the processes and methodologies to identify and assess impacts, risks, and opportunities related to affected communities issues (see **Chapter 3: Single and Double Materiality Assessment and 13.1 Affected Communities Assessment Methodology**).

TIPS AND TRICKS

Social Impact Assessments

The methodology outlined in this chapter is ideal for organisations that have little or sporadic engagement with communities, which will be the case for the majority of organisations. However, should you have a more substantial, in-depth relationship, or dependency on a single or multiple communities, then you should consider a detailed social impact assessment system. This would be the case for organisations where they are the principal employer in a community, or a tourism and hospitality business, or involved with a community development project and/or NGO that has a direct link with operations. Similarly, if your organisation is somehow involved with a community benefit project, you should also consider a more detailed impact assessment approach.

A peer-reviewed, detailed community impact assessment tool has been developed by the author and applied in a tourism scenario, and can be easily adapted – details can be found at:

https://www.degruyter.com/document/doi/10.1515/tw-2021-0010/html

For more information, contact the author at info@esgreportpro.com

Table 85 Affected community assessment indicators relating to policies and stakeholder engagement.

ESRS	DR	Paragraph	Name	Data type	Tips and tricks	Related CDR	VSMEs	GRI
S3	SBM-3	9	All affected communities who can be materially impacted by undertaking are included in scope of disclosure under ESRS 2	Semi-narrative	Description of how affected communities (including indigenous peoples) material impacts relate to risks and opportunities and their interaction with strategy and business model			GRI 413-2
S3	SBM-3	9 a)	Description of types of affected communities subject to material impacts	Narrative				GRI 413-2
S3	SBM-3	9 a i)-v)	Type of communities subject to material impacts by own operations or through value chain	Semi-narrative				GRI 413-2
S3	SBM-3	9 b	Material negative impacts occurrence on affected communities	Semi-narrative				GRI 410-3-3, GRI 411-3-3, GRI 413-3-3, GRI 413-2
S3	SBM-3	9 c	Description of activities that result in positive impacts and types of affected communities that are positively affected or could be positively affected	Narrative				
S3	SBM-3	9 d	Description of material risks and opportunities arising from impacts and dependencies on affected communities	Narrative				
S3	SBM-3	10	Disclosure of whether and how the undertaking has developed an understanding of how affected communities with particular characteristics or those living in particular contexts, or those undertaking particular activities may be at greater risk of harm	Narrative				

S3	SBM-3	11	Disclosure of which of material risks and opportunities arising from impacts and dependencies on affected communities are impacts on specific groups	Narrative				
ESRS 2	S3		Disclosures to be reported in case the undertaking has not adopted policies and/or actions and/or targets	Narrative				
S3	S3-1	14	Policies to manage material impacts, risks, and opportunities related to affected communities, including specific affected communities or all affected communities	MDR-P	Policies, process, mitigation, and reporting channels relating to affected communities – use stakeholder engagement assessment with materiality assessment, 5.1.1 MDR-P and:		V	GRI 2-23, GRI 410-3-3, GRI 411-3-3, GRI 413-3-3
S3	S3-1	15	Disclosure of any particular policy provisions for preventing and addressing impacts on indigenous peoples	Narrative				GRI 411-3-3
S3	S3-1	16	Description of relevant human rights policy commitments relevant to affected communities	Narrative	13.2 Community engagement and consultation	SFDR		GRI 2-23
S3	S3-1	16 a	Disclosure of general approach in relation to respect for human rights	Narrative	13.3 Community investment	SFDR		GRI 2-23
S3	S3-1	16 b	Disclosure of general approach in relation to engagement with affected communities	Narrative	13.4 Investments in community development projects	SFDR		GRI 2-23, GRI 2-29
S3	S3-1	16 c	Disclosure of general approach in relation to measures to provide and/or enable remedy for human rights impacts	Narrative	13.5 Local procurement and supplier diversity	SFDR		GRI 2-23, GRI 2-25, GRI 410-3-3, GRI 411-3-3, GRI 411-1, GRI 413-3-3

(Continued)

Table 85 (*Continued*).

ESRS	DR	Paragraph	Name	Data type	Tips and tricks	Related CDR	VSMEs	GRI
S3	S3-1	17	Description of how policies are aligned with relevant internationally recognised instruments	Narrative	13.6 Percentage of local procurement 13.7 Community satisfaction with the organisation's presence and activities 13.8 Stakeholder grievance mechanism effectiveness 13.9 Social and economic impacts on local communities See also: – UN Guiding Principles on Business & Human Rights[1] – UN Global Compact Principles[2] – OECD Guidelines for Multinational Enterprises[3] – ILO Declaration of Fundamental Rights and Principles at Work[4]			GRI 2-23
S3	S3-1	17	Disclosure of extent and indication of nature of cases of non-respect of the UN Guiding Principles on Business and Human Rights, ILO Declaration on Fundamental Principles and Rights at Work or OECD Guidelines for Multinational Enterprises that involve affected communities	Narrative				GRI 2-23
S3	S3-1	AR 9	Disclosure of explanations of significant changes to policies adopted during reporting year	Narrative			V	
S3	S3-1	AR 11	Disclosure on an illustration of the types of communication of its policies to those individuals, group of individuals, or entities for whom they are relevant	Narrative			V	GRI 2-23
S3	S3-2	21	Disclosure of how perspectives of affected communities inform decisions or activities aimed at managing actual and potential impacts	Narrative				GRI 3-3, GRI 2-29, GRI 410-3-3, GRI 411-3-3, GRI 413-3-3, GRI 413-1
S3	S3-2	21 a	Engagement occurs with affected communities or their legitimate representatives directly, or with credible proxies	Semi-narrative				GRI 2-29, GRI 3-3

S3	S3-2	21 b	Disclosure of stage at which engagement occurs, type of engagement, and frequency of engagement	Narrative				GRI 2-29, GRI 3-3
S3	S3-2	21 c	Disclosure of function and the most senior role within undertaking that has operational responsibility for ensuring that engagement happens and that results inform undertakings approach	Narrative				GRI 3-3
S3	S3-2	21 d	Disclosure of how the undertaking assesses the effectiveness of its engagement with affected communities	Narrative				GRI 2-29, GRI 3-3
S3	S3-2	22	Disclosure of how effectiveness of engagement with affected communities is assessed	Narrative				GRI 2-29
S3	S3-2	23	Disclosure of steps taken to gain insight into perspectives of affected communities that may be particularly vulnerable to impacts and/or marginalised	Narrative				
S3	S3-2	24	Disclosure of timeframe for adoption of general process to engage with affected communities in case the undertaking has not adopted a general process for engagement	Narrative			V	
S3	S3-2	AR 14	Type of role or function handling with engagement	Semi-narrative			V	GRI 413-1

[1] https://www.ohchr.org/sites/default/files/documents/publications/guidingprinciplesbusinesshr_en.pdf
[2] https://unglobalcompact.org/what-is-gc/mission/principles
[3] https://mneguidelines.oecd.org/mneguidelines/
[4] https://www.ilo.org/ilo-declaration-fundamental-principles-and-rights-work

(Continued)

Table 85 (*Continued*).

ESRS	DR	Paragraph	Name	Data type	Tips and tricks	Related CDR	VSMEs	GRI
S3	S3-3	27 a	Disclosure of general approach to and processes for providing or contributing to remedy where undertaking has identified that it connected with a material negative impact on affected communities	Narrative				GRI 2-25
S3	S3-3	27 b	Disclosure of specific channels in place for affected communities to raise concerns or needs directly with undertaking and have them addressed	Narrative				GRI 2-25
S3	S3-3	27 c	Disclosure of processes through which undertaking supports or requires availability of channels	Narrative				
S3	S3-3	27 d	Disclosure of how issues raised and addressed are tracked and monitored and how effectiveness of channels is ensured	Narrative				GRI 2-26
S3	S3-3	28	Disclosure of how it is assessed that affected communities are aware of and trust structures or processes as a way to raise their concerns or needs and have them addressed	Narrative				GRI 2-25

S3	S3-3	28	Policies regarding protection against retaliation for individuals that use channels to raise concerns or needs are in place	Semi-narrative				
S3	S3-3	29	Statement in case the undertaking has not adopted a general process to engage with its own workforce	Narrative				
S3	S3-3	29	Disclosure of timeframe for channel or processes for raising concerns to be in place and	Narrative			V	
S3	S3-3	AR 19	Disclosure of whether and how affected communities are able to access channels at level of undertaking they are affected by	Narrative			V	
S3	S3-3	AR 20	Third-party mechanisms are accessible to all affected communities	Narrative			V	
S3	S3-3	AR 21	Grievances are treated confidentially and with respect to rights of privacy and data protection	Semi-narrative			V	
S3	S3-3	AR 21	Affected communities are allowed to use anonymously channels to raise concerns or needs	Semi-narrative			V	

Source: CDR (EU) 2023/2772 supplementing Directive 2013/34/EU, *GRI 1, 2, 3: 11, 12, 13, 14, 306, 101, 201, 202, 203, 204, 205, 206, 207,301, 302, 303, 304, 305, 306, 308,401, 402, 403, 404, 405, 406, 407, 408, 409, 410, 411, 413, 414, 415, 416, 417, 418*, and the author.

13.1 Affected Communities Assessment Methodology

Due to a lack or proximity and direct authority, affected communities can be difficult to engage with, so the focus of disclosures is how your organisation reaches out to them and harnesses their inputs, including collaborations and/or cooperation with other organisations, as well as indicators for critical issues.

The methodology provides a framework for organisations to assess the material impacts on communities that have been caused or contributed to by its operations. It also considers the material impacts directly linked to the organisation's own operations, products, or services through its business relationships. The methodology includes the following key components:

1. Identification and assessment of impacts:

Conduct a comprehensive assessment to identify material positive and negative actual or potential impacts on affected communities. Focus on areas where risks are most likely to be present and severe. Consider social, environmental, and economic dimensions to gain a holistic understanding of community impacts.

2. Actions and results:

Implement actions to prevent, mitigate, or remediate actual or potential negative impacts on affected communities. Report on the results of these actions to demonstrate their effectiveness in addressing community-related impacts. Foster transparency by sharing information on efforts to minimise adverse effects and contribute positively to the well-being of affected communities.

3. Management of risks and opportunities:

Identify the nature, type, and extent of material risks and opportunities related to impacts and dependencies on affected communities. Develop strategies and mechanisms to manage these risks and leverage the identified opportunities. Embrace robust risk management practices and seize opportunities for community development.

4. Financial effects:

Evaluate the short-, medium-, and long-term financial effects of material risks and opportunities arising from impacts and dependencies on affected communities. Assess the financial implications of managing and mitigating risks. Consider the potential returns associated with seizing opportunities for community development.

By integrating these components, the methodology ensures a comprehensive approach to disclosing information on material impacts, actions, risks, and opportunities related to affected communities. It emphasises the identification and assessment of impacts, the implementation of actions and their outcomes, the management of risks and opportunities, and the financial implications of the organisation's community-related activities. Through transparent reporting, the undertaking aims to foster understanding, accountability, and positive change in its interactions with affected communities.

13.2 Community Engagement and Consultation (S3-1)

Relevant SDG: SDG 16 – Peace, Justice, and Strong Institutions.

Description: This indicator evaluates the organisation's efforts to engage and consult with the communities affected by its operations. It assesses the extent to which the organisation involves community members in decision-making processes and considers their perspectives and needs.

1. Identify the key stakeholders and communities affected by the organisation's operations – see Chapter 2, Stakeholder Engagement.
2. Co-create with the community, a community engagement plan outlining the objectives, methods, and frequency of engagement activities.
3. Conduct regular community meetings, consultations, or focus groups to gather community input and feedback.
4. Document the engagement process, including the number of community members involved, topics discussed, and outcomes.
5. Evaluate the effectiveness of community engagement efforts through feedback surveys or assessments.

13.3 Community Investment (S3-1, S3-2)

Relevant SDG: SDG 16 – Peace, Justice, and Strong Institutions.

The community investment measurement assesses the monetary value of a company's investments in the community. It includes both capital and operating expenditures incurred during the reporting period. The measurement involves two indicators:

1. Total community investments: This indicator calculates the overall amount of community investments during a specific reporting period.

Table 86 Affected community assessment indicators relating to actions and targets.

ESRS	DR	Paragraph	Name	Data type	Tips and tricks	Related CDR	VSMEs	GRI
S3	S3-4		Action plans and resources to manage its material impacts, risks, and opportunities related to affected communities	MDR-A	Disclose all planned or underway actions to prevent, mitigate, or remediate material negative impacts by reviewing stakeholder engagement with own workforce assessment methodology, 5.1.2 MDR-A 13.9 Social and economic impacts on local communities 13.10 Community health and safety incidents related to the organisation's operations 13.11 Indigenous and cultural heritage preservation 13.12 Contribution to local employment and job creation 13.13 Community investment and social infrastructure development 13.14 Local economic impact			GRI 410-3-3, GRI 411-3-3, GRI 411-1
S3	S3-4	32 a	Description of action taken, planned or underway to prevent, mitigate or remediate material negative impacts on affected communities, including training programs	Narrative				GRI 201-2, GRI 410-3-3, GRI 411-3-3, GRI 413-3-3
S3	S3-4	32 b	Description of whether and how the undertaking has taken action to provide or enable remedy in relation to an actual material impact	Narrative				GRI 410-3-3, GRI 411-3-3, GRI 411-1, GRI 413-3-3
S3	S3-4	32 c	Description of additional initiatives or processes with primary purpose of delivering positive impacts for affected communities	Narrative				GRI 410-3-3, GRI 411-3-3, GRI 413-3-3
S3	S3-4	32 d	Description of how effectiveness of actions or initiatives in delivering outcomes for affected communities is tracked and assessed	Narrative				GRI 410-3-3, GRI 411-3-3, GRI 413-3-3
S3	S3-4	33 a	Description of approach to identifying what action is needed and appropriate in response to particular actual or potential material negative impact on affected communities	Narrative				GRI 3-3, GRI 410-3-3, GRI 411-3-3, GRI 413-3-3
S3	S3-4	33 b	Description of approach to taking action in relation to specific material impacts on affected communities	Narrative				GRI 3-3, GRI 410-3-3, GRI 411-3-3, GRI 411-1, GRI 413-3-3

S3	S3-4	33 c	Description of approach to ensuring that processes to provide or enable remedy in event of material negative impacts on affected communities are available and effective in their implementation and outcomes	Narrative				GRI 3-3 GRI 2-25
S3	S3-4	34 a	Description of what action is planned or underway to mitigate material risks arising from impacts and dependencies on affected communities and how effectiveness is tracked	Narrative				
S3	S3-4	34 b	Description of what action is planned or underway to pursue material opportunities in relation to affected communities	Narrative				
S3	S3-4	35	Disclosure of how it is ensured that own practices do not cause or contribute to material negative impacts on affected communities	Narrative				GRI 410-3-3, GRI 411-3-3, GRI 413-3-3
S3	S3-4	36	Disclosure of severe human rights issues and incidents connected to upstream and downstream value chain	Narrative		SFDR		GRI 410-3-3, GRI 411-3-3, GRI 411-1, GRI 413-3-3
S3	S3-4	38	Disclosure of resources allocated to management of material impacts	Narrative				
S3	S3-4	AR 25 a	Disclosure of general and specific approaches to addressing material negative impacts	Narrative			V	
S3	S3-4	AR 25 b	Disclosure of initiatives aimed at contributing to additional material positive impacts	Narrative			V	
S3	S3-4	AR 25 c	Disclosure of how far undertaking has progressed in efforts during reporting period	Narrative			V	
S3	S3-4	AR 25 d	Disclosure of aims for continued improvement	Narrative			V	

(Continued)

Table 86 (*Continued*).

ESRS	DR	Paragraph	Name	Data type	Tips and tricks	Related CDR	VSMEs	GRI
S3	S3-4	AR 27	Disclosure of how undertaking seeks to use leverage with relevant business relationships to manage material negative impacts affecting affected communities	Narrative			V	GRI 2-24
S3	S3-4	AR 28	Disclosure of how participation in industry or multi-stakeholder initiative and undertaking's own involvement is aiming to address material impacts	Narrative			V	
S3	S3-4	AR 34 a)	Disclosure of how affected communities play a role in decisions regarding design and implementation of programmes or processes	Narrative			V	GRI 3-3, GRI 410-3-3, GRI 411-3-3, GRI 413-3-3
S3	S3-4	AR 34 b)	Information about intended or achieved positive outcomes of programmes or processes for affected communities	Narrative			V	
S3	S3-4	AR 34 c)	Explanation of the approximate scope of affected communities covered by the described social investment or development programmes, and, where applicable, the rationale for why selected communities were chosen	Narrative			V	GRI 413-1
S3	S3-4	AR 35	Initiatives or processes whose primary aim is to deliver positive impacts for affected communities are designed also to support achievement of one or more of Sustainable Development Goals	Semi-narrative			V	

S3	S3-4	AR 43	Description of internal functions that are involved in managing impacts and types of action taken by internal functions to address negative and advance positive impacts	Narrative			V	
S3	S3-5	41	Targets set to manage material impacts, risks, and opportunities related to affected communities [see ESRS 2–MDR-T]	MDR-T	Disclose all planned or underway targets and specific metrics to prevent, mitigate, or remediate material negative impacts by reviewing stakeholder engagement with own workforce assessment methodology, 5.1.3 MDR-M, 5.1.4 MDR-T and that related to above actions metrics			GRI 410-3-3, GRI 411-3-3, GRI 413-3-3
S3	S3-5	42 a	Disclosure of how affected communities were engaged directly in setting targets	Narrative				
S3	S3-5	42 b	Disclosure of how affected communities were engaged directly in tracking performance against targets	Narrative				GRI 3-3
S3	S3-5	42 c	Disclosure of how affected communities were engaged directly in identifying lessons or improvements as a result of undertaking's performance	Narrative				GRI 3-3
S3	S3-5	AR 44 a	Disclosure of intended outcomes to be achieved in lives of affected communities	Narrative			V	
S3	S3-5	AR 44 b	Information about stability over time of target in terms of definitions and methodologies to enable comparability	Narrative			V	
S3	S3-5	AR 44 c	Disclosure of references to standards or commitments on which target is based	Narrative			V	

Source: CDR (EU) 2023/2772 supplementing Directive 2013/34/EU, *GRI 1, 2, 3: 11, 12, 13, 14, 306, 101, 201, 202, 203, 204, 205, 206, 207,301, 302, 303, 304, 305, 306, 308,401, 402, 403, 404, 405, 406, 407, 408, 409, 410, 411, 413, 414, 415, 416, 417, 418*, and the author.

Community investments should be expressed in monetary terms and encompass the expenditures related to capital and operating costs.

2. Community investment ratio: This indicator expresses a company's community investments in period *t* as a percentage of its total assets (and/or revenue). It provides insights into the proportion of resources allocated to community investments. The following formulae can be referred to:

   Community Investment Ratio = (Community Investments in Period *t*/ Total Assets or Revenue in Period *t*) × 100

where:

To calculate the first indicator and the numerator of the second indicator, the following classification can be used to track community investments over the reporting period:

a. Contributions to charities, non-governmental organisations (NGOs), and research institutes not associated with your organisation's commercial research and development activities.

b. Funding community infrastructures such as education, medical, and recreational facilities, even if they are outside the primary business activities. For example, constructing a school or hospital for employees and their families.

c. Direct costs of social programs (e.g., arts and educational events) or emergency relief provided during natural disasters.

When supporting community infrastructures, the calculation depends on the situation:

- If your organisation purchases an existing infrastructure, the calculation should consider the expenditures incurred.
- If your organisation contributes to the construction of a facility, the calculation should include costs of materials, labour, and all construction expenses specific to that facility.
- If your organisation funds the daily operations of a community facility, the reported amount should include the associated operating costs.

Regarding support for social programs, the calculation should focus on the specific operating costs related to the programs financed by your organisation.

The calculation should also encompass non-monetary contributions, such as workers volunteering their time and capabilities for community projects and in-kind donations (valued at fair market value).

13.4 Investments in Community Development Projects (S3-1, S3-5)

Relevant SDG: SDG 1 – No Poverty and SDG 11 – Sustainable Cities and Communities.

This indicator assesses the organisation's investments in community development projects aimed at improving the quality of life, social well-being, and economic opportunities of the communities affected by its operations.

1. Identify and document community development projects supported or initiated by the organisation.
2. Quantify the investments made in terms of financial contributions, in-kind support, or volunteer hours.
3. Assess the scope and impact of each community development project, considering factors such as the number of beneficiaries, project outcomes, and sustainability.
4. Conduct regular evaluations to measure the effectiveness and long-term impact of community development investments.

13.5 Local Procurement and Supplier Diversity (S3-1)

Relevant SDG: SDG 8 – Decent Work and Economic Growth.

This indicator assesses the organisation's commitment to local procurement and supplier diversity in the communities affected by its operations. It focuses on supporting local businesses, fostering economic resilience, and promoting diversity in the supply chain.

1. Identify the percentage of total procurement spent on local suppliers within the affected communities.
2. Evaluate the diversity of the local supplier base, considering factors such as the representation of women-owned businesses, minority-owned businesses, and small enterprises.
3. Monitor and report on changes in local procurement percentages and supplier diversity over time.

4. Implement initiatives to support capacity building and supplier development programs for local businesses.

13.6 Percentage of Local Procurement (S3-1, S3-2, E5-2)

Relevant SDG: SDG 1 – No Poverty and SDG 11 – Sustainable Cities and Communities.

The percentage of local procurement indicator assesses the extent to which a reporting organisation engages in procurement from local suppliers within a specific geographical market. It measures the proportion of invoices or commitments made to local suppliers during the reporting period, based on the accrual accounting principle. Local suppliers refer to organisations or individuals providing products or services within the same geographical market as the reporting organisation. The definition of the geographical market may vary, encompassing the community surrounding operations, a region within a country, or the entire country itself.

To calculate the indicator, invoices or commitments made to local suppliers are considered. Purchasing is deemed local when it involves products or services produced within the same country as the reporting entity, provided by an entity incorporated in the same country, or fulfilling the local content or entity requirements set by the government of that country.

Considering the challenges of defining and tracking local purchases, it is recommended to use the country as a distinguishing criterion. Checking whether transnational payments have been made to suppliers can serve as an initial step to identify local purchases for inclusion in the calculation. By examining invoices, reporting organisations can determine the local purchasing items within the reporting period and compute the costs on an accrual basis. The indicator is presented as both an absolute figure (in monetary terms) and as a percentage of the reporting entity's total purchasing.

The following formulae can be referred to:

Percentage of Local Procurement = (Total Value of Invoices or Commitments to Local Suppliers/Total Value of Invoices or Commitments for Procurement) × 100

13.7 Community Satisfaction with the Organisation's Presence and Activities (SBM-3, S3-3)

Relevant SDG: SDG 16 – Peace, Justice, and Strong Institutions.

This indicator measures the satisfaction levels of the communities affected by the organisation's operations regarding the organisation's presence, activities, and contributions to community well-being. It focuses on understanding and addressing community concerns and expectations.

1. Conduct community satisfaction surveys to gather feedback and perceptions of community members regarding the organisation's presence and activities.
2. Develop a standardised survey instrument or custom questionnaire based on community-specific concerns and expectations.
3. Analyse survey responses quantitatively to calculate community satisfaction scores.
4. Consider additional qualitative feedback through focus groups or interviews to gain deeper insights into community satisfaction and concerns.

13.8 Stakeholder Grievance Mechanism Effectiveness (SBM-3, S3-3)

Relevant SDG: SDG 16 – Peace, Justice, and Strong Institutions.

This indicator assesses the effectiveness of the organisation's grievance mechanism in addressing and resolving community stakeholders' concerns and complaints. It focuses on ensuring timely and fair resolution of grievances to maintain positive relationships with affected communities.

1. Establish a formal grievance mechanism that is accessible, transparent, and culturally appropriate for the affected communities.
2. Document all grievances received, including the nature of the complaint, dates, and actions taken.
3. Assign a dedicated team to review and investigate grievances, ensuring impartiality and adherence to defined procedures.
4. Measure the resolution time for each grievance, from the time of receipt to closure, to assess the timeliness of the process.
5. Conduct periodic reviews and evaluations of the grievance mechanism's effectiveness, incorporating feedback from community stakeholders.

13.9 Social and Economic Impacts on Local Communities (S3-4)

Relevant SDG: SDG 1 – No Poverty and SDG 8 – Decent Work and Economic Growth.

This indicator assesses the social and economic impacts of the organisation's operations on local communities. It focuses on understanding how the organisation contributes to community development, livelihoods, and overall well-being.

1. Identify key social and economic indicators relevant to the local communities, such as employment, income, education, healthcare, and infrastructure development.
2. Collect quantitative data on these indicators through surveys, interviews, or government records.
3. Analyse the data to assess the organisation's contributions to local employment, income generation, skill development, and community infrastructure.
4. Compare the data against baseline indicators or regional averages to understand the organisation's impact on the local communities.

13.10 Community Health and Safety Incidents Related to the Organisation's Operations (S3-4, SBM-3)

Relevant SDG: SDG 3 – Good Health and Well-being.

This indicator measures the number and severity of health and safety incidents that occur within the communities affected by the organisation's operations. It focuses on understanding and addressing the health and safety risks faced by community members.

1. Collect data on community health and safety incidents related to the organisation's operations, such as accidents, pollution events, or public health issues.
2. Calculate the incident rate per community member or per household using the formula:

   Incident rate = (Total number of incidents/Total number of community members or households) × 1000
3. Analyse incident trends over time and assess the severity and impact of each incident.

13.11 Indigenous and Cultural Heritage Preservation (S3-4)

Relevant SDG: SDG 11 – Sustainable Cities and Communities and SDG 16 – Peace, Justice, and Strong Institutions.

This indicator assesses the organisation's efforts to preserve and protect indigenous rights and cultural heritage in the communities affected by its operations. It focuses on respecting and promoting the cultural diversity and heritage of indigenous communities.

1. Identify the indigenous communities affected by the organisation's operations and the cultural heritage elements that require preservation.
2. Collaborate with indigenous community leaders and cultural experts to develop culturally sensitive preservation plans.
3. Implement strategies such as capacity building, community participation, documentation, and awareness programs to safeguard cultural heritage.
4. Monitor and evaluate the implementation of preservation plans, including the maintenance of cultural sites, documentation of traditional practices, and involvement of indigenous community members.

13.12 Contribution to Local Employment and Job Creation (S3-4)

Relevant SDG: SDG 8 – Decent Work and Economic Growth.

This indicator assesses the organisation's contribution to local employment and job creation in the communities affected by its operations. It focuses on understanding the organisation's role in stimulating economic growth and enhancing livelihood opportunities.

1. Identify the number of local employees directly employed by the organisation and their job roles.
2. Analyse the organisation's supply chain and identify the number of indirect employment opportunities created for local communities.
3. Calculate the total number of direct and indirect local employment positions.

4. Monitor and report on changes in local employment numbers over time.
5. Evaluate the organisation's efforts to enhance local workforce skills through training and capacity-building programs.

13.13 Community Investment and Social Infrastructure Development (S3-4)

Relevant SDG: SDG 1 – No Poverty and SDG 11 – Sustainable Cities and Communities.

This indicator assesses the organisation's investment in community development and the development of social infrastructure in the communities affected by its operations. It focuses on enhancing community well-being, access to basic services, and infrastructure development.

1. Identify community development projects and social infrastructure initiatives supported or initiated by the organisation.
2. Quantify the financial contributions, in-kind support, or volunteer hours dedicated to community investment and infrastructure development.
3. Assess the scope and impact of each project, considering factors such as the number of beneficiaries, project outcomes, and long-term sustainability.
4. Monitor and evaluate the effectiveness and impact of community investment and infrastructure development initiatives.

13.14 Local Economic Impact S3-4

Relevant SDGs: (DEPENDS ON THE EVENT BEING ASSESSED).

Local economic impact refers to the effects of a hazard or event on the economic conditions of a specific local area or region. It encompasses various metrics that assess the implications on local economic factors, such as gross domestic product (GDP) losses, changes in employment levels, shifts in business activity, and impacts on property values. Understanding the local economic impact helps to evaluate the consequences of hazards and guide decision-making for resilience and recovery efforts. The methodology for assessing local economic impact typically involves the following considerations:

- **Identify relevant economic indicators**: Determine the key economic indicators that are most pertinent to the specific local area or region under analysis. This can include metrics such as GDP, employment rates, business revenues, tax revenues, property values, and other relevant economic data.
- **Quantify changes in economic indicators**: Evaluate the impact of the hazard or event on the identified economic indicators. This involves comparing pre-event and post-event data to quantify the magnitude of the changes. For example, calculating the decrease in GDP, the increase in unemployment rate, or the decline in property values.
- **Analyse multiplier effects**: Consider the multiplier effects that can amplify the initial impacts on the local economy. This includes evaluating how changes in one economic sector can ripple through other sectors and affect overall economic activity. It can involve assessing indirect and induced impacts on employment, businesses, and investment in the local area.
- **Estimate financial losses**: Assess the financial losses incurred as a result of the local economic impact. This involves quantifying the monetary value associated with GDP reductions, decreased tax revenues, business closures, job losses, and other relevant factors. These estimates help in understanding the magnitude of the economic consequences.

14

S4 Consumers and End-Users

In this chapter: Assessment of consumer and end-user impacts and actions to prevent and mitigate negative impacts. Management of consumer and end-user dependencies and their financial effects. Risks and opportunities related to consumer and end-user impacts and their financial effects. Common indicators and metrics.

In today's dynamic business landscape, organisations are increasingly recognising the importance of understanding and addressing their impacts on consumers and end-users. As consumers become more conscious of their purchasing decisions, they are seeking products and services that align with their values and expectations. This shift in consumer behaviour has elevated the significance of considering the material impacts an organisation has on its consumers and end-users.

The objective of this methodology is to provide a framework for your organisation to comprehensively evaluate and manage your impacts on consumers and end-users. By doing so, organisations can gain a deeper understanding of the positive and negative effects their products and services may have on individuals and communities. Moreover, this methodology aims to enable organisations to take proactive measures to prevent or mitigate any potential negative impacts, while maximising the opportunities to create positive outcomes for consumers and end-users.

Through a systematic analysis of impacts, your organisation can identify areas for improvement, develop action plans, and implement strategies that enhance consumer satisfaction, trust, and loyalty. Additionally, by considering the financial implications of their impacts and dependencies on consumers and end-users, your organisation can align business strategies with long-term sustainability and financial success.

By adopting this methodology, your organisation demonstrates commitment to responsible and sustainable practices, gaining a competitive edge in the marketplace and fostering positive relationships with their customers. Ultimately, the effective management of impacts on consumers and end-users not only benefits the organisation but also contributes to the overall well-being and satisfaction of individuals and communities.

In the following chapter:

- None of the data points are applicable to your ESG report if the topic is deemed not material, with the exception of IRO1 that are to be disclosed irrespective of the outcome of your materiality assessment.
- Not all DPs are to be reported if the undertaking has not adopted the respective policies, implemented the respective actions, or set the respective targets in relation to a sustainability matter that has been assessed to be material.
- Organisations with less than 750 employees during the financial year of their reporting may omit all data points in this chapter for the first two years of preparation, unless deemed to be material.

ESRS S4: CONSUMERS AND END-USERS		
GOVERNANCE & STRATEGY	**IMPACTS, RISKS, & OPPORTUNITIES**	**METRICS & TARGETS**
INCLUDES CROSS-CUTTING DISCLOSURES: • SBM-2: Views and interests of *stakeholders* • SBM-3: *Material* impacts, risks, and opportunities and their interaction with strategy and business model(s) • Information on *violations of UNGC principles and OECD guidelines, monitoring processes, and compliance*	• S4-1: *Policies* related to consumers and end-users • S4-2: Processes for *engaging* with consumers and end-users about impacts • S4-3: Processes to *remediate negative impacts* and channels for consumers and end-users to raise concerns • S4-4: *Taking action* on material impacts on consumers and end-users, and approaches to *mitigating* material risks and pursuing material opportunities related to consumers and end-users, and effectiveness of those actions	• S4-5: *Targets* and *tracking effectiveness* of policies, actions, and targets related to managing material negative impacts, advancing positive impacts, and managing material risk and opportunities

Before diving into detailed metrics and targets, begin with describing your materiality assessment and organisational governance initiatives that directly interact with consumer and end-user goals, most importantly, the processes and methodologies to identify and assess impacts, risks, and opportunities related to consumer and end-user issues (see **Chapter 3: Single and Double Materiality Assessment and 14.1 Consumer and End-User Assessment Methodology**).

Table 87 Consumers and end-users assessment indicators relating to policies and stakeholder engagement.

ESRS	DR	Paragraph	Name	Data type	Tips and tricks	Related CDR	VSMEs	GRI
S4	SBM-3	10	All consumers and end-users who can be materially impacted by undertaking are included in scope of disclosure	Semi-narrative	Description of how consumers and end-users' material impacts relate to risks and opportunities and their interaction with strategy and business model			
S4	SBM-3	10 a)	Description of types of consumers and end-users subject to material impacts	Narrative				
S4	SBM-3	10 a i)-v)	Type of consumers and end-users subject to material impacts by own operations or through value chain	Semi-narrative				
S4	SBM-3	10 b	Material negative impacts occurrence on consumers and end-users: individual incidents (e.g., a defect linked to a particular product) or to specific business relationships (e.g., a business partner uses marketing that inappropriately targets young consumers)	Semi-narrative				GRI 406-3-3, GRI 416-3-3, GRI 417-3-3, GRI 418-3-3
S4	SBM-3	10 c	Description of activities that result in positive impacts and types of consumers and end-users that are positively affected or could be positively affected	Narrative				
S4	SBM-3	10 d	Description of material risks and opportunities arising from impacts and dependencies on consumers and end-users	Narrative				
S4	SBM-3	11	Disclosure of how understanding of how own workers/consumers and end-users with particular characteristics, working in particular contexts, or undertaking particular activities may be at greater risk of harm has been developed	Narrative				

S4	SBM-3	12	Disclosure of which material risks and opportunities arising from impacts and dependencies on consumers and end-users are impacts on specific groups	Narrative				
ESRS 2	S4		Disclosures to be reported in case the undertaking has not adopted policies and/or actions and/or targets	Narrative				
S4	S4-1	15	Policies to manage material impacts, risks, and opportunities related to consumers and end-users	MDR-P	Policies, process, mitigation, and reporting channels relating to consumer and end-user stakeholder engagement assessment with materiality assessment, 5.1.1 MDR-P 14.2 Product safety and quality assurance 14.3 Consumer data privacy and security 14.4 Product labelling and transparency 14.5 Customer satisfaction and feedback		V	GRI 2-23, GRI 416-3-3 , GRI 417-3-3, GRI 418-3-3
S4	S4-1	16	Description of relevant human rights policy commitments relevant to consumers and/or end-users	Narrative		SFDR		GRI 2-23
S4	S4-1	16 a	Disclosure of general approach in relation to respect for human rights of consumers and end-users	Narrative		SFDR		GRI 2-23
S4	S4-1	16 b	Disclosure of general approach in relation to engagement with consumers and/or end-users	Narrative		SFDR		GRI 2-23, GRI 2-29
S4	S4-1	16 c	Disclosure of general approach in relation to measures to provide and/or enable remedy for human rights impacts	Narrative		SFDR		GRI 2-23, GRI 2-25, GRI 406-3-3, GRI 416-3-3, GRI 417-3-3, GRI 418-3-3

(Continued)

Table 87 (*Continued*).

ESRS	DR	Paragraph	Name	Data type	Tips and tricks	Related CDR	VSMEs	GRI
S4	S4-1	17	Description of how policies are aligned with relevant internationally recognised instruments	Narrative	14.6 Customer education and awareness 14.7 Responsible marketing and advertising practices See also: – UN Guiding Principles on Business & Human Rights[1] – UN Global Compact Principles[2] – OECD Guidelines for Multinational Enterprises[3] – ILO Declaration of Fundamental Rights and Principles at Work[4]			GRI 2-23
S4	S4-1	17	Disclosure of extent and indication of nature of cases of non-respect of the UN Guiding Principles on Business and Human Rights, ILO Declaration on Fundamental Principles and Rights at Work, or OECD Guidelines for Multinational Enterprises that involve consumers and/or end-users					GRI 2-23
S4	S4-1	AR 9	Disclosure of explanations of significant changes to policies adopted during reporting year	Narrative			V	
S4	S4-1	AR 13	Disclosure on an illustration of the types of communication of its policies to those individuals, group of individuals, or entities for whom they are relevant	Narrative			V	GRI 2-23
S4	S4-2	20	Disclosure of how perspectives of consumers and end-users inform decisions or activities aimed at managing actual and potential impacts	Narrative				GRI 3-3, GRI 2-29, GRI 406-3-3, GRI 416-3-3, GRI 417-3-3, GRI 418-3-3
S4	S4-2	20 a	Engagement occurs with consumers and end-users or their legitimate representatives directly, or with credible proxies	Semi-narrative				GRI 2-29, GRI 3-3
S4	S4-2	20 b	Disclosure of stage at which engagement occurs, type of engagement, and frequency of engagement	Narrative				GRI 2-29, GRI 3-3

S4	S4-2	20 c	Disclosure of function and most senior role within undertaking that has operational responsibility for ensuring that engagement happens and that results inform undertakings approach	Narrative			GRI 3-3
S4	S4-2	20 d	Disclosure of how effectiveness of engagement with consumers and end-users is assessed	Narrative			GRI 2-29, GRI 3-3
S4	S4-2	21	Disclosure of steps taken to gain insight into perspectives of consumers and end-users/consumers and end-users who may be particularly vulnerable to impacts and (or) marginalised	Narrative			GRI 2-29
S4	S4-2	22	Disclosure of timeframe for adoption of general process to engage with consumers and end-users in case the undertaking has not adopted a general process for engagement	Narrative		V	
S4	S4-2	AR 15	Type of role or function handling with engagement	Semi-narrative		V	
S4	S4-3	25 a	Disclosure of general approach to and processes for providing or contributing to remedy where undertaking has identified that it connected with a material negative impact on consumers and end-users	Narrative			GRI 2-25
S4	S4-3	25 b	Disclosure of specific channels in place for consumers and end-users to raise concerns or needs directly with undertaking and have them addressed	Narrative			GRI 2-25
S4	S4-3	25 c	Disclosure of processes through which undertaking supports or requires availability of channels	Narrative			

[1] https://www.ohchr.org/sites/default/files/documents/publications/guidingprinciplesbusinesshr_en.pdf
[2] https://unglobalcompact.org/what-is-gc/mission/principles
[3] https://mneguidelines.oecd.org/mneguidelines/
[4] https://www.ilo.org/ilo-declaration-fundamental-principles-and-rights-work

(Continued)

Table 87 (*Continued*).

ESRS	DR	Paragraph	Name	Data type	Tips and tricks	Related CDR	VSMEs	GRI
S4	S4-3	25 d	Disclosure of how issues raised and addressed are tracked and monitored and how effectiveness of channels is ensured	Narrative				GRI 2-26
S4	S4-3	26	Disclosure of how it is assessed that consumers and end-users are aware of and trust structures or processes as a way to raise their concerns or needs and have them addressed	Narrative				
S4	S4-3	26	Policies regarding protection against retaliation for individuals that use channels to raise concerns or needs are in place	Narrative				
S4	S4-3	27	Statement in case the undertaking has not adopted a general process to engage with its own workforce	Narrative				
S4	S4-3	27	Disclosure of timeframe for channel or processes for raising concerns to be in place	Narrative			V	
S4	S4-3	AR 20	Disclosure of whether and how consumers and/or end-users are able to access channels at level of undertaking they are affected by	Narrative			V	
S4	S4-3	AR 21	Third-party mechanisms are accessible to all consumers and/or end-users	Narrative			V	
S4	S4-3	AR 22	Grievances are treated confidentially and with respect to rights of privacy and data protection	Semi-narrative			V	
S4	S4-3	AR 22	Consumers and end-users are allowed to use anonymously channels to raise concerns or needs	Semi-narrative			V	
S4	S4-3	AR 23	Number of complaints received from consumers and/or end-users during the reporting period	Integer			V	GRI 2-25, GRI 418-1

Source: CDR (EU) 2023/2772 supplementing Directive 2013/34/EU, *GRI 1, 2, 3: 11, 12, 13, 14, 306, 101, 201, 202, 203, 204, 205, 206, 207,301, 302, 303, 304, 305, 306, 308,401, 402, 403, 404 405, 406, 407, 408, 409, 410, 411, 413, 414, 415, 416, 417, 418*, and the author.

14.1 Consumer and End-User Assessment Methodology

Compared to value chain workers and affected communities, organisations are much more familiar with engaging with consumers and end-users through customer service and support functions. However, for your ESG report, we need to dig deeper, and to effectively assess and address the material impacts of organisations on consumers and end-users, a comprehensive methodology is essential. The following steps outline the key components of the methodology:

1. Impact analysis:
Identify the key aspects of your organisation's products and services that directly impact consumers and end-users. Conduct a thorough lifecycle analysis (see 10.2, Life Cycle Analysis) of the material positive and negative actual or potential impacts on consumers and end-users. Evaluate the extent to which these impacts align with sustainable and responsible business practices.

2. Action plan:
Develop and implement strategies and actions to prevent, mitigate, or remediate any actual or potential negative impacts on consumers and end-users. Monitor and evaluate the effectiveness of these actions in addressing the identified impacts. Continuously improve and update the action plan based on feedback, emerging trends, and evolving consumer needs.

3. Risk and opportunity assessment:
Assess the nature, type, and extent of material risks and opportunities associated with your organisation's impacts and dependencies on consumers and end-users. Consider factors such as changing consumer preferences, market trends, and regulatory requirements. Develop strategies to manage and capitalise on these risks and opportunities, ensuring the long-term sustainability and competitiveness of the organisation.

4. Financial implications:
Evaluate the financial effects of material risks and opportunities arising from your organisation's impacts and dependencies on consumers and end-users. Consider short-term, medium-term, and long-term time horizons to assess the financial implications. Incorporate these financial considerations into your organisation's overall financial planning, budgeting, and decision-making processes.

By following this methodology, your organisation can gain valuable insights into its impacts on consumers and end-users, take appropriate actions to address any negative effects, and identify opportunities to create positive outcomes for their customers. This methodology enables your organisation to foster consumer trust, enhance brand reputation, and contribute to the well-being and satisfaction of their consumers and end-users.

14.2 Product Safety and Quality Assurance (S4-1)

Relevant SDG: SDG 3 (Good Health and Well-being).

This indicator assesses your organisation's commitment to product safety and quality assurance for its consumers and end-users. It focuses on ensuring that products meet:

1. Implement rigorous quality assurance processes, including testing, inspection, and certification, to ensure product safety and quality.
2. Conduct regular product testing and analysis to verify compliance with relevant safety regulations and standards.
3. Monitor customer feedback, complaints, and product return data to identify potential safety or quality issues.
4. Establish corrective actions and continuous improvement processes to address any identified safety or quality concerns.

14.3 Consumer Data Privacy and Security (S4-1)

Relevant SDG: SDG 16 (Peace, Justice, and Strong Institutions).

This indicator assesses your organisation's commitment to protecting consumer data privacy and ensuring the security of personal information collected through products or services. It focuses on maintaining consumer trust and complying with relevant data protection regulations.

1. Implement robust data privacy policies and procedures aligned with applicable data protection laws and regulations.
2. Conduct regular audits of data collection, storage, and processing practices to ensure compliance and identify potential vulnerabilities.
3. Provide transparent information to consumers about data collection practices, usage, and security measures.
4. Monitor and respond to data breach incidents promptly, including incident investigation, notification to affected consumers, and remediation actions.

Table 88 Consumers and end-users assessment indicators relating to actions and targets.

ESRS	DR	Paragraph	Name	Data type	Tips and tricks	Related CDR	VSMEs	GRI
S4	S4-4		Action plans and resources to manage its material impacts, risks, and opportunities related to consumers and end-users	MDR-A	Disclose all planned or underway actions to prevent, mitigate, or remediate material negative impacts by reviewing stakeholder engagement with own workforce assessment methodology, 5.1.2 MDR-A 14.8 Product recalls and responsiveness 14.9 Product lifecycle management and end-of-life considerations 14.10 Product innovation and sustainable design 14.11 Product accessibility and affordability			GRI 3-3, GRI 416-3-3, GRI 417-3-3, GRI 418-3-3
S4	S4-4	31 a	Description of action planned or underway to prevent, mitigate, or remediate material negative impacts on consumers and end-users	Narrative				GRI 3-3 , GRI 417-3-3, GRI 418-3-3
S4	S4-4	31 b	Description of action to provide or enable remedy in relation to an actual material impact	Narrative				GRI 3-3 , GRI 417-3-3, GRI 418-3-3
S4	S4-4	31 c	Description of additional initiatives or processes with primary purpose of delivering positive impacts for consumers and end-users	Narrative				GRI 3-3 , GRI 417-3-3, GRI 418-3-3
S4	S4-4	31 d	Description of how effectiveness of actions or initiatives in delivering outcomes for consumers and end-users is tracked and assessed	Narrative				GRI 3-3 , GRI 417-3-3, GRI 418-3-3
S4	S4-4	32 a	Description of approach to identifying what action is needed and appropriate in response to particular actual or potential material negative impact on consumers and end-users	Narrative				GRI 406-3-3, GRI 416-3-3, GRI 417-3-3, GRI 418-3-3
S4	S4-4	32 b	Description of approach to taking action in relation to specific material impacts on consumers and end-users	Narrative				GRI 406-3-3, GRI 416-3-3, GRI 417-3-3, GRI 418-3-3

(*Continued*)

Table 88 *(Continued).*

ESRS	DR	Paragraph	Name	Data type	Tips and tricks	Related CDR	VSMEs	GRI
S4	S4-4	32 c	Description of approach to ensuring that processes to provide or enable remedy in event of material negative impacts on consumers and end-users are available and effective in their implementation and outcomes	Narrative				GRI 2-25
S4	S4-4	33 a	Description of what action is planned or underway to mitigate material risks arising from impacts and dependencies on consumers and end-users and how effectiveness is tracked	Narrative				
S4	S4-4	33 b	Description of what action is planned or underway to pursue material opportunities in relation to consumers and end-users	Narrative				
S4	S4-4	34	Disclosure of how it is ensured that own practices do not cause or contribute to material negative impacts on consumers and end-users	Narrative				

S4	S4-4	35	Disclosure of severe human rights issues and incidents connected to upstream and downstream value chain	Narrative		SFDR		GRI 406-3-3, GRI 416-3-3, GRI 416-2, GRI 417-3-3, GRI 417-2, GRI 417-3, GRI 418-3-3, GRI 418-1
S4	S4-4	37	Disclosure of resources allocated to management of material impacts	Narrative				
S4	S4-4	AR 25 a	Disclosure of general and specific approaches to addressing material negative impacts	Narrative			V	
S4	S4-4	AR 25 b	Disclosure of initiatives aimed at contributing to additional material positive impacts	Narrative			V	
S4	S4-4	AR 25 c	Disclosure of how far undertaking has progressed in efforts during reporting period	Narrative			V	
S4	S4-4	AR 25 d	Disclosure of aims for continued improvement	Narrative			V	
S4	S4-4	AR 27	Disclosure of how undertaking seeks to use leverage with relevant business relationships to manage material negative impacts affecting consumers and end-users	Narrative			V	GRI 2-24
S4	S4-4	AR 28	Disclosure of how participation in industry or multi-stakeholder initiative and undertaking's own involvement is aiming to address material impacts	Narrative			V	
S4	S4-4	AR 33 a)	Disclosure of how consumers and end-users play a role in decisions regarding design and implementation of programmes or processes	Narrative			V	GRI 3-3, GRI 416-3-3, GRI 417-3-3, GRI 418-3-3

(Continued)

Table 88 (*Continued*).

ESRS	DR	Paragraph	Name	Data type	Tips and tricks	Related CDR	VSMEs	GRI
S4	S4-4	AR 33 b)	Information about intended or achieved positive outcomes of programmes or processes for consumers and end-users	Narrative			V	
S4	S4-4	AR 34	Initiatives or processes whose primary aim is to deliver positive impacts for consumers and/or end-users are designed also to support achievement of one or more of Sustainable Development Goals	Semi-narrative			V	
S4	S4-4	AR 41	Description of internal functions that are involved in managing impacts and types of action taken by internal functions to address negative and advance positive impacts	Narrative			V	
S4	S4-5	41	Targets set to manage material impacts, risks, and opportunities related to consumers and end-users (see ESRS 2–MDR-T)	MDR-T	Disclose all planned or underway targets and specific metrics to prevent, mitigate, or remediate material negative impacts by reviewing stakeholder engagement with own			
S4	S4-5	41	Disclosure of process for setting time-bound and outcome-oriented targets	Narrative				
S4	S4-5	41 a	Disclosure of how consumers and end-users were engaged directly in setting targets	Narrative				

S4	S4-5	41 b	Disclosure of how consumers and end-users were engaged directly in tracking performance against targets	Narrative	workforce assessment methodology, 5.1.3 MDR-M, 5.1.4 MDR-T and that relate to above actions metrics			GRI 3-3, GRI 406-3-3, GRI 416-3-3, GRI 417-3-3, GRI 418-3-3
S4	S4-5	41 c	Disclosure of how consumers and end-users were engaged directly in identifying lessons or improvements as result of undertaking's performance	Narrative				GRI 3-3, GRI 406-3-3, GRI 416-3-3, GRI 417-3-3, GRI 418-3-3
S4	S4-5	AR 42 a	Disclosure of intended outcomes to be achieved in lives of consumers and end-users	Narrative			V	
S4	S4-5	AR 42 b	Information about stability over time of target in terms of definitions and methodologies to enable comparability	Narrative			V	
S4	S4-5	AR 42 c	Disclosure of references to standards or commitments on which target is based	Narrative			V	

Source: CDR (EU) 2023/2772 supplementing Directive 2013/34/EU, *GRI 1, 2, 3: 11, 12, 13, 14, 306, 101, 201, 202, 203, 204, 205, 206, 207,301, 302, 303, 304, 305, 306, 308,401, 402, 403, 404, 405, 406, 407, 408, 409, 410, 411, 413, 414, 415, 416, 417, 418*, and the author.

14.4 Product Labelling and Transparency (S4-1, S4-3)

Relevant SDG: SDG 12 (Responsible Consumption and Production).

This indicator evaluates the organisation's transparency in product labelling and provides consumers and end-users with accurate and comprehensive information about your product's ingredients, origin, manufacturing processes, and potential risks or allergens.

1. Develop clear and standardised product labelling guidelines that comply with relevant regulations and best practices.
2. Conduct regular audits to ensure accurate and consistent product labelling across all products and markets.
3. Include relevant information such as ingredient lists, allergen warnings, country of origin, manufacturing dates, and expiration dates on product labels.
4. Provide additional transparency through online platforms, websites, or mobile applications where consumers can access detailed product information.

14.5 Customer Satisfaction and Feedback (S4-2)

Relevant SDG: SDG 8 (Decent Work and Economic Growth) and SDG 12 (Responsible Consumption and Production).

This indicator measures customer satisfaction and your organisation's responsiveness to customer feedback. It focuses on understanding and improving customer experiences, addressing concerns, and building long-term customer relationships.

1. Conduct customer satisfaction surveys to gather feedback on product satisfaction, customer service experiences, and overall satisfaction levels.
2. Use a standardised survey instrument or develop a custom survey based on industry best practices.
3. Analyse survey responses quantitatively to calculate customer satisfaction scores and identify areas for improvement.
4. Implement mechanisms to capture and address customer feedback and complaints promptly.

14.6 Customer Education and Awareness (S4-2)

Relevant SDG: SDG 12 (Responsible Consumption and Production).

This indicator evaluates your organisation's efforts to educate and raise awareness among consumers and end-users about product usage, safety precautions, and responsible consumption practices. It aims to empower consumers with knowledge and promote sustainable consumption behaviour.

1. Develop educational materials, user manuals, or online resources that provide clear instructions and safety guidelines for product usage.
2. Conduct consumer awareness campaigns, workshops, or training sessions on responsible consumption, recycling, and sustainable product use.
3. Monitor and assess the effectiveness of educational initiatives through feedback surveys or quizzes to measure consumer knowledge and behaviour change.
4. Collaborate with relevant organisations or community groups to expand the reach of consumer education programs.

14.7 Responsible Marketing and Advertising Practices (S4-2, S4-4)

Relevant SDG: SDG 16 (Peace, Justice, and Strong Institutions).

This indicator evaluates your organisation's adherence to responsible marketing and advertising practices, including ethical advertising, accurate product claims, and protection of vulnerable consumers.

1. Develop marketing and advertising guidelines that align with industry standards, regulations, and ethical principles.
2. Regularly review marketing materials and advertisements to ensure compliance with guidelines and legal requirements.
3. Monitor consumer feedback, complaints, and advertising monitoring reports to identify potential issues or violations.
4. Implement corrective actions and continuous improvement processes to address any identified marketing or advertising concerns.

14.8 Product Recalls and Responsiveness (S4-3)

Relevant SDG: SDG 3 (Good Health and Well-being) and SDG 16 (Peace, Justice, and Strong Institutions).

This indicator assesses your organisation's responsiveness to product recalls and the effectiveness of communication and remedial actions when product defects or safety issues arise. It focuses on protecting consumer safety and maintaining trust in the organisation's products.

1. Establish a product recall protocol that outlines the process for identifying, communicating, and addressing product defects or safety concerns.
2. Implement mechanisms to promptly identify and track product issues, including consumer complaints, quality control checks, and monitoring of safety reports.
3. Communicate product recalls clearly and effectively to consumers through various channels, including traditional media, online platforms, and direct notifications.
4. Analyse the effectiveness of product recall efforts by monitoring the response rate, the number of returned products, and customer satisfaction with the recall process.

14.9 Product Lifecycle Management and End-of-Life Considerations (S4-4)

Relevant SDG: SDG 12 (Responsible Consumption and Production) and SDG 13 (Climate Action).

This indicator evaluates your organisation's practices related to product lifecycle management, including considerations for end-of-life disposal, recycling, and responsible waste management. It focuses on minimising the environmental impact of products throughout their lifecycle.

1. Conduct a comprehensive assessment of the product lifecycle, from sourcing raw materials to end-of-life disposal or recycling.
2. Implement measures to reduce waste generation, promote recycling, and facilitate proper disposal of products.
3. Establish partnerships with recycling facilities, waste management companies, or extended producer responsibility programs.
4. Track and report on key performance indicators, such as the percentage of products recycled or the amount of waste diverted from landfills.

14.10 Product Innovation and Sustainable Design (S4-4)

Relevant SDG: SDG 12 (Responsible Consumption and Production).

This indicator assesses your organisation's commitment to product innovation and sustainable design, aiming to create products that minimise environmental impact, promote resource efficiency, and meet evolving consumer needs and expectations.

1. Establish a product development process that integrates sustainability criteria, such as lifecycle assessment, eco-design principles, and material sourcing guidelines.
2. Conduct regular assessments of product sustainability performance, considering factors such as energy efficiency, recyclability, and the use of renewable materials.
3. Encourage collaboration with suppliers and external partners to foster innovation and share best practices in sustainable design.
4. Track and report on the percentage of products with sustainable design features or eco-friendly certifications.

14.11 Product Accessibility and Affordability (S4-4, S4-5)

Relevant SDG: SDG 10 (Reduced Inequalities) and SDG 12 (Responsible Consumption and Production).

This indicator assesses your organisation's efforts to ensure product accessibility and affordability for a diverse range of consumers and end-users. It focuses on making products available and affordable to underserved populations and promoting inclusive consumption.

1. Identify target populations or segments that may face accessibility or affordability barriers.
2. Analyse pricing strategies, distribution channels, and product availability in different regions or markets.
3. Conduct surveys or market research to understand consumer affordability concerns and identify potential pricing or accessibility improvements.
4. Implement initiatives such as pricing adjustments, discounts, or partnerships to enhance product accessibility and affordability.

15

G1: Business Conduct

In this chapter: Assessment of business conduct impacts and actions to prevent and mitigate negative impacts. Management of business conduct dependencies and their financial effects. Risks and opportunities related to business conduct impacts and their financial effects. Common indicators and metrics.

This methodology focuses on your organisation's business conduct and governance, and aims to guide your organisation in adopting responsible and sustainable practices by embracing several key aspects essential for fostering transparency, integrity, and ethical behaviour. Positive impacts are created by inclusive and values-driven corporate cultures that promote integrity, diversity, and inclusivity. In turn, this will lead to your organisation championing ethical practices, fair trade principles, and social and environmental sustainability with peers and through relationships, especially your value chain. Additionally, combatting corruption and bribery requires robust measures to prevent and address misconduct. By complying with laws and regulations, fostering transparency, and establishing a culture of accountability, your organisation can mitigate risks and enhance ethical business practices.

Transparency is also important when disclosing political influence, including lobbying activities as well as the protection of whistle-blowers, whose role in exposing wrongdoing and promoting organisational accountability is crucial. Effective whistle-blower protection mechanisms encourage a culture of openness and trust, enabling individuals to report misconduct without fear of retaliation. Respect and care for biodiversity, including animal welfare, is another aspect covered by this methodology, and emphasises your organisation's responsibilities towards animal rights and promoting sustainable and humane practices throughout their operations. Finally, this methodology addresses payment practices, specifically with regard to late payment to small and medium enterprises (SMEs). Timely and fair payment to SMEs is encouraged to support their financial stability and contribute to a healthy business ecosystem.

Integrating these principles into your organisation's business conduct and governance will enhance reputation, strengthen stakeholder trust, and contribute to a more sustainable and ethical business environment.

Remember that there are more details on the support website https://esgreportpro.com along with examples of some key metrics.

In the following chapter:

- Not all DPs are to be reported if the undertaking has not adopted the respective policies, implemented the respective actions, or set the respective targets in relation to a sustainability matter that has been assessed to be material.

ESRS G1: BUSINESS CONDUCT		
GOVERNANCE & STRATEGY	**IMPACTS, RISKS, & OPPORTUNITIES**	**METRICS & TARGETS**

INCLUDES CROSS-CUTTING DISCLOSURES:
• ESRS 2: Governance (GOV), Strategy (SBM), and Management of impacts, risks, and opportunities (IRO)
• ESRS 2 GOV-1: The role of the administrative, supervisory, and management bodies
• ESRS 2 IRO-1: Description of the processes to identify and assess material impacts, risks, and opportunities
	• G1-1: Corporate culture and business conduct *policies*
	• G1-2: Management of relationships with *suppliers*
	• G1-3: Prevention and detection of *corruption/bribery*
	INCLUDES CROSS-CUTTING DISCLOSURES:
	• Information on *violations of UNGC principles and OECD guidelines, monitoring processes and compliance* with focus on:
	  • Anti-corruption
	  • Competition
	  • Tax
	  • Science and technology
		• G1-4: **Confirmed incidents** of corruption or bribery
		• G1-5: **political influence and lobbying activities**
		• G1-6: *Payment practices*
		INCLUDES CROSS-CUTTING DISCLOSURES:
		• *Tracking effectiveness* of policies, actions, and targets related to managing material negative impacts, advancing positive impacts, and managing material risk and opportunities

Before diving into detailed metrics and targets, begin with describing your materiality assessment and organisational governance initiatives that directly interact with business conduct, most importantly, the processes and methodologies to identify and assess impacts, risks, and opportunities related to business conduct issues (see **Chapter 3: Single and Double Materiality Assessment and 15.1 Business Conduct Assessment Methodology**).

TIPS AND TRICKS

Transparency Begins at the Top.

An EU-based private tobacco company (cannot be named for legal reasons) must comply with the CSRD/ESRS and produce an ESG report.

Problem: During their governance audit by an external consultant, it was found that the CEO also holds a permanent tenure as President of the BoD. So how is integrity assured? Neither EU nor national law demands that these positions must be held by separate individuals, unless they are a listed company.

Solution: Since there is no legal demand to adhere to, the same person continues to hold both positions, and the ESG report discloses this. However, the report explains how integrity is ensured through an extensive internal audit process. Will this be enough to satisfy investors? Only time will tell, but there is no doubt that transparency could be improved with a comprehensive external audit process of organisational governance.

Table 89 Business conduct assessment indicators relating to policies and stakeholder engagement.

ESRS	DR	Paragraph	Name	Data type	Tips and tricks	Related CDR	VSMEs	GRI
G1	GOV-1	5 a	Disclosure of role of administrative, management, and supervisory bodies related to business conduct	Narrative	See ESRS GOV-1			GRI 2-12
G1	GOV-1	5 b	Disclosure of expertise of administrative, management, and supervisory bodies on business conduct matters	Narrative				GRI 2-9
G1	G1-1	7	Policies in place to manage its material impacts, risks, and opportunities related to climate change mitigation and adaptation	MDR-P	Policies, process, mitigation, and reporting channels relating to business conduct and corporate culture using stakeholder engagement assessment with materiality assessment, 5.1.1 MDR-P, and policy audit of business conduct, protection of whistle-blowers, anti-corruption and anti-bribery, and animal welfare. Include: reference to 15.1 Stakeholder participation 15.2 Legitimation of management			GRI 2-23 GRI 205-3-3
G1	G1-1	9	Description of how the undertaking establishes, develops, promotes, and evaluates its corporate culture	Narrative				GRI 2-16, GRI 2-23, GRI 2-24
G1	G1-1	10a	Description of the mechanisms for identifying, reporting, and investigating concerns about unlawful behaviour or behaviour in contradiction of its code of conduct or similar internal rules	Narrative				GRI 2-26
G1	G1-1	10 b	If there are no policies on anti-corruption or anti-bribery consistent with United Nations Convention against Corruption are in place	Semi-narrative		SFDR		
G1	G1-1	10 b	If no policies, then timetable for implementation of policies on anti-corruption or anti-bribery consistent with United Nations Convention against Corruption	Narrative				
G1	G1-1	10 c	Disclosure of safeguards for reporting irregularities including whistle-blowing protection	Narrative				

(*Continued*)

Table 89 (*Continued*).

ESRS	DR	Paragraph	Name	Data type	Tips and tricks	Related CDR	VSMEs	GRI
G1	G1-1	10 d	State whether there are policies on the protection of whistle-blowers in place	Semi-narrative	15.3 Board members by age range 15.4 Number of board meetings and attendance rate 15.5 Term limits for board of directors 15.6 Attendance at annual general meetings 15.7 Democratic elections 15.8 Number of meetings of audit committee and attendance rate 15.9 Access to remedy 15.10 Employee turnover rate 15.11 Responsible and ethical sourcing	SFDR		
G1	G1-1	10 d	If no policies, then timetable for implementation of policies on protection of whistle-blowers	Narrative				
G1	G1-1	10 e	Undertaking is committed to investigate business conduct incidents promptly, independently, and objectively	Semi-narrative				
G1	G1-1	10 f	Where applicable, policies with respect to animal welfare are in place	Semi-narrative				
G1	G1-1	10 g	Information about policy for training within the organisation on business conduct, including target audience, frequency, and depth of coverage	Narrative				GRI 2-24
G1	G1-1	10 h	Disclosure of the functions that are most at risk in respect of corruption and bribery	Narrative				
G1	G1-1	11	Entity is subject to legal requirements with regard to protection of whistle-blowers	Semi-narrative			V	
G1	G1-2	14	Description of policy to prevent late payments, especially to SMEs	Narrative				GRI 308-3-3
G1	G1-2	15 a	Description of approaches in regard to relationships with suppliers, taking into account risks related to supply chain and impacts on sustainability matters	Narrative				GRI 204-3-3 , GRI 308-3-3, GRI 414-3-3
G1	G1-2	15 b	Disclosure of how social and environmental criteria are taken into account for selection of supply-side contractual partners	Narrative				GRI 308-1, GRI 414-1
ESRS 2	G1		Disclosures to be reported if the undertaking has not adopted policies and/or actions and/or targets	Narrative				

G1	G1-3	18 a	Information about procedures in place to prevent, detect, and address allegations or incidents of corruption or bribery	Narrative	Management of relationships with suppliers – assessment of policy implementation, includes reported incidents and outcomes relating to above plus the following: 15.12 Average hours of training on anti-corruption issues per year per employee 15.13 Corruption-related fines paid or payable due to settlements 15.14 Corporate political influence 15.15 Days sales outstanding 15.16 Collection costs 15.17 Payment terms compliance 15.18 Days payable outstanding		V	GRI 2-26, GRI 205-1, GRI 205-3-3
G1	G1-3	18 b	Investigators or investigating committee are separate from the chain of management involved in prevention and detection of corruption or bribery	Semi-narrative				
G1	G1-3	18 c	Information about the process to report outcomes to administrative, management, and supervisory bodies	Narrative				GRI 2-13, GRI 2-16
G1	G1-3	19	Disclosure of plans to adopt procedures to prevent, detect, and address allegations or incidents of corruption or bribery in case of no procedure	Narrative			V	
G1	G1-3	20	Information about how policies are communicated to those for whom they are relevant (prevention and detection of corruption or bribery)	Narrative				GRI 205-2
G1	G1-3	21 a	Information about nature, scope, and depth of anti-corruption or anti-bribery training programmes offered or required	Narrative				
G1	G1-3	21 b	Percentage of functions-at-risk covered by training programmes	Per cent				GRI 205-2
G1	G1-3	21 c	Information about members of administrative, supervisory, and management bodies relating to anti-corruption or anti-bribery training	Narrative				GRI 205-2
G1	G1-3	AR 7	Disclosure of an analysis of its training activities by, for example, region of training or category	Narrative			V	GRI 205-2
G1	G1-3	AR 8	Prevention and detection of corruption or bribery – anti-corruption and bribery training table	Table				GRI 205-2
G1	G1-4	24 a	Number of convictions for violation of anti-corruption and anti-bribery laws	Integer		SFDR	V	GRI 2-27

(Continued)

Table 89 (*Continued*).

ESRS	DR	Paragraph	Name	Data type	Tips and tricks	Related CDR	VSMEs	GRI
G1	G1-4	24 a	Amount of fines for violation of anti-corruption and anti-bribery laws	Monetary	15.19 Supplier payment history 15.20 Tax gap	SFDR	V	GRI 2-27
G1	G1-4	25 a	Number of confirmed incidents of corruption or bribery and action taken	Integer			V	GRI 205-1, GRI 205-3
G1	G1-4	25 a	Information about the nature of confirmed incidents of corruption or bribery	Narrative			V	GRI 205-3
G1	G1-4	25 b	Number of confirmed incidents in which own workers were dismissed or disciplined for corruption or bribery-related incidents	Integer			V	GRI 205-3
G1	G1-4	25 c	Number of confirmed incidents relating to contracts with business partners that were terminated or not renewed due to violations related to corruption or bribery	Integer			V	GRI 205-3
G1	G1-4	25 d	Information about details of public legal cases regarding corruption or bribery brought against undertaking and own workers and about outcomes of such cases				V	GRI 205-3
G1	G1-5	29 a	Information about representative(s) responsible in administrative, management, and supervisory bodies for oversight of political influence and lobbying activities	Narrative				GRI 415-3-3
G1	G1-5	29 b	Information about financial or in-kind political contributions	Narrative				GRI 415-1
G1	G1-5	29 b i	Financial political contributions made	Monetary				
G1	G1-5	AR 12 a	Amount of internal and external lobbying expenses	Monetary			V	
G1	G1-5	AR 12 b	Amount paid for membership to lobbying associations	Monetary			V	

G1	G1-5	29 b i	In-kind political contributions made	Monetary			
G1	G1-5	29 b (ii)	Disclosure of how monetary value of in-kind contributions is estimated	Narrative			
G1	G1-5	29 b (ii)	Financial and in-kind political contributions made [table]	Table			
G1	G1-5	29 c	Disclosure of main topics covered by lobbying activities and undertaking's main positions on these topics	Narrative			
G1	G1-5	29 d	Undertaking is registered in EU Transparency Register or in equivalent transparency register in member states	Semi-narrative			
G1	G1-5	30	Information about appointment of any members of administrative, management, and supervisory bodies who held comparable position in public administration in two years preceding such appointment	Narrative			
G1	G1-5	AR13	The entity is legally obliged to be a member of a chamber of commerce or other organisation that represents its interests	Semi-narrative		V	GRI general disclosure
G1	G1-6	33 a	Average number of days to pay invoice from the date when contractual or statutory term of payment starts to be calculated	Integer			
G1	G1-6	33 b	Description of undertakings standard payment terms in number of days by main category of suppliers	Narrative			
G1	G1-6	33 b	Percentage of payments aligned with standard payment terms	Per cent			
G1	G1-6	33 c	Number of outstanding legal proceedings for late payments	Integer			
G1	G1-6	33 d	Disclosure of contextual information regarding payment practices	Narrative			

Source: CDR (EU) 2023/2772 supplementing Directive 2013/34/EU, *GRI 1, 2, 3: 11, 12, 13, 14, 306, 101, 201, 202, 203, 204, 205, 206, 207,301, 302, 303, 304, 305, 306, 308,401, 402, 403, 404, 405, 406, 407, 408, 409, 410, 411, 413, 414, 415, 416, 417, 418*, and the author.

15.1 Stakeholder Participation (G1-1)

Relevant SDGs: SDG 8 (Decent Work and Economic Growth), SDG 9 (Industry, Innovation, and Infrastructure), SDG 12 (Responsible Consumption and Production).

This indicator assesses whether your organisation has formal mechanisms in place for non-employee stakeholders, such as members, consumers, and communities, to participate in decision-making on strategic issues. Determine and disclose the presence of formal mechanisms for non-employee stakeholder participation in strategic decision-making, for example, feedback systems, community relations, and media. This can include processes that allow stakeholders to provide input, express opinions, or collaborate with the organisation on strategic matters.

15.2 Legitimation of Management (G1-1)

Relevant SDGs: SDG 8 (Decent Work and Economic Growth), SDG 12 (Responsible Consumption and Production).

This indicator assesses the extent to which staff members are involved in the selection process of managers within the organisation. It consists of two disclosures: (i) the proportion of managers selected by their own staff, and (ii) the specific processes or mechanisms through which staff participate in the selection.

1. **Proportion of managers selected by staff**: The organisation determines and discloses the percentage of managers who are selected by their own staff. This is calculated by dividing the total number of managers selected by staff (MSS) by the total number of managers employed (TME) and multiplying by 100. The result represents the percentage of managers selected by staff in a specific year (*t*).

   PMH = (MSS/TME) × 100

2. **Disclosure of specific participation process(es) or mechanism(s)**: Disclose the specific processes or mechanisms by which staff participate in the selection of their own managers. This can include formal consultation processes, participation in selection committees, elections, co-determination, or other forms of staff involvement.

15.3 Board Members by Age Range (G1-1)

Relevant SDGs: SDG 8 (Decent Work and Economic Growth), SDG 12 (Responsible Consumption and Production).

This indicator assesses the age diversity of your organisation's board members by categorising them into different age ranges. It provides insights into the distribution of board members across various age groups.

To calculate this indicator, organisations should establish specific age ranges, such as 'Under 30 years', '30–50 years', and 'Over 50 years', for classifying board members. Accurate and current data on the age of each board member should be collected from sources such as the investor relations office, company secretary, or human resources manager. Board members are then categorised into their respective age ranges based on the defined categories. The number of board members in each age range is counted to determine the distribution.

15.4 Number of Board Meetings and Attendance Rate (G1-1)

Relevant SDGs: SDG 8 (Decent Work and Economic Growth), SDG 12 (Responsible Consumption and Production).

This indicator assesses the frequency of board meetings and the attendance rate of board members at these meetings. It provides insights into the level of board engagement and participation in key decision-making processes. The number of board meetings conducted within a specified period is identified by reviewing the records maintained by the organisation, such as those kept by the investor relations office, company secretary, or human resources manager. Additionally, the attendance rate at each meeting is determined by dividing the number of board members present by the total number of board members, using the ratio of attendees to the overall board size. To calculate the overall annual attendance rate, the attendance rates from each meeting are averaged.

15.5 Term Limits for Board of Directors (G1-1)

Relevant SDGs: SDG 8 (Decent Work and Economic Growth), SDG 12 (Responsible Consumption and Production).

This indicator examines whether there are term limits in place for members of the board of directors. It assesses the organisation's governance practices regarding board member tenure. Your organisation determines and discloses whether there is a term limit for members of the board of directors. This can be done by reviewing the organisation's policies, bylaws, or other governance documents to identify the existence of specific term limits for board members. The disclosure should indicate whether term limits are implemented (Yes) or not (No).

15.6 Attendance at Annual General Meetings (G1-1)

Relevant SDGs: SDG 8 (Decent Work and Economic Growth), SDG 12 (Responsible Consumption and Production).

This indicator assesses the average level of attendance at annual general meetings (AGMs) or equivalent meetings/mechanisms by members of the organisation over the past five years. It provides insights into member engagement and participation in these important organisational events.

Determine and disclose the average level of attendance at AGMs or equivalent meetings/mechanisms by members over the past five years. The calculation involves tracking the rate of attendance at each AGM or equivalent mechanism using the equation:

$$ROA = (NOM/TSM) \times 100$$

where:

ROA represents the rate of attendance.

NOM represents the number of members attending the AGM or equivalent mechanism.

TSM represents the total standing members at the time of the specific AGM or equivalent mechanism. The five-year trend in the rate of attendance is reported, showcasing the level of member participation over time.

15.7 Democratic Elections (G1-1)

Relevant SDGs: SDG 8 (Decent Work and Economic Growth), SDG 12 (Responsible Consumption and Production).

This indicator assesses whether your organisation uses a 'one person, one vote' system, with or without delegation of votes, for electing individuals to managerial, executive, and organisational governance roles. It focuses on the democratic process of electing officers and board members within the organisation.

Determine and disclose whether it utilises a 'one person, one vote' system for elections. This system ensures that each member has an equal voting right, either directly or through delegated votes. The organisation indicates whether such a system is in place by reporting 'Yes' or 'No' accordingly.

15.8 Number of Meetings of Audit Committee and Attendance Rate (G1-1, ESRS 2 GOV-1)

Relevant SDGs: SDG 8 (Decent Work and Economic Growth), SDG 12 (Responsible Consumption and Production).

The indicator measures the number of meetings conducted by the audit committee within a specified period and the attendance rate of committee members at each meeting. It provides insights into the effectiveness of the audit committee's governance and oversight functions. The indicator includes the calculation of the overall annual attendance rate by averaging the attendance rates over the year.

1. Number of meetings: Identify the total number of audit committee meetings conducted during the reporting period.
2. Attendance rate calculation: For each meeting, record the number of audit committee members in attendance.
   - Calculate the attendance rate at each meeting by dividing the number of members present by the total number of committee members.
   - To determine the overall annual attendance rate, average the attendance rates of all meetings conducted throughout the year.
3. Reporting format:
   - Report the total number of audit committee meetings held during the reporting period.
   - Report the attendance rate for each meeting as a percentage.
   - Report the overall annual attendance rate as a percentage.

15.9 Access to Remedy (G1-1, S1-3)

Relevant SDGs: SDG 8 (Decent Work and Economic Growth), SDG 12 (Responsible Consumption and Production).

This indicator measures whether your organisation has mechanisms for access to remedy (i.e., non-state-based grievance mechanisms) for any issues related to labour rights and whether there are clear and known procedures with an indicative timeframe of the grievance process, and whether there are any cases where access to remedy has been demanded.

Your organisation should disclose whether they have mechanisms for access to remedy for labour rights issues, the existence of clear and known procedures for the grievance process, and whether there are any cases where access to remedy has been demanded.

15.10 Employee Turnover Rate (G1-1, S1-15)

Relevant SDGs: SDG 8 (Decent Work and Economic Growth), SDG 12 (Responsible Consumption and Production).

This indicator measures the rate at which employees leave your organisation within a given year. Determine and disclose the number of employees who left your organisation during the year, divided by the average number of employees during that same year:

Employee turnover rate = (Number of employees who left during the year/Average number of employees during the year) × 100

15.11 Responsible and Ethical Sourcing (G1-2, GOV-3)

Relevant SDGs: SDG 8 (Decent Work and Economic Growth), SDG 12 (Responsible Consumption and Production).

This indicator assesses whether your organisation engages in responsible and ethical sourcing and purchasing practices. You must disclose whether there are policies and programs that align with commercial and sustainability goals, and whether they guard against negative social and environmental outcomes for your suppliers. You should also indicate whether your organisation supports suppliers' efforts to upgrade their labour standards and social and environmental impacts, and whether it provides financial incentives to its suppliers.

1. **Policies and programs**: Your organisation must disclose whether it has policies and programs in place that ensure consistency and alignment between its commercial and sustainability goals and outcomes.

2. **Supplier protection**: Your organisation must disclose whether it has policies in place to guard against negative social and environmental outcomes for its suppliers. This includes policies against aggressive pricing, short-term or insecure contractual relationships, and excessive and unplanned overtime.

3. **Capacity of workers**: Your organisation must disclose whether it provides workers in its supply chains with the capacity to contest and shape the upgrading of supply chains.

4. **Financial support and incentives**: Your organisation must disclose whether it provides financial support and incentives to its suppliers to upgrade their labour standards and social and environmental impacts, and the percentage of suppliers and/or facilities receiving such incentives and support.

15.12 Average Hours of Training on Anti-corruption Issues per Year per Employee (G1-3)

Relevant SDGs: SDG 8 (Decent Work and Economic Growth), SDG 12 (Responsible Consumption and Production).

The indicator measures the average number of hours of anti-corruption training provided to employees and other workers within your organisation each year. It helps assess your organisation's commitment to preventing corruption and promoting ethical behaviour among its workforce. The indicator can be reported as the annual figure and the trend over the past five years.

1. **Calculation of total hours**: Calculate the aggregate number of hours each year in which employees and other workers within the organisation are engaged in anti-corruption training. This includes both mandatory and voluntary training sessions related to anti-corruption issues.

2. **Reporting format**: Report the annual number of hours of anti-corruption training for the most recent year (t). Report the trend in hours of training for the past five years (t, $t-1$, $t-2$, $t-3$, $t-4$).

3. **Total hours calculation**: THT (total hours of anti-corruption training) represents the sum of all hours of anti-corruption training provided by the organisation.

Table 90 Anti-corruption and bribery training reporting template.

During the financial year 20XX, [ABC Organisation] provided training to its at-risk own workers in terms of its policy (see note Z). For those at-risk functions the training was mandatory, but [ABC organisation] also made available voluntary training for other own workers. Details of the training during the year are:

	At-risk functions	Managers	Administrative, management, and supervisory bodies	Other own workers
Training coverage				
Total	20,000	200	16	70,000
Total received training	19,500	150	8	5,000
Delivery method and duration				
Classroom training	5 hours			
Computer-based training	1 hour	2 hours	1 hour	
Voluntary computer-based training				1 hour
Frequency				
How often is training required	Annually	Annually	Bi-annually	–
Topics covered				
Definition of corruption	X	X	X	X
Policy	X	X	X	X
Procedures on suspicion/detection	X	X		
Etc.	X			

Source: CDR (EU) 2023/2772 supplementing Directive 2013/34/EU.

15.13 Corruption-Related Fines Paid or Payable Due to Settlements (G1-4)

Relevant SDGs: SDG 8 (Decent Work and Economic Growth), SDG 12 (Responsible Consumption and Production).

This indicator measures the amount of fines paid or payable by your organisation as a result of settlements related to corruption-related violations of laws or regulations.

1. **Identify convictions and settlements**: Identify all convictions and settlements pertaining to violations of corruption-related laws or regulations, which can normally be found in the income statement as an expense.

2. **Determine fine amounts**: For each conviction or settlement, determine the amount paid or payable in fines.

3. **Summation**: Total all the fine amounts identified during the reporting period to calculate the total amount of fines paid or payable.
4. **Disclosure**: Disclose the total number of relevant convictions and settlements, as well as the total amount of fines paid or payable.

15.14 Corporate Political Influence (G1-5)

Relevant SDGs: SDG 8 (Decent Work and Economic Growth), SDG 12 (Responsible Consumption and Production).

Your organisation must disclose its policies, programs, and practices related to corporate political influence. It must include a description of its expenditure on political campaigns, candidates, and advocacy/lobbying in the last five years, the jurisdictions covered, the top five issues, and recipients of expenditure. Additionally, it must disclose whether it has a policy aligned with CSR, ESG, SDGs, or other sustainability performance frameworks.

Group or company-wide expenditure divided by revenue related to political campaigns/candidates and advocacy/lobbying in the last five years. Indirect payments to lobbying firms/organisations and business/trade associations/industry groups are also included.

Table 91 Corporate political funding reporting template.

	20XX	20xx+1 (Est.)
Political funding provided:	100	
Funding to EFG	75	
Funding to QRS	25	

Source: CDR (EU) 2023/2772 supplementing Directive 2013/34/EU.

15.15 Days Sales Outstanding (G1-6)

Relevant SDGs: SDG 8 (Decent Work and Economic Growth), SDG 12 (Responsible Consumption and Production).

Days sales outstanding (DSO) measures the average number of days it takes for your organisation to collect payment from its customers after a sale. It indicates the efficiency of your organisation's credit and collection processes.

1. **Calculate average daily sales**: Divide the total sales revenue by the number of days in the period under consideration.

Average daily sales = Total Sales Revenue/Number of Days.

2. **Determine accounts receivable**: Obtain the total amount of outstanding customer invoices or accounts receivable at the end of the period.

3. **Calculate DSO**: Divide the accounts receivable by the average daily sales and multiply the result by the number of days in the period.

   DSO = (Accounts Receivable/Average Daily Sales) × Number of Days.

15.16 Collection Costs (G1-6)

Relevant SDGs: SDG 8 (Decent Work and Economic Growth), SDG 12 (Responsible Consumption and Production).

Collection costs refer to the total expenses incurred by your company in the process of collecting overdue payments from customers. It includes costs such as personnel, legal fees, and collection agency fees. Total the various costs associated with debt collection activities, including salaries of collection staff, legal expenses, third-party collection agency fees, and other related expenses.

15.17 Payment Terms Compliance (G1-6)

Relevant SDGs: SDG 8 (Decent Work and Economic Growth), SDG 12 (Responsible Consumption and Production).

Payment terms compliance measures the percentage of customers who adhere to the agreed-upon payment terms, such as paying within 30 days or making instalment payments as agreed.

1. **Determine the number of payments made within payment terms**: Count the number of payments made by customers within the agreed-upon payment terms.

2. **Calculate payment terms compliance**:

   Payment Terms Compliance = (Number of Payments Made Within Payment Terms/Total Number of Payments) × 100

15.18 Days Payable Outstanding (G1-6)

Relevant SDGs: SDG 8 (Decent Work and Economic Growth), SDG 12 (Responsible Consumption and Production).

Days payable outstanding (DPO) measures the average number of days it takes for your organisation to pay its suppliers after receiving an invoice. It provides insights into your organisation's efficiency in managing accounts payable:

1. **Determine accounts payable**: Obtain the total amount of outstanding supplier invoices or accounts payable at the end of the period.
2. **Calculate average daily purchases**:

   Average Daily Purchases = Total Purchases/Number of Days
3. **Calculate DPO**: Divide the accounts payable by the average daily purchases and multiply the result by the number of days in the period:

   DPO = (Accounts Payable/Average Daily Purchases) × Number of Days

15.19 Supplier Payment History (G1-6)

Relevant SDGs: SDG 8 (Decent Work and Economic Growth), SDG 12 (Responsible Consumption and Production).

Supplier payment history tracks the payment history and track record of your organisation with its suppliers. It provides insights into your organisation's reliability and consistency in meeting its payment obligations.

1. **Review payment records**: Assess the payment records to track the history of payments made to suppliers, including the frequency of on-time payments and any instances of late or missed payments.
2. **Analyse payment performance**: Calculate the percentage of on-time payments made to suppliers:

   Payment Performance = (Number of On-Time Payments/Total Number of Payments) × 100

15.20 Tax Gap

Relevant SDGs: SDG 8 (Decent Work and Economic Growth), SDG 12 (Responsible Consumption and Production.

The tax gap indicator measures the difference between the actual taxes paid by your organisation and the statutory taxes due. This provides an assessment of the extent to which your organisation is avoiding taxes. It is calculated by comparing the effective tax rate (ETR) to the statutory tax rate (STR). The STR is determined by taking the weighted average of tax rates from at least three countries or regions of operation with the highest revenue. The ETR is calculated by taking the weighted average of the annual income tax payable to the annual earnings before tax (EBT) as disclosed by companies from each respective country or region:

$$\text{Annual Tax Gap} = [(\text{STR} - \text{ETR})/\text{STR}] \times 100$$

To assess sustainability, the tax gap should be reported over a five-year period, including the current year (t), as well as the previous four years ($t - 1$, $t - 2$, $t - 3$, $t - 4$). The five-year average tax gap (ATG) is calculated using the following formula:

$$\text{ATG} = (\text{Annual Tax Gap } t + \text{Annual Tax Gap } t - 1 + \text{Annual Tax Gap } t - 2 + \text{Annual Tax Gap } t - 3 + \text{Annual Tax Gap } t - 4)/5$$

A score of $\text{ATG} \geq 1.0$ is considered sustainable, while a score of $\text{ATG} < 1.0$ is considered unsustainable.

16

Report Formation and Publication

In this chapter: Importance of transparency in ESG reporting and disclosure practices for ESG data. Formats, structure, and layout. Compliance with ESG regulations, legal considerations, and statements.

You are now in possession of an enormous amount of information that needs to be structured into a report that readers will be able to use to make informed decisions about your organisation. Sustainability reports are not only useful tools for organisations to address and communicate sustainability information and manage sustainability-related risks, they are also highly relevant for investors, regulators, and stakeholders. So when creating your report, you need to remain mindful of the audience's (see Chapter 1.4, 'Understanding Your ESG Report Audience') needs in scope, format, and content accessibility. But first, let us review the purpose of producing an ESG report beyond the simple reason of compliance.

16.1 The Purpose of Your ESG Report

An environmental, social, and governance (ESG) report serves as a crucial tool for organisations to communicate their sustainable behaviours, impacts, risks, and opportunities to stakeholders. ESG factors have become increasingly important for investors, customers, employees, and other stakeholders in evaluating the overall performance and long-term sustainability of a company.

ESG reporting is a mechanism through which companies disclose non-financial information (but always cross-referenced to the financial statements) related to their environmental, social, and governance practices. These reports are designed to provide stakeholders with a comprehensive view of how an organisation is managing key sustainability issues that could impact its performance and reputation over time. The primary purpose of an ESG report is to enhance transparency, accountability, and trust by sharing information on the company's sustainability efforts and their outcomes. But an ESG report can go further than making factual statements about an organisation's performance; there should also be nuanced explanation of the context affecting decision-making, and the vision driving transitioning and adaptation so that stakeholders (especially investors) can gain better, more accurate insight.

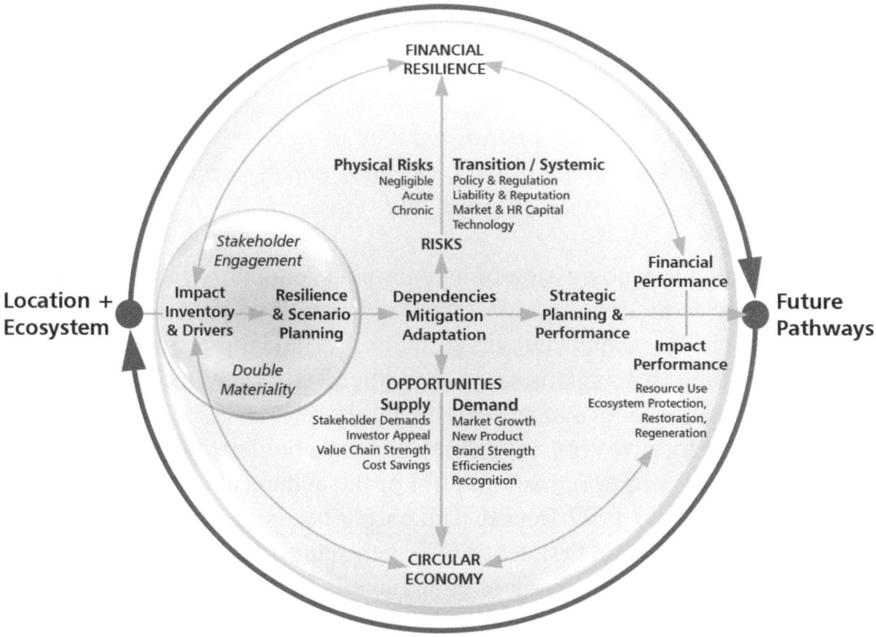

Graphic 24 The ESG reporting system. Source: Author.

The ESG report is therefore a description of the 'systems thinking' approach that the organisation has towards its impacts, risks, and opportunities, at the heart of which are organisational dependencies, impact mitigation, and operational adaptation. The vision and mission, the articulation of forward-looking pathways to achieve short- and long-term goals, should clearly encompass the entire system. The disclosure of ESG information is therefore inherently different from financial disclosures; in ESG reports, we need to be reading about the current to future landscape and performance, whereas in a financial report we are reviewing past to current activities. It is this shift in perspective that will likely cause some of the greatest challenges to report preparation, and it represents a change in mindset, and that is never easy.

An additional challenge is the level of transparency concerning impacts and risks, which human nature tends to want to avoid discussing. Faithful representation and extensive due diligence are the first steps towards disclosure openness, especially regarding the methodologies employed to assess materiality. This is likely to be a major focus of ESG report scrutiny in the coming years.

16.2 Drafting Your ESG Report

The CSRD/ESRS uses the term 'sustainability statement' to refer to a section of an organisation's management report that presents sustainability information. The sustainability statement's structure should include general, environmental, social, and governance sections containing quantitative and qualitative information relevant for stakeholder decision-making. Sustainability statements should therefore:

- include board statements that link sustainability strategy, management approach, and sustainability-related governance processes;
- be aligned with your organisation's fiscal year and business ownership model;
- apply the same consolidation scope used for financial reporting (including links and references) and explain reporting boundaries and list the entities included;
- contain specific disclosures presented in a similar structure, including information about impacts, metrics, targets, management approach, and other relevant information;
- for all metrics disclosed in the current period, include one year of comparative information, and when relevant to understanding a narrative disclosure, include one year of comparative information – except in the year of adoption, when an organisation may defer the presentation of comparative information by one year;
- explain the extent of the report's coverage of the organisation's value chain; and
- define time intervals as of the end of the reporting period, whereby short-term is the period adopted in the financial statements, medium-term is from the end of the short-term reporting period to five years, and long-term is more than five years.

General disclosures begin with the presentation of your organisation's current and upcoming opportunities and challenges, and explanation of how it intends to generate value in a responsible way. This is best achieved with a principal message that is reinforced through the report while considering the audience and main stakeholder groups. This should include an analysis of gathered data, averages, trends, materiality assessment, and other results that support the principal message. It is crucial to remain transparent and avoid greenwashing.

Table 92 Sustainability statement narrative development.

Introduction to your organisation	Governance and management	Report contents	Support information
Understanding the What? Why? How?		Mandatory disclosures and references	
Organisation activities, products, services, performance, locations, value chain, stakeholder engagement, materiality topics, impacts, risks opportunities, and challenges	Oversight, control, strategy, policies, management approach to actions, and targets related to material topics and relevant areas	Material topics, quantitative and qualitative data, and trends	Brief case studies, best practice examples, success and human interest stories, and pictures

Source: Author.

The CSRD/ESRS requires organisations to show stakeholders that they understand and address all relevant impacts using neutral language and legitimate scientific approaches, which will improve trust and minimise potential reputational risks. So, data collection, normalisation methods, and calculations should remain consistent throughout all reporting periods. Any changes, data gaps (including missing data), estimations, or inaccurate data require appropriate explanations. All non-material information should be excluded rather than disclosed and potential 'other reasons' for not disclosing certain information in the report include confidentiality constraints and legal prohibitions. Accordingly, it is advisable to have your ESG report reviewed by a legal specialist to check both inclusions and rationale for exclusions.

TIPS AND TRICKS

- Define roles for content approval at the beginning of the reporting process. Ensure engagement with board representatives as well as investor relations, legal, and communication departments (not only Green Team).
- Ensure representatives of the board and top management review key data trends periodically, not just once a year for the purpose of publication.
- Ensure strategic understanding and the opportunity to add comments on data trends.
- Benchmark main data values with peers, sector averages.
- Define the principal message, structure, key information, and KPIs before drafting the report.
- Early in the process, send out internal requests for supporting information (e.g., case studies).
- Identify relevant people who will provide feedback on the report narrative to avoid the process becoming protracted.
- When discussing achievements, avoid all hyperbole and exaggeration, and include challenges and mistakes — be transparent and avoid greenwashing.

16.3 Report Structure

Section 8 of ESRS 1 (General requirements) defines the structure of the sustainability statement into four parts: general information, environmental information, social information, and governance information, which follow this structure and with any sector-specific disclosures grouped by cross-cutting reporting area and, where applicable, by sustainability topic (see Graphic 25).

Article 8 of the Regulation includes a disclosure obligation of turnover, CapEx and OpEx, which applies to entities that are already required to include a non-financial statement in their management report under the Non-Financial Reporting Directive (NFRD). For more information, consult with your financial disclosure assurer and/or refer to https://www.eurosif.org/policies/eutaxonomy/

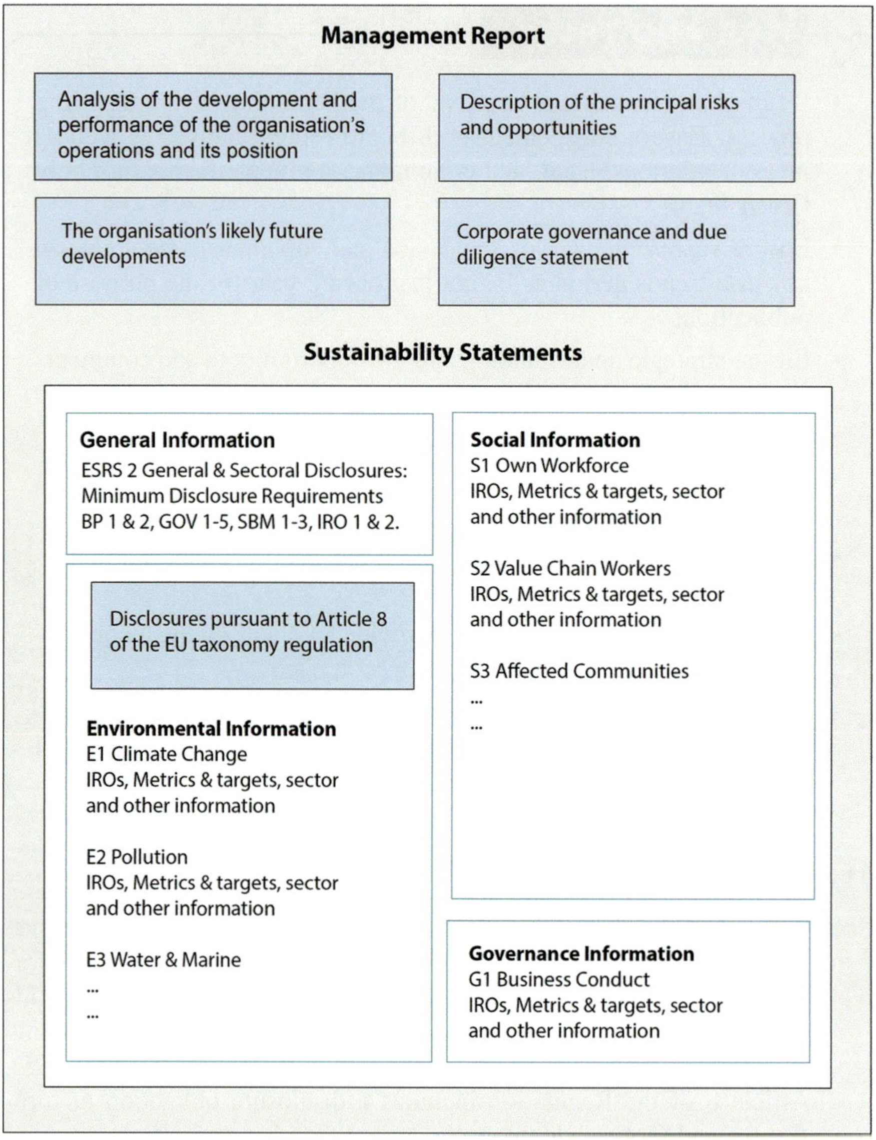

Graphic 25 Structure of a sustainability statement. Source: CDR (EU) 2023/2772 supplementing Directive 2013/34/EU.

Where organisations are incorporated by reference, then it is possible to cascade sustainability statement information as separate elements, but coherence of information must be retained.

The overall narrative and relationships between different pieces of information is vital in sustainability statements and may involve connecting governance, strategy, risk, and related metrics and targets to provide a holistic perspective. For example, to allow users to assess connections between pieces of information, your organisation might need to explain the effect or likely effect of the sustainability strategy on financial statements or financial plans, or on metrics and targets used to measure progress against performance. Furthermore, the explanation might need to delve into how the use of natural resources and changes within the supply chain could amplify, change, or reduce material impacts, risks, and opportunities. This may in turn link to the potential or actual effect(s) on production costs, strategic responses to mitigate such impacts, or risks and related investment in new assets, which are outlined in the financial statements.

16.4 Format and Accessibility

In accordance with the CSRD/ESRS and the guidelines of the European Securities and Markets Authority (ESMA), your sustainability statement should be easily accessible from your organisation's website and be machine readable and available through other channels, depending on audience preferences. Under Article 8 XBRL Taxonomy,[1] your sustainability statement must be machine readable using XBRL tagging so that it can be uploaded to the proposed European Single Access Point (ESAP) due in 2027/2028 as well as facilitate automated database management. Once finalised, the XBRL Taxonomy released by EFRAG will be the basis for the European Securities and Market Authority (ESMA) to develop draft Regulatory Technical Standards (RTS) for tagging the ESRS sustainability statement. The tagging rules will finally be adopted by the EC by way of a Delegated Act (DA) amending Commission Delegated Regulation (EU) 2019/815 on the European Single Electronic Format sometime during 2025.

Each reference table throughout this manual is based on the CSRD/ESRS reporting structure, which is then cross-referenced in the XBRL taxonomy

[1] For details of XBRL Taxonomy entry points, see:

Article 8 All Entry Point: https://xbrl.efrag.org/taxonomy/article8/2023-06-27/art8_all.xsd

Article 8 Disclosures for non-financial undertakings: https://xbrl.efrag.org/taxonomy/article8/2023-06-27/art8_nonfinancial_undertakings.xsd

Article 8 Disclosures for asset managers: https://xbrl.efrag.org/taxonomy/article8/2023-06-27/art8_asset_managers.xsd

Article 8 Disclosures for credit institutions: https://xbrl.efrag.org/taxonomy/article8/2023-06-27/art8_credit_institutions.xsd

Article 8 Disclosures for investment firms: https://xbrl.efrag.org/taxonomy/article8/2023-06-27/art8_investment_firms.xsd

Article 8 Disclosures for insurance and reinsurance undertakings: https://xbrl.efrag.org/taxonomy/article8/2023-06-27/art8_insurance_and_reinsurance_undertakings.xsd

recommended by EFRAG whereby each disclosure requirement has been broken down into elemental, separable (narrative) data points.

Graphic 26 Human readable to machine readable formats requires XBRL software. Source: Author.

XBRL software is freely available and can be downloaded from software.xbrl.org or it may be included in ESG reporting software, although at the time of publishing this book there were no examples to cite. The CSRD/ESRS machine readable XBRL structure is currently available from www.efrag.org/Lab4 and follows the same Topic > Sub-Topic > Sub-Sub-Topic structure (expressed as Levels 1, 2, and 3 in XBRL) that we covered in 3.5, Impact Materiality. The XBRL taxonomy can also be used by loading the following entry points in compliant XBRL software:

- ESRS All (importing ESRS Core with all topics and disclosure requirements, including presentation and definition link-bases and validation rules):

  https://xbrl.efrag.org/taxonomy/draft-esrs/2023-07-31/esrs_all.xsd

- ESRS Core (concepts, labels, and references only):

  https://xbrl.efrag.org/taxonomy/draft-esrs/2023-07-31/common/esrs_cor.xsd

XBRL allows tagging of narrative, qualitative, tabular, Boolean (yes/no variations), soft links between topics, and other information that may add context or additional meaning. As the XBRL Taxonomy is still in development, there are no examples of actual reports that can be cited, but a good explanation of how to complete XBRL tagging is available at https://www.youtube.com/watch?v=-fG08QtNF7E and there is extensive supporting documentation available at www.efrag.org/Lab4

It is expected that during the course of the first quarter of 2025, software developers will produce pre-tagged XBRL report structures that link with ESG reporting systems that will also be printable as human-readable pre-formatted documents.

16.5 National Implementation and Penalties

The CSRD/ESRS provides a harmonised framework for companies on transparency, accounting, and auditing, while allowing flexibility for EU Member States to implement provisions in their national legal systems. The CSRD directive[2] is explicit in that member states must transpose it into their national law by 6 July 2024. In fact, member states do not have much flexibility regarding the content of the reporting standards, as they are set at EU level; however, they are still required to incorporate the directive's provisions into the national legal framework, and to publish the administrative provisions necessary to comply with the CSRD. When it comes to penalties for non-compliance, the directive is not very prescriptive and leaves it to member states to define enforcing measures. However, the CSRD/ESRS does not introduce any new penalties to the existing provisions of the Accounting Directive (article 51), which requires member states to establish effective, proportionate, and dissuasive penalties for infringements of the directive, which now includes sustainability disclosures.

Member states are also responsible for providing adequate resources for supervisory authorities, National Competent Authorities (NCAs), that may control corporate sustainability reporting practices. Examples are, France which has introduced criminal sanctions for company directors that would fail to appoint an auditor of sustainability information and for the obstruction of a sustainability audit; and, Hungary that has legislated that should undertakings fail to comply with the ESG reporting obligations, the national Supervisory Authority for Regulated Activities may impose financial penalties.

Additionally, member states are responsible for implementing rules on the statutory audit of annual and consolidated accounts, including the appointment of auditors, and the independence of auditors and independent assurance service providers (IASPs). So, each member state must empower their NCA to monitor possible breaches of the CSRD (e.g., in case of failure or ethical misconduct), and to impose appropriate penalties including fines or other disciplinary measures. In addition to, or instead of, financial penalties, NCAs may require audit firms and independent assurance service providers to take corrective measures to address identified deficiencies in their audit processes, quality control, or compliance with professional standards. Public oversight and transparency of the auditing process is handled

[2] Annex I to the Commission Delegated Regulation (EU) 2023/2772 supplementing Directive 2013/34/EU of the European Parliament and of the Council as regards sustainability reporting standards and as published in the Official Journal of the European Union on 22 December 2023.

by the Committee of European Auditing Oversight Bodies (CEAOB) who is in charge of coordinating national audit oversight bodies at EU level, including on certain aspects of cross-border enforcement and penalties – the formal deadline for Limited Assurance standards from CEAOB is October 2026. The CSRD also mandates that ESMA issue guidelines on the supervision of sustainability reporting by national competent authorities.

16.6 ESG Report Examples

The CSRD/ESRS is a new standard that applies from 2024 for reports to be published by the end of March 2025, so there are currently no examples of compliant reports. However, some members of the EFRAG Financial Reporting Technical Expert Group[3] have included many of the approaches included in the CSRD/ESRS, such that their reports provide an excellent insight into sustainability statement publications, some examples are given in Table 93 below:

Table 93 Recommend examples of ESG reports from various sectors.

Primary sector	Company and HQ country	ESG report location
Banking:	Danske Bank, Denmark	danskebank.com/sustainability
	Société Générale, France	www.societegenerale.com/sites/default/files/documents/2023-05/integrated-report-2022-2023.pdf
Energy sector:	Siemens Energy, Germany	www.siemens-energy.com/global/en/home/company/sustainability.html
Insurance:	Assicurazioni Generali SpA, Italy	www.generali.com/info/download-center/sustainability/bilanci
	Allianz Group, Switzerland	www.allianz.com/content/dam/onemarketing/azcom/Allianz_com/sustainability/documents/Allianz_Group_Sustainability_Report_2022-web.pdf
Pharmaceutical and healthcare:	Philips, The Netherlands	www.philips.com/a-w/about/environmental-social-governance/downloads.html
Residential real estate:	Vonovia SE, Germany, Austria and Sweden.	report.vonovia.com/2022/nachhaltigkeitsbericht/en/about-this-report/
Manufacturing and retail:	Inditex Group, Spain	static.inditex.com/annual_report_2022/en/committed/environment/
Services:	Deutsche Post DHL Group, Germany	group.dhl.com/en/sustainability/sustainability-roadmap/sustainability-reports.html

Source: Author.

[3] https://www.efrag.org/About/Governance/1/EFRAG-Financial-Reporting-Technical-Expert-Group-EFRAG-FR-TEG

European legal references available on *https://eur-lex.europa.eu:*

- Directive (EU) 2021/2139 of 4 June 2021 supplementing Regulation (EU) 2020/852 of the European Parliament and of the Council by establishing the technical screening criteria for determining the conditions under which an economic activity qualifies as contributing substantially to climate change mitigation or climate change adaptation and for determining whether that economic activity causes no significant harm to any of the other environmental objectives.
- Commission Recommendation of 16.12.2021 on the use of the Environmental Footprint methods to measure and communicate the life cycle environmental performance of products and organisations with Annexes 1–4.
- Additionally, the We Mean Business Colation produce a CSRD Early Adopters report of anticipated best practices and examples, see EARLY ADOPTERS' CSRD REPORTING – Inspiring reporting practice from reporting year 2023[4].
- Directive 2013/34/EU of the European Parliament and of the Council of 26 June 2013 on the annual financial statements, consolidated financial statements and related reports of certain types of companies, amending Directive 2006/43/EC of the European Parliament and of the Council and repealing Council Directives 78/660/ EEC and 83/349/EE.
- Council Directive 91/674/EEC on the annual accounts and consolidated accounts.
- Regulation (EU) 2019/2088 of the European Parliament and of the Council of 27 November 2019 on sustainability-related disclosures in the financial services sector.
- Regulation 575/2013 of the European Parliament and of the Council of 26 June 2013 on prudential requirements for credit institutions and amending Regulation (EU) No 648/2012.
- Directive 2013/36/EU of the European Parliament and of the Council of 26 June 2013 on access to the activity of credit institutions and the prudential supervision of credit institutions, amending Directive 2002/87/EC and repealing Directives 2006/48/EC and 2006/49/EC.

---

[4] https://www.wemeanbusinesscoalition.org/wp-content/uploads/2024/07/WMBC_Early_Adopters_CSRD_reporting.pdf

- Council Directive 86/635/EEC of 8 December 1986 on the annual accounts and consolidated accounts of banks and other financial institutions.
- Regulation 765/2008 of the European Parliament and of the Council of 9 July 2008 setting out the requirements for accreditation and market surveillance relating to the marketing of products and repealing Regulation (EEC) No 339/93.
- Regulation (EC) No 1606/2002 of the European Parliament and of the Council of 19 July 2002 on the application of international accounting standards.
- Regulation (EU) 2021/1119 of the European Parliament and of the Council of 30 June 2021 establishing the framework for achieving climate neutrality and amending Regulations (EC) No 401/2009 and (EU) 2018/1999.
- Directive (EU) 2022/2464 of the European Parliament and of the Council of 14 December 2022 amending Regulation (EU) No 537/2014, Directive 2004/109/EC, Directive 2006/43/EC and Directive 2013/34/EU, as regards corporate sustainability reporting.
- Directive 2009/138/EC of the European Parliament and of the Council of 25 November 2009 on the taking-up and pursuit of the business of insurance and reinsurance.
- Directive 2004/109/EC of the European Parliament and of the Council of 15 December 2004 on the harmonisation of transparency requirements in relation to information about issuers whose securities are admitted to trading on a regulated market and amending Directive 2001/34/EC.
- Regulation (EU) 2020/852 of the European Parliament and of the Council of 18 June 2020 on the establishment of a framework to facilitate sustainable investment, and amending Regulation (EU) 2019/2088.
- Commission Delegated Regulation (EU) 2020/1816 of 17 July 2020 supplementing Regulation (EU) 2016/1011 of the European Parliament and of the Council as regards the explanation in the benchmark statement of how environmental, social and governance factors are reflected in each benchmark provided and published.
- Commission Delegated Regulation (EU) 2020/1817 of 17 July 2020 supplementing Regulation (EU) 2016/1011 of the European Parliament and of the Council as regards the minimum content of the explanation

on how environmental, social and governance factors are reflected in the benchmark methodology.

- Commission Delegated Regulation (EU) 2020/1818 of 17 July 2020 supplementing Regulation (EU) 2016/1011 of the European Parliament and of the Council as regards minimum standards for EU Climate Transition Benchmarks and EU Paris-aligned Benchmarks.
- Directive 2003/87/EC of the European Parliament and of the Council of 13 October 2003 establishing a scheme for greenhouse gas emission allowance trading within the Community and amending Council Directive 96/61/EC.
- Regulation (EC) No 1221/2009 of the European Parliament and of the Council of 25 November 2009 on the voluntary participation by organisations in a Community eco-management and audit scheme (EMAS), repealing Regulation (EC) No 761/2001 and Commission Decisions 2001/681/EC and 2006/193/EC.
- Directive (EU) 2019/1937 of the European Parliament and of the Council of 23 October 2019 on the protection of persons who report breaches of Union law?
- Commission Delegated Regulation (EU) 2018/815 of 17 December 2018 supplementing Directive 2004/109/ EC of the European Parliament and of the Council with regard to regulatory technical standards on the specification of a single electronic reporting format (ESEF).
- Directive (EU) 2017/1132 of the European Parliament and of the Council of 14 June 2017 relating to certain aspects of company law (codification).
- Directive 2006/43/EC of the European Parliament and of the Council of 17 May 2006 on statutory audits of annual accounts and consolidated accounts, amending Council Directives 78/660/EEC and 83/349/EEC and repealing Council Directive 84/253/EEC.
- Regulation (EU) No 1095/2010 of the European Parliament and of the Council of 24 November 2010 establishing a European Supervisory Authority (European Securities and Markets Authority), amending Decision No 716/2009/EC and repealing Commission Decision 2009/77/EC.
- Regulation (EU) No 537/2014 of the European Parliament and of the Council of 16 April 2014 on specific requirements regarding statutory audit of public-interest entities and repealing Commission Decision 2005/909/EC.

- Directive 2014/65/EU of the European Parliament and of the Council of 15 May 2014 on markets in financial instruments and amending Directive 2002/92/EC and Directive 2011/61/EU (recast).
- Regulation (EU) No 575/2013 (CRR), Definition of small and non-complex institutions and its applicability to SSM significant institutions (consolidation 2022).
- Commission Delegated Regulation (EU) supplementing Directive 2013/34/EU of the European Parliament and of the Council as regards sustainability reporting standards.

Other References

- A. Gilbert-D'Halluin, *CSRD Essentials*, 2024.
- Climate Disclosure Standards Board (CDSB), 2020, *CDSB Framework: Application guidance for water-related disclosures.*
- European Financial Reporting Advisory Group (EFRAG):
    - May 2024, *ESRS–ISSB Standards Interoperability Guidance.*
    - April 2024, *ESRS Q&A Platform*, https://www.efrag.org/lab7
    - January 2024, *Voluntary ESRS for Non-Listed Small- and Medium-Sized Enterprises (VSMEs) (draft)*
    - December 2023, *Implementation Guidance 2: Value Chain (IG 2) (draft)*
    - December 2023, *Explanatory note: List of ESRS Datapoints (IG 3) (draft)*
    - December 2023, *Guidelines on Enforcement of Sustainability Information, (draft)*
    - November 2023, *Draft ESRS Disclosures: ESRS 1, ESRS 2, E1, E2, E3, E4, E5, S1, S2, S3, S4, G1, and Annexes.*
    - March 2023, *Implementation Guidance 1: Materiality assessment (IG 1), (draft).*
- The European Economic and Social Committee (EESC), October 2023, *Opinion of the European Economic and Social Committee on the proposal for a Directive of the European Parliament and of the Council amending Directive 2008/98/EC on waste.*

- Eurostat. 2023. *https://ec.europa.eu/eurostat/statistics-explained/index.php?title=SDG_12_-_Responsible_consumption_and_production#-Responsible_consumption_and_production_in_the_EU:_overview_and_key_trends*
- Financial Conduct Authority (FCA), 2023, *Policy Statement PS23-16: Sustainability Disclosure Requirements (SDR) and investment labels.*
- Forico Future Fibre, September 2023, *Illustrative Example of Integrated TCFD + TNFD Disclosures.*
- The Global Reporting Initiative (GRI), *GRI Standards System: https://www.globalreporting.org/standards/*
- Green Finance Institute, April 2024, *Assessing the Materiality of Nature-Related Financial Risks for the UK.*
- Gullison, R.E., J. Hardner, S. Anstee, M. Meyer. 2015, *Good Practices for the Collection of Biodiversity Baseline Data*
- Jonker J and Faber N, 2020, *Organizing for Sustainability: A Guide to Developing New Business Models*, Management Impact, ISBN 978-3-030-78156-9, ISBN 978-3-030-78157-6 (eBook), https://doi.org/10.1007/978-3-030-78157-6
- Lammerant J. et al., 2022, *Assessment of Biodiversity Measurement Approaches for Businesses and Financial Institutions, Update Report 4* on behalf of the EU Business @ Biodiversity Platform.
- Mark Wayne McElroy, 2008, *Social footprints. Measuring the social sustainability performance of organizations*, Thesis Rijksuniversiteit Groningen
- Network for Greening the Financial System (NGFS), 2023, Technical document: *Recommendations toward the development of scenarios for assessing nature-related economic and financial risks.*
- OECD, 2023, *ACHIEVING ALIGNMENT: Syncing EU due diligence legislation with the updated OECD Guidelines.*
- OECD, 2018, *Due Diligence Guidance for Responsible Business Conduct.*
- Securities and Exchange Commission (SEC), March 2024, *The Enhancement and Standardization of Climate-Related Disclosures for*

Investors - 17 CFR 210, 229, 230, 232, 239, and 249, [Release Nos. 33-11280; 34-99908; File No. S7-10-22], RIN 3235-AM87.

- Sustainable Fitch, March 2024, *ESG Regulations and Reporting Standards Tracker.*
- Task Force on Climate-related Financial Disclosures (TCFD), *Guidance on Metrics, Targets and Transition Plans*, October 2021.
- Task Force on Climate-related Financial Disclosures (TCFD). 2020, *Guidance on Scenario Analysis for Non-Financial Companies.*
- Task Force on Climate-related Financial Disclosures (TCFD), June 2017, *Final Report: Recommendations of the Task Force on Climate-related Financial Disclosures.*
- Task Force on Nature-related Financial Disclosures (TNFD), October 2023, *Conducting advanced scenario analysis.* Cited under: Creative Commons Attribution 4.0 International License
- Task Force on Nature-related Financial Disclosures (TNFD), September 2023, *Guidance on scenario analysis.*
- Task Force on Nature-related Financial Disclosures (TNFD), March 2022, *The TNFD Nature-related Risk & Opportunity Management and Disclosure Framework.*
- Transition Plan Taskforce (TPT), October 2023, *Disclosure Framework.*
- World Business Council for Sustainable Development (WBCSD), 2023, *Circular Transition Indicators V4.0 Metrics for business, by business.*
- World Resources Institute (WRI) & World Business Council for Sustainable Development (WBCSD), 2013, *Greenhouse Gas Protocol: Technical Guidance for Calculating Scope 3 Emissions.*
- United Nations Environment Programme (UNEP), 2020, *Guidelines for Social Life Cycle Assessment of Products and Organizations*, Benoît Norris, C., Traverso, M., Neugebauer, S., Ekener, E., Schaubroeck, T., Russo Garrido, S., Berger, M., Valdivia, S., Lehmann, A., Finkbeiner, M., Arcese, G. (eds.). United Nations Environment Programme (UNEP).
- United Nations Environment Programme (UNEP), 2016, *Global Guidance for life cycle impact assessment indicators*, ISBN: 978-92-807-3630-4

- Ilcheong Yi, Samuel Bruelisauer, Peter Utting, Mark McElroy, Marguerite Mendell, Sonja Novkovic and Zhen Lee. 2022. *Authentic Sustainability Assessment: A User Manual for the Sustainable Development Performance Indicators*. Geneva, UNRISD. ISBN 978-92-9085-131-8
- Vidal-Legaz B, Sala S, Antón A, Maia De Souza D, Nocita M, Putman B, Teixeira RFM, 2016, *Land-use related environmental indicators for Life Cycle Assessment*. JRC Technical report. Luxembourg: Publications Office of the European Union, pp.44 doi: 10.2788/905478

Annex 1

List of data points in cross-cutting and topical standards that are required by EU law

The table below illustrates the data points in ESRS 2 and topical ESRS that emanate from other EU legislation. They are to be reported irrespective of the outcome of the materiality assessment.

Source: ANNEX I to Commission Delegated Regulation (EU) 2023/2772 supplementing Directive 2013/34/EU of the European Parliament and of the Council as regards sustainability reporting standards, as published in the Official Journal of the European Union on 22 December 2023.

Disclosure Requirement and related data point	SFDR reference	Pillar 3 reference	Benchmark regulation reference	EU Climate Law reference
ESRS 2 GOV-1 Board's gender diversity paragraph 21 (d)	Indicator n.13 of Table #1 of Annex 1		CDR (EU) 2020/1816, Annex II	
ESRS 2 GOV-1 Percentage of board members who are independent paragraph 21 (e)			CDR (EU) 2020/1816, Annex II	
ESRS 2 GOV-4 Statement on due diligence paragraph 30	Indicator n. 10 Table #3 of Annex 1			
ESRS 2 SBM-1 Involvement in activities related to fossil fuel activities paragraph 40 (d) i	Indicators n. 4 Table #1 of Annex 1	Article 449a Regulation (EU) No 575/2013; Commission Implementing Regulation (EU) 2022/2453 (6) Table 1: Qualitative information on environmental risk and Table 2: Qualitative information on social risk	CDR (EU) 2020/1816, Annex II	
ESRS 2 SBM-1 Involvement in activities related to chemical production paragraph 40 (d) ii	Indicator n. 9 Table #2 of Annex 1		CDR (EU) 2020/1816, Annex II	
ESRS 2 SBM-1 Involvement in activities related to controversial weapons paragraph 40 (d) iii	Indicator n. 14 Table #1 of Annex 1		CDR (EU) 2020/1818, Article 12 (1) CDR (EU) 2020/1816, Annex II	
ESRS 2 SBM-1 Involvement in activities related to cultivation and production of tobacco paragraph 40 (d) iv			CDR (EU) 2020/1818, Article 12 (1) CDR (EU) 2020/1816, Annex II	
ESRS E1-1 Transition plan to reach climate neutrality by 2050 paragraph 14				Regulation (EU) 2021-1119 Article 2 (1)

Disclosure Requirement and related data point	SFDR reference	Pillar 3 reference	Benchmark regulation reference	EU Climate Law reference
ESRS E1-1 Undertakings excluded from Paris-aligned Benchmarks paragraph 16 (g)		Article 449a Capital Requirements Regulation – CRR; Template 1: Banking book-Climate Change transition risk: Credit quality of exposures by sector, emissions, and residual maturity	CDR (EU) 2020/1818, Article 12.1 (d) to (g), and Article 12.2	
ESRS E1-4 GHG emission reduction targets paragraph 34	Indicator n. 4 Table #2 of Annex 1	Article 449a Regulation (EU) No 575/2013; Commission Implementing Regulation (EU) 2022/2453 Template 3: Banking book – Climate change transition risk: alignment metrics	CDR (EU) 2020/1818, Article 6	
ESRS E1-5 Energy consumption from fossil sources disaggregated by sources (only high climate impact sectors) paragraph 38	Indicator n. 5 Table #1 and Indicator n. 5 Table #2 of Annex 1			
ESRS E1-5 Energy consumption and mix paragraph 37	Indicator n. 5 Table #1 of Annex 1			
ESRS E1-5 Energy intensity associated with activities in high climate impact sectors paragraphs 40–43	Indicator n. 6 Table #1 of Annex 1			
ESRS E1-6 Gross Scope 1, 2, and 3 and total GHG emissions paragraph 44	Indicators n. 1 and 2 Table #1 of Annex 1	Article 449a; Regulation (EU) No 575/2013; Commission Implementing Regulation (EU) 2022/2453 Template 1: Banking book – Climate change transition risk: Credit quality of exposures by sector, emissions, and residual maturity	CDR (EU) 2020/1818, Article 5(1), 6 and 8(1)	

(Continued)

Disclosure Requirement and related data point	SFDR reference	Pillar 3 reference	Benchmark regulation reference	EU Climate Law reference
ESRS E1-6 Gross GHG emissions intensity paragraphs 53–55	Indicators n. 3 Table #1 of Annex 1	Article 449a Regulation (EU) No 575/2013; Commission Implementing Regulation (EU) 2022/2453 Template 3: Banking book – Climate change transition risk: alignment metrics	CDR (EU) 2020/1818, Article 8(1)	
ESRS E1-7 GHG removals and carbon credits paragraph 56				Regulation (EU) 2021-1119 Article 2 (1)
ESRS E1-9 Exposure of the benchmark portfolio to climate-related physical risks paragraph 66			CDR (EU) 2020/1818, Annex II CDR (EU) 2020/1816, Annex II	

Disclosure Requirement and related data point	SFDR reference	Pillar 3 reference	Benchmark regulation reference	EU Climate Law reference
ESRS E1-9 Disaggregation of monetary amounts by acute and chronic physical risk paragraph 66 (a) ESRS E1-9 Location of significant assets at material physical risk paragraph 66 (c)		Article 449a Regulation (EU) No 575/2013; Commission Implementing Regulation (EU) 2022/2453 paragraphs 46 and 47; Template 5: Banking book-Climate change physical risk: Exposures subject to physical risk		
ESRS E1-9 Breakdown of the carrying value of its real estate assets by energy-efficiency classes paragraph 67 (c)		Regulation (EU) No 575/2013; Commission Implementing Regulation (EU) 2022/2453 paragraph 34; Template 2:Banking book-Climate change transition risk: Loans collateralised by immovable property - Energy efficiency of the collateral		

ESRS E1-9 Degree of exposure of the portfolio to climate-related opportunities paragraph 69			CDR (EU) 2020/1818, Annex II	
ESRS E2-4 Amount of each pollutant listed in Annex II of the E-PRTR Regulation (European Pollutant Release and Transfer Register) emitted to air, water, and soil, paragraph 28	Indicator number 8 Table #1 of Annex 1 Indicator number 2 Table #2 of Annex 1 Indicator number 1 Table #2 of Annex 1 Indicator number 3 Table #2 of Annex 1			

Disclosure Requirement and related data point	**SFDR reference**	**Pillar 3 reference**	**Benchmark regulation reference**	**EU Climate Law reference**
ESRS E3-1 Water and marine resources paragraph 9	Indicator n. 7 Table #2 of Annex 1			
ESRS E3-1 Dedicated policy paragraph 13	Indicator n.8 Table 2 of Annex 1			
ESRS E3-1 Sustainable oceans and seas paragraph 14	Indicator n. 12 Table #2 of Annex 1			
ESRS E3-4 Total water recycled and reused paragraph 28 (c)	Indicator n. 6.2 Table #2 of Annex 1			
ESRS E3-4 Total water consumption in m3 per net revenue on own operations paragraph 29	Indicator n. 6.1 Table #2 of Annex 1			
ESRS 2- IRO 1 – E4 paragraph 16 (a) i	Indicator n. 7 Table #1 of Annex 1			
ESRS 2- IRO 1 – E4 paragraph 16 (b)	Indicator n. 10 Table #2 of Annex 1			
ESRS 2- IRO 1 – E4 paragraph 16 (c)	Indicator n.14 Table #2 of Annex 1			

(Continued)

Disclosure Requirement and related data point	SFDR reference	Pillar 3 reference	Benchmark regulation reference	EU Climate Law reference
ESRS E4-2 Sustainable land / agri-culture practices or policies paragraph 24 (b)	Indicator n.11 Table #2 of Annex 1			
ESRS E4-2 Sustainable oceans/seas practices or policies paragraph 24 (c)	Indicator n. 12 Table #2 of Annex 1			
ESRS E4-2 Policies to address deforestation paragraph 24 (d)	Indicator n. 15 Table #2 of Annex 1			
ESRS E5-5 Non-recycled waste paragraph 37 (d)	Indicator n. 13 Table #2 of Annex 1			
ESRS E5-5 Hazardous waste and radioactive waste paragraph 39	Indicator n.9 Table #1 of Annex 1			
ESRS 2- SBM3 – S1 Risk of incidents of forced labour paragraph 164(f)	Indicator 13 Table #3 of Annex I			
ESRS 2- SBM3 – S1 Risk of incidents of child labour paragraph 14 (g)	Indicator 12 Table #3 of Annex I			

Disclosure Requirement and related data point	SFDR reference	Pillar 3 reference	Benchmark regulation reference	EU Climate Law reference
ESRS S1-1 Human rights policy commitments paragraph 20	Indicator n. 9 Table #3 and Indicator n. 11 Table #1 of Annex I			
ESRS S1-1 Due diligence policies on issues addressed by the fundamental International Labour Organisation Conventions 1–8, paragraph 21			CDR (EU) 2020/1816, Annex II	
ESRS S1-1 Processes and measures for preventing trafficking in human beings paragraph 22	Indicator n.11 Table #3 of Annex I			

ESRS S1-1 Workplace accident prevention policy or management system paragraph 23	Indicator n.1 Table #3 of Annex I			
ESRS S1-3 Grievance/complaints handling mechanisms paragraph 33 (c)	Indicator n. 5 Table #3 of Annex I			
ESRS S1-14 Number of fatalities and number and rate of work-related accidents paragraph 88 (b) and (c)	Indicator n. 2 Table #3 of Annex I		CDR (EU) 2020/1816, Annex II	
ESRS S1-14 Number of days lost to injuries, accidents, fatalities, or illness paragraph 88 (e)	Indicator n. 3 Table #3 of Annex I			
ESRS S1-16 Unadjusted gender pay gap and weighted average gender pay gap paragraph 97 (a)	Indicator n. 12 Table #1 of Annex I		CDR (EU) 2020/1816, Annex II	
ESRS S1-16 Excessive CEO pay ratio paragraph 97 (b)	Indicator n. 8 Table #3 of Annex I			
ESRS S1-17 Incidents of discrimination paragraph 103 (a)	Indicator n. 7 Table #3 of Annex I			
ESRS S1-17 Violations of UNGC principles and OECD paragraph 104 (a)	Indicator n. 10 Table #1 and Indicator n. 14 Table #3 of Annex I		CDR (EU) 2020/1816, Annex II CDR (EU) 2020/1818 Art 12 (1)	
ESRS 2-SBM3 – S2 Significant risk of child labour or forced labour in the value chain paragraph 11 (b)	Indicators n. 12 and n. 13 Table #3 of Annex I			
ESRS S2-1 Human rights policy commitments paragraph 17	Indicator n. 9 Table #3 and Indicator n. 11 Table #1 of Annex 1			
ESRS S2-1 Policies related to value chain workers paragraph 18	Indicator n. 11 and n. 4 Table #3 of Annex 1			
ESRS S2-1 Violations of UNGC principles and OECD guidelines paragraph 19	Indicator n. 10 Table #1 of Annex 1		CDR (EU) 2020/1816, Annex II CDR (EU) 2020/1818, Art 12 (1)	

(Continued)

Disclosure Requirement and related data point	SFDR reference	Pillar 3 reference	Benchmark regulation reference	EU Climate Law reference
ESRS S2-1 Due diligence policies on issues addressed by the fundamental International Labour Organisation Conventions 1–8, paragraph 19			CDR (EU) 2020/1816, Annex II	
ESRS S2-4 Human rights issues and incidents connected to its upstream and downstream value chainparagraph 36	Indicator n. 14 Table #3 of Annex 1			
ESRS S3-1 Human policy commitments paragraph 16	Indicator n. 9 Table #3 of Annex 1 and Indicator n. 11 Table #1 of Annex 1			
ESRS S3-1 Violations of UNGC principles and OECD guidelines paragraph 17	Indicator n. 10 Table #1 Annex 1		CDR (EU) 2020/1816, Annex II CDR (EU) 2020/1818, Art 12 (1)	
ESRS S3-4 Human rights issues and incidents paragraph 36	Indicator 14 Table #3 of Annex 1			
ESRS S4-1 Policies related to consumers and end-users paragraph 16	Indicator n. 9 Table #3 and Indicator n. 11 Table #1 of Annex 1			
ESRS S4-1 Violations of UNGC principles and OECD guidelines paragraph 17	Indicator 10 Table #1 of Annex 1		CDR (EU) 2020/1816, Annex II CDR (EU) 2020/1818, Art 12 (1)	
ESRS S4-4 Human rights issues and incidents paragraph 35	Indicator n. 14 Table #3 of Annex 1			

Disclosure Requirement and related data point	SFDR reference	Pillar 3 reference	Benchmark regulation reference	EU Climate Law reference
ESRS G1-1 United Nations Convention against Corruption paragraph 10 (b)	Indicator n. 15 Table #3 of Annex 1			
ESRS G1-1 Protection of whistle- blowers paragraph 10 (d)	Indicator n. 6 Table #3 of Annex 1			
ESRS G1-4 Fines for violation of anti-corruption and anti-bribery laws paragraph 24 (a)	Indicator n. 17 Table #3 of Annex 1		CDR (EU) 2020/1816, Annex II)	
ESRS G1-4 Standards of anti-corruption and anti-bribery paragraph 24 (b)	Indicator 16 Table #3 of Annex 1			

Annex 2

List of phased-in Disclosure Requirements.

Source: CDR (EU) 2023/2772 supplementing Directive 2013/34/EU.

[draft] ESRS	Disclosure Requirement	Full name of the Disclosure Requirement	Phase-in or effective date (including the first year)
ESRS 2	SBM-1	Market position, strategy, business model(s), and value chain	The undertaking shall report the information prescribed by ESRS 2 SBM-1 paragraph 40(b) (breakdown of total revenue by significant ESRS sector) and 40(c) (list of additional significant ESRS sectors) starting from the application date specified in a Commission Delegated Act to be adopted pursuant to article 29b(1) third subparagraph, point (ii), of Directive 2013/34/EU.
ESRS 2	SBM-3	Material impacts, risks, and opportunities and their interaction with strategy and business model	The undertaking may omit the information prescribed by ESRS 2 SBM-3 paragraph 48(e) (anticipated financial effects) for the first year of preparation of its sustainability statement. The undertaking may comply with ESRS 2 SBM-3 paragraph 48(e) by reporting only qualitative disclosures for the first three years of preparation of its sustainability statement, if it is impracticable to prepare quantitative disclosures.
ESRS E1	E1-6	Gross Scopes 1, 2, and 3 and total GHG emissions	Undertakings or groups not exceeding on their balance sheet dates the average number of 750 employees during the financial year (on a consolidated basis where applicable) may omit the data points on Scope 3 emissions and total GHG emissions for the first year of preparation of their sustainability statement.

(Continued)

[draft] ESRS	Disclosure Requirement	Full name of the Disclosure Requirement	Phase-in or effective date (including the first year)
ESRS E1	E1-9	Anticipated financial effects from material physical and transition risks and potential climate-related opportunities	The undertaking may omit the information prescribed by ESRS E1-9 for the first year of preparation of its sustainability statement. The undertaking may comply with ESRS E1-9 by reporting only qualitative disclosures for the first three years of preparation of its sustainability statement, if it is impracticable to prepare quantitative disclosures.
ESRS E2	E2-6	Anticipated financial effects from pollution-related impacts, risks, and opportunities	The undertaking may omit the information prescribed by ESRS E2-6 for the first year of preparation of its sustainability statement. Except for the information prescribed by paragraph 40 (b) on the operating and capital expenditures occurred in the reporting period in conjunction with major incidents and deposits, the undertaking may comply with ESRS E2-6 by reporting only qualitative disclosures, for the first three years of preparation of its sustainability statement.
ESRS E3	E3-5	Anticipated financial effects from water and marine resources-related impacts, risks, and opportunities	The undertaking may omit the information prescribed by ESRS E3-5 for the first year of preparation of its sustainability statement. The undertaking may comply with ESRS E3-5 by reporting only qualitative disclosures, for the first three years of preparation of its sustainability statement.
ESRS E4	E4-6	Anticipated financial effects from biodiversity and ecosystem-related impacts, risks, and opportunities	Undertakings or groups not exceeding on their balance sheet dates the average number of 750 employees during the financial year (on a consolidated basis where applicable) may omit the information specified in the disclosure requirements of ESRS E4 for the first two years of preparation of their sustainability statement.
ESRS E5	E5-6	Anticipated financial effects from resource use and circular economy-related impacts, risks, and opportunities	The undertaking may omit the information prescribed by ESRS E4-6 for the first year of preparation of its sustainability statement. The undertaking may comply with ESRS E4-6 by reporting only qualitative disclosures, for the first three years of preparation of its sustainability statement.
ESRS S1	All disclosure requirements	All disclosure requirements	Undertakings or groups not exceeding on their balance sheet dates the average number of 750 employees during the financial year (on a consolidated basis where applicable) may omit the information specified in the disclosure requirements of ESRS S1 for the first year of preparation of their sustainability statement.
ESRS S1	S1-7	Characteristics of non-employee workers in the undertaking's own workforce	The undertaking may omit reporting for all data points in this Disclosure Requirement for the first year of preparation of its sustainability statement.

ESRS S1	S1-8	Collective bargaining coverage and social dialogue	The undertaking may omit this Disclosure Requirement with regard to its own employees in non-EEA countries for the first year of preparation of its sustainability statement.
ESRS S1	S1-11	Social protection	The undertaking may omit the information prescribed by ESRS S1-11 for the first year of preparation of its sustainability statement.
ESRS S1	S1-12	Percentage of employees with disabilities	The undertaking may omit the information prescribed by ESRS S1-12 for the first year of preparation of its sustainability statement.
ESRS S1	S1-13	Training and skills development indicators	The undertaking may omit the information prescribed by ESRS S1-13 for the first year of preparation of its sustainability statement.
ESRS S1	S1-14	Health and safety	The undertaking may omit the data points on cases of work-related ill-health and on number of days lost to injuries, accidents, fatalities, and work-related ill health for the first year of preparation of its sustainability statement.
ESRS S1	S1-14	Health and safety	The undertaking may omit reporting on non-employees for the first year of preparation of its sustainability statement.
ESRS S1	S1-15	Work–life balance	The undertaking may omit the information prescribed by ESRS S1-15 for the first year of preparation of its sustainability statement.
ESRS S2	All disclosure requirements	All disclosure requirements	Undertakings or groups not exceeding on their balance sheet dates the average number of 750 employees during the financial year (on a consolidated basis where applicable) may omit the information specified in the disclosure requirements of ESRS S2 for the first two years of preparation of their sustainability statement.
ESRS S3	All disclosure requirements	All disclosure requirements	Undertakings or groups not exceeding on their balance sheet dates the average number of 750 employees during the financial year (on a consolidated basis where applicable) may omit the information specified in the disclosure requirements of ESRS S3 for the first two years of preparation of their sustainability statement.
ESRS S4	All disclosure requirements	All disclosure requirements	Undertakings or groups not exceeding on their balance sheet dates the average number of 750 employees during the financial year (on a consolidated basis where applicable) may omit the information specified in the disclosure requirements of ESRS S4 for the first two years of preparation of their sustainability statement.

(*Continued*)

Annex 3

Scope 3 categories based on GHG corporate protocol.

Category	Description	Minimum boundary
Upstream activities		
1. Purchased goods and services	• Extraction, production, and transportation of goods and services purchased or acquired by the reporting company in the reporting year, not otherwise included in Categories 2–8	• All upstream (cradle-to-gate) emissions of purchased goods and services
2. Capital goods	• Extraction, production, and transportation of capital goods purchased or acquired by the reporting company in the reporting year	• All upstream (cradle-to-gate) emissions of purchased capital goods
3. Fuel- and energy-related activities (not included in Scope 1 or Scope 2)	• Extraction, production, and transportation of fuels and energy purchased or acquired by the reporting company in the reporting year, not already accounted for in Scope 1 or Scope 2, including: a. Upstream emissions of purchased fuels (extraction, production, and transportation of fuels consumed by the reporting company) b. Upstream emissions of purchased electricity (extraction, production, and transportation of fuels consumed in the generation of electricity, steam, heating, and cooling consumed by the reporting company) c. Transmission and distribution (T&D) losses (generation of electricity, steam, heating, and cooling that is consumed (i.e., lost) in a T&D system) – reported by end user	a. For upstream emissions of purchased fuels: All upstream (cradle-to-gate) emissions of purchased fuels (from raw material extraction up to the point of, but excluding combustion) b. For upstream emissions of purchased electricity: All upstream (cradle-to-gate) emissions of purchased fuels (from raw material extraction up to the point of, but excluding, combustion by a power generator) c. For T&D losses: All upstream (cradle-to-gate) emissions of energy consumed in a T&D system, including emissions from combustion d. For generation of purchased electricity that is sold to end users: Emissions from the generation of purchased energy

	d Generation of purchased electricity that is sold to end users (generation of electricity, steam, heating, and cooling that is purchased by the reporting company and sold to end users) – reported by utility	
4. Upstream transportation and distribution	• Transportation and distribution of products purchased by the reporting company in the reporting year between a company's tier 1 suppliers and its own operations (in vehicles and facilities not owned or controlled by the reporting company) • Transportation and distribution services purchased by the reporting company in the reporting year, including inbound logistics, outbound logistics (e.g., of sold products), and transportation and distribution between a company's own facilities (in vehicles and facilities not owned or controlled by the reporting company)	• The Scope 1 and Scope 2 emissions of transportation and distribution providers that occur during the use of vehicles and facilities (e.g., from energy use) • Optional: The lifecycle emissions associated with manufacturing vehicles, facilities, or infrastructure
5. Waste generated in operations	• Disposal and treatment of waste generated in the reporting company's operations in the reporting year (in facilities not owned or controlled by the reporting company)	• The Scope 1 and Scope 2 emissions of waste management suppliers that occur during disposal or treatment • Optional: Emissions from transportation of waste
6. Business travel	• Transportation of employees for business-related activities during the reporting year (in vehicles not owned or operated by the reporting company)	• The Scope 1 and Scope 2 emissions of transportation carriers that occur during the use of vehicles (e.g., from energy use) • Optional: The lifecycle emissions associated with manufacturing vehicles or infrastructure
7. Employee commuting	• Transportation of employees between their homes and their worksites during the reporting year (in vehicles not owned or operated by the reporting company)	• The Scope 1 and Scope 2 emissions of employees and transportation providers that occur during the use of vehicles (e.g., from energy use) • Optional: Emissions from employee teleworking
8. Upstream leased assets	• Operation of assets leased by the reporting company (lessee) in the reporting year and not included in Scope 1 and Scope 2 – reported by lessee	• The Scope 1 and Scope 2 emissions of lessors that occur during the reporting company's operation of leased assets (e.g., from energy use) • Optional: The lifecycle emissions associated with manufacturing or constructing leased assets

(*Continued*)

Category	Description	Minimum boundary
Downstream activities		
9. Downstream transportation and distribution	• Transportation and distribution of products sold by the reporting company in the reporting year between the reporting company's operations and the end consumer (if not paid for by the reporting company), including retail and storage (in vehicles and facilities not owned or controlled by the reporting company)	• The Scope 1 and Scope 2 emissions of transportation providers, distributors, and retailers that occur during the use of vehicles and facilities (e.g., from energy use) • Optional: The lifecycle emissions associated with manufacturing vehicles, facilities, or infrastructure
10. Processing of sold products	• Processing of intermediate products sold in the reporting year by downstream companies (e.g., manufacturers)	• The Scope 1 and Scope 2 emissions of downstream companies that occur during processing (e.g., from energy use)
11. Use of sold products	• End use of goods and services sold by the reporting company in the reporting year	• The direct use-phase emissions of sold products over their expected lifetime (i.e., Scope 1 and Scope 2 emissions of end users that occur from the use of: products that directly consume energy (fuels or electricity) during use; fuels and feedstocks; and GHGs and products that contain or form GHGs that are emitted during use) • Optional: The indirect use-phase emissions of sold products over their expected lifetime (i.e., emissions from the use of products that indirectly consume energy (fuels or electricity) during use)

12. End-of-life treatment of sold product	• Waste disposal and treatment of products sold by the reporting company (in the reporting year) at the end of their life	• Scope 1 and Scope 2 emissions of waste management companies that occur during disposal or treatment of sold products
13. Downstream leased assets	• Operation of assets owned by the reporting company (lessor) and leased to other organisations in the reporting year, not included in Scope 1 and Scope 2 – reported by lessor	• Scope 1 and Scope 2 emissions of lessees that occur during the operation of leased assets (e.g., from energy use). • Optional: The lifecycle emissions associated with manufacturing or constructing leased assets
14. Franchises	• Operation of franchises in the reporting year, not included in Scope 1 and Scope 2 – reported by franchisor	• Scope 1 and Scope 2 emissions of franchisees that occur during operation of franchises (e.g., from energy use) • Optional: The lifecycle emissions associated with manufacturing or constructing franchises
15. Investments	• Operation of investments (including equity and debt investments and project finance) in the reporting year, not included in Scope 1 or Scope 2	• See the description of category 15 (Investments) in Section 5.5 for the required and optional boundaries

Source: GHG Protocol.

Index

About the Author

Robin Boustead has combined his love for nature with his decades of experience as an entrepreneur and sustainability consultant in business development, carbon calculation and reporting solutions, strategic marketing and cross-cultural management. His inspiring portfolio includes growing a multi-million-dollar adventure travel and retail business, launching a handcrafted interior design and fashion accessories brand, sustainable agricultural infrastructure development, enhancing Bhutan's status as the greenest country on earth, coaching more than 170 businesses in voluntary sustainability standards and creating the world's highest and hardest adventure trekking route, the Great Himalaya Trail. He has written 5 books, makes maps of Nepal and Bhutan, writes short stories and has published numerous research papers and articles about sustainability.

For Product Safety Concerns and Information please contact our EU representative GPSR@taylorandfrancis.com Taylor & Francis Verlag GmbH, Kaufingerstraße 24, 80331 München, Germany